G. A. Sekon

**The Evolution of the Steam Locomotive**

(1803 to 1898)

G. A. Sekon

**The Evolution of the Steam Locomotive**
*(1803 to 1898)*

ISBN/EAN: 9783743465176

Printed in Europe, USA, Canada, Australia, Japan

Cover: Foto ©ninafisch / pixelio.de

Manufactured and distributed by brebook publishing software (www.brebook.com)

G. A. Sekon

**The Evolution of the Steam Locomotive**

THE

# EVOLUTION

OF THE

# STEAM LOCOMOTIVE.

(1803 TO 1898.)

BY

## G. A. SEKON

*Editor of the "Railway Magazine" and "Railway Year Book,"*
*Author of "A History of the Great Western Railway,"*
*&c., &c.*

---

London:

THE RAILWAY PUBLISHING CO., LTD.,

79 TO 83, TEMPLE CHAMBERS, TEMPLE AVENUE, E.C.

1899.

# PREFACE.

In connection with the marvellous growth of our railway system there is nothing of so paramount importance and interest as the evolution of the locomotive steam engine.

At the present time it is most important to place on record the actual facts, seeing that attempts have been made to disprove the correctness of the known and accepted details relative to several interesting, we might almost write historical, locomotives.

In this work most diligent endeavours have been made to chronicle only such statements as are actually correct, without reference to personal opinions.

In a broad sense, and taken as a whole, the old works on locomotive history may be accepted as substantially correct.

From these, therefore, and from authentic documents provided by the various railways, locomotive builders, and designers, together with the result of much original research, has the earlier portion of this account of the evolution of the locomotive steam engine been constructed. The various particulars of modern locomotive practice have been kindly supplied by the locomotive superintendents of the different British railways, so that no question can arise as to the strict accuracy of this portion of the work.

Nearly forty years ago it was authoritatively stated: "That kind of knowledge of the locomotive engine which answers the purpose of a well-informed man has already become so popular that it almost amounts to ignorance to be without it. Locomotive mechanism is very simple in its elementary nature, and the mind is naturally disposed to receive and retain any adequate explanation of striking phenomena, whether mechanical or otherwise; and hence it is that there are thousands of persons who, although in no way concerned in the construction or working of railway engines, are nevertheless competent to give a fair general explanation of their structure and mode of working."

If such were true at that time it is abundantly evident that it is more so at the threshold of the 20th century, considering the growth of inquiry into, and appreciation of, scientific and mechanical knowledge by an ever widening and increasing circle of general readers, which has been one of the marked signs of intellectual development during recent years. Under such circumstances it is not surprising that the locomotive and its history have received a large share of public attention. Whilst railway officers, with the intelligence for which they are justly distinguished, have always evinced a proper desire to be acquainted with the evolution of the "steam horse," the spread of education has increased and quickened a desire for knowledge concerning the locomotive amongst all classes in a remarkable manner. Many of the numerous illustrations that embellish the book have been specially collected for the purpose, and several will be quite new to the majority of readers. Special pains have been taken to admit only such illustrations the authenticity of which was known to the author, and for the same reason many otherwise interesting pictures, upon the accuracy of which suspicion rested, were excluded from the collection.

Despite these exclusions, we believe that no other book on locomotive history in the English language is so fully illustrated.

As it is proposed to deal with the railway locomotive only, it is not necessary to make more than a passing reference to the more or less crude proposals of Sir Isaac Newton, the Marquess of Worcester, Savery, Dr. Robinson, Leupold, and other writers and scientists, who hinted at the possibility of steam locomotion. Nor does the writer propose to discuss the alleged use of railways and steam locomotives in Germany at a date prior to their general introduction into England. The claims of Cugnot, Symington, Evans, Murdoch, and others as builders or designers of actual or model steam road locomotives will also be passed without discussion.

We take this opportunity of expressing our sincere thanks to the locomotive superintendents of British railways, who have all been so willing to assist the author, not only in supplying accurate data concerning the locomotives of their own design, but also for so kindly revising the portions of the volume that relate to the locomotive history of the particular railway with which each one of these gentlemen is connected.

In conclusion, we leave the "Evolution of the Steam Locomotive" to the kindly consideration of our readers, hoping that from a perusal of it they may derive both information and pleasure.

G. A. SEKON.

December, 1898.

# CONTENTS.

|  | PAGE |
|---|---|
| PREFACE | iii. |
| LIST OF ILLUSTRATIONS | vi. |
| CHAPTER I. | 1 |
| „ II. | 10 |
| „ III. | 28 |
| „ IV. | 40 |
| „ V. | 56 |
| „ VI. | 66 |
| „ VII. | 82 |
| „ VIII. | 103 |
| „ IX. | 130 |
| „ X. | 156 |
| „ XI. | 185 |
| „ XII. | 205 |
| „ XIII. | 231 |
| „ XIV. | 260 |
| „ XV. | 294 |
| INDEX | 321 |

# LIST OF ILLUSTRATIONS.

|  | PAGE |
|---|---|
| "990," the latest type of Great Northern Railway express engine | Frontispiece |
| The First Railway Locomotive of which authentic particulars are known | 3 |
| Locomotive built by Murray for Blenkinsopp's Railway | 6 |
| Brunton's "Mechanical Traveller" Locomotive | 8 |
| Hackworth's "Wylam Dilly," generally known as Hedley's "Puffing Billy" | 11 |
| Hackworth's or Hedley's Second Design, used on the Wylam Rwy. in 1815 | 13 |
| Stephenson's Initial Driving Gear for Locomotives | 15 |
| Stephenson and Dodd's Patent Engine, built in 1815 | 16 |
| Stephenson's Improved Engine, as altered, fitted with Steel Springs | 17 |
| "Locomotion," the First Engine to Run on a Public Railway | 20 |
| The First Successful Locomotive, Hackworth's "Royal George" | 23 |
| Hackworth's Blast Pipe in the "Royal George" | 24 |
| Waste Steam Pipe in Stephenson's "Rocket" | 25 |
| The "Novelty," entered by Braithwaite and Ericsson for the Rainhill Prize | 29 |
| Hackworth's "Sanspareil," one of the Competitors at Rainhill | 32 |
| Stephenson's "Rocket," the Winner of the Rainhill Prize of £500 | 35 |
| Winan's "Cycloped" Horse Locomotive | 38 |
| Bury's Original "Liverpool," the First Engine with Inside Cylinders, etc. | 41 |
| The "Invicta," Canterbury and Whitstable Railway, 1830 | 45 |
| The "Northumbrian," the Engine that Opened the Liverpool and Manchester Rwy. | 46 |
| Hackworth's "Globe" for the Stockton and Darlington Railway | 48 |
| Stephenson's "Planet," Liverpool and Manchester Railway | 49 |
| "Wilberforce," a Stockton and Darlington Railway Locomotive | 53 |
| Galloway's "Caledonian," built for the Liverpool & Manchester Rwy. in 1832 | 54 |
| Roberts's "Experiment," with Verticle Cylinders, Bell Cranks, etc. | 57 |
| Hawthorn's "Comet," First Engine of the Newcastle & Carlisle Rwy., 1835 | 59 |
| "Sunbeam," built by Hawthorn for the Stockton and Darlington Railway | 64 |
| The "Grasshopper," with 10ft. driving wheels, built by Mather, Dixon & Co., for the G.W. Rwy. | 73 |
| The "Hurricane," with 10ft. driving wheels, a Broad-Guage Engine, built on Harrison's System | 76 |
| The "Thunderer," a geared-up Broad-Guage Engine, built on Harrison's Plan | 78 |
| Bury's Standard Passenger Engine for the London and Birmingham Railway | 83 |
| "Garnet," one of the First Engines of the London and Southampton Rwy. | 85 |
| "Harpy," one of Gooch's "Firefly" Class of Broad-Gauge Engines | 90 |
| Interior of Paddington Engine House, showing the Broad-Guage Locomotives of 1840 | 92 |
| "Jason," one of Gooch's First Type of Goods Engines for the G.W. Rwy. | 95 |
| Paton & Millar's Tank Engine, for working off the Cowlairs Incline, Glasgow | 98 |
| Stephenson's "Long Boiler" Goods Engine, Eastern Counties Railway | 104 |
| Gray's Prototype of the "Jenny Lind," No. 49, London & Brighton Rwy. | 104 |
| "Hero," a Great Western Railway Six-Coupled Broad-Gauge Goods Engine | 106 |
| The "Great Western," Broad-Gauge Engine as originally Constructed | 107 |
| The Original "Great Western," as Rebuilt with Two Pairs of Leading Wheels | 109 |
| The "Namur," the First Engine built on Crampton's Principle | 112 |
| Crampton's "London," First Engine with a Name, L. & N.W. Rwy. | 113 |
| "Great Britain," one of Gooch's Famous 8ft. "Singles," G.W. Rwy. | 114 |
| "No. 61," London and Brighton Railway | 115 |
| The "Jenny Lind," a Famous Locomotive, built by Wilson and Co. | 119 |
| Trevithick's "Cornwall," with 8ft. 6in. Driving Wheels, and Boiler below the Driving Axle | 120 |

## LIST OF ILLUSTRATIONS

| | PAGE |
|---|---|
| Trevithick's "Cornwall," as now Running between Liverpool and Manchester | 121 |
| "Old Copper Nob," No. 3, Furness Rwy., Oldest Locomotive now at work | 123 |
| The "Albion," a Locomotive built on the "Cambrian" System | 127 |
| The "Fairfield," Adams's Combined Broad-Gauge Engine and Train | 132 |
| The "Enfield," Combined Engine and Train for the Eastern Counties Railway | 134 |
| "Red Star," a 7ft. Single Broad-Gauge Saddle Tank Engine | 136 |
| "No. 148," L. & N.W. Rwy.; Example of Stephenson's "Long Boiler" Engines | 137 |
| Adams's "Light" Locomotive for the Londonderry and Enniskillen Railway | 139 |
| England & Co.'s "Little England," Locomotive—Exhibition, London, 1851 | 142 |
| Crampton's "Liverpool," London and North Western Railway | 145 |
| Timothy Hackworth's "Sanspareil, No. 2" | 149 |
| Caledonian Railway Engine, "No. 15" | 153 |
| "Mac's Mangle," No. 227, London and North Western Railway | 154 |
| "President," one of McConnell's "Bloomers," as originally built | 155 |
| One of McConnell's "Bloomers," as Rebuilt by Ramsbottom | 155 |
| The "Folkestone," a Locomotive on Crampton's System, built for the S.E.R., 1851 | 159 |
| One of J. V. Gooch's "Single" Tank Engines, Eastern Counties Railway | 161 |
| "Ely," a Taff Vale Railway Engine, built in 1851 | 163 |
| McConnell's "300," London and North Western Railway | 165 |
| Pasey's Compressed Air Locomotive, Tried on the E.C. Rwy., 1852 | 170 |
| The First Type of Great Northern Railway Passenger Engine, one of the "Little Sharps" | 171 |
| Sturrock's Masterpiece, the Famous Great Northern Railway, "215" | 172 |
| Pearson's 9ft. "Single" Tank Engine, Bristol and Exeter Railway | 174 |
| One of Pearson's 9ft. "Single" Tanks, taken over by the Great Western Railway | 176 |
| A Bristol and Exeter Railway Tank Engine, as Rebuilt (with Tender) by the G.W.R. | 178 |
| "Ovid," a South Devon Railway Saddle Tank Engine, with Leading Bogie | 180 |
| "Plato," a Six-Coupled Saddle Tank Banking Engine, South Devon Railway | 181 |
| The First Type of Narrow-Gauge Passenger Engines, Great Western Rwy. | 182 |
| "Robin Hood," a Broad-Gauge Express Engine, with Coupled Wheels 7ft. in diameter | 183 |
| North British Railway Inspection Engine, No. 879 | 184 |
| The "Dane," L. and S.W.R., fitted with Beattie's Patent Apparatus for Burning Coal | 187 |
| Cudworth's Sloping Fire Grate, for Burning Coal, as fitted to S.E.R. Locomotives | 189 |
| "Nunthorpe," a Stockton and Darlington Railway Passenger Engine, 1856 | 193 |
| Beattie's Four-Coupled Tank Engine, London & South Western Rwy., 1857 | 194 |
| Sinclair's Outside Cylinder, Four-Coupled Goods Engine, Eastern Counties Railway (Rebuilt) | 196 |
| Six-Coupled Mineral Engine, Taff Vale Railway, built 1860 | 202 |
| "Brougham," No. 160, Stockton and Darlington Railway | 206 |
| Conner's 8ft. 2in. "Single" Engine, Caledonian Railway (Rebuilt) | 208 |
| "Albion," Cambrian Railways, 1863 | 210 |
| A Great Northern Railway Engine, fitted with Sturrock's Patent Steam Tender | 218 |
| Sinclair's Design of Tank Engine for the Eastern Counties Railway | 219 |
| Beattie's Standard Goods Engine, London and South Western Railway, 1866 | 226 |
| Beattie's Goods Engine, London and South Western Railway (Rebuilt) | 227 |
| Adams's Passenger Tank Engine, N.L. Rwy., as Rebuilt by Mr. Pryce | 228 |
| Pryce's Six-Coupled Tank Goods Engine, North London Railway | 229 |
| Locomotive and Travelling Crane, North London Railway | 230 |
| "Python," a 7ft. 1in. Coupled Express Engine, L. and S.W. Rwy. | 232 |
| 8ft. 1in. "Single" Express Engine, Great Northern Railway | 237 |
| "John Ramsbottom," one of Webb's "Precedent" Class, L. & N.W. Rwy. | 238 |
| "Firefly," a London and South Western Outside Cylinder Tank Engine | 239 |
| "Kensington," a Four-Coupled Passenger Engine, London, Brighton and South Coast Railway | 240 |
| "Teutonic," a London and North Western Railway "Compound" Locomotive on Webb's System | 244 |
| "Queen Empress," one of Webb's Compound Locomotives, L. & N.W. Rwy. | 245 |

## LIST OF ILLUSTRATIONS

| | PAGE |
|---|---|
| "Black Prince," L. & N.W. Railway, a Four-Coupled Four-Cylinder Compound Engine | 248 |
| Johnson's 7ft. 9in. "Single" Engine, Midland Railway | 251 |
| "George A. Wallis," an Engine of the "Gladstone" Class, L., B. and S.C. Railway | 252 |
| "1463," North Eastern Railway, one of the "Tennant" Locomotives | 253 |
| Holmes's Type of Express Engines for the North British Railway | 254 |
| 7ft. "Single" Engine, Great Eastern Railway, fitted with Holden's Liquid Fuel Apparatus | 256 |
| "No. 10," the Latest Type of Great Eastern Railway Express Engine, Fired with Liquid Fuel | 258 |
| "Goldsmid," one of the new London, Brighton and South Coast Railway Express Passenger Engines | 261 |
| "Inspector," London, Brighton and South Coast Railway | 262 |
| "No. 192," a Standard Express Passenger Locomotive, L.C. & D.Rwy. | 263 |
| Standard Express Passenger Engine, Cambrian Railways | 264 |
| Standard Passenger Tank Engine, Cambrian Railways | 265 |
| "No. 240," the S.E. Railway Engine that obtained the Gold Medal, Paris Exhibition, 1889 | 267 |
| Standard Goods Engine, South Eastern Railway | 268 |
| Standard Passenger Tank Locomotive, South Eastern Railway | 269 |
| Latest Type of Express Passenger Engine, South Eastern Railway | 271 |
| Adams's Standard Express Engine, London and South Western Railway | 273 |
| A "Windcutter" Locomotive, "No. 136," L. and S.W. Railway, fitted with Convex Smoke-Box Door | 274 |
| Drummond's Four-Cylinder Engine, London and South Western Railway | 275 |
| Four-Coupled Passenger Engine with Leading Bogie, North British Railway | 277 |
| Holmes's Latest Type of Express Engine, North British Railway | 279 |
| Four-Wheels-Coupled Saddle Tank Engine, London & North Western Rwy. | 281 |
| Standard Express Passenger Locomotive, Lancashire and Yorkshire Railway | 282 |
| Standard Eight-Wheel Passenger Tank Engine, Lancashire & Yorkshire Rwy. | 283 |
| Oil-Fired Saddle Tank Shunting Engine, Lancashire and Yorkshire Railway | 284 |
| "Dunalastair," Caledonian Railway | 285 |
| One of McIntosh's "Dunalastair 2nd" Caledonian Express Locomotives | 287 |
| Six-Wheels-Coupled Condensing Engine, Caledonian Railway | 288 |
| "Carbrook," one of Drummond's Express Engines for the Caledonian Railway | 289 |
| McIntosh's 5ft. 9in. Condensing Tank Engine, Caledonian Railway | 290 |
| "No. 143," Taff Vale Railway Tank Locomotive, for working on incline | 292 |
| A favourite Locomotive of the Isle of Wight Central Railway | 293 |
| 7ft. 8in. "Single" Convertible Engine, Great Western Railway | 295 |
| "Empress of India." Standard G.W. 7ft. 8in. "Single" Express Locomotive | 296 |
| "Gooch," a Four-Coupled Express Engine, Great Western Railway | 297 |
| "Pendennis Castle," one of the Great Western "Hill Climbers" | 298 |
| "Single" Express Engine, Six-Wheel Type, Great Western Railway | 300 |
| 6ft. 6in. Four-Coupled Passenger Locomotive, Great Western Railway | 300 |
| 6ft. Four-Coupled Passenger Engine, Great Western Railway | 301 |
| "Barrington," New Type of Four-Coupled Engine, Great Western Railway | 301 |
| Four-Coupled-in-Front Passenger Tank Engine, Great Western Railway | 302 |
| "No. 1312," one of Mr. Ivatt's (1073) Smaller Class of Four-Coupled Bogie Engines, Great Northern Railway | 304 |
| The Latest Type of 6ft. 6in. Coupled Engine, Great Northern Railway | 305 |
| Latest Type of G.N.R. Express Locomotive; 7ft. 6in. "Single," with Inside Cylinders, etc. | 308 |
| "No. 100," one of the "T" Class Four-Coupled Passenger Engines, Great North of Scotland Railway | 311 |
| Pettigrew's New Goods Engine for the Furness Railway | 315 |
| Six-Wheels-Coupled Bogie Engine, with Outside Cylinders, Highland Railway | 316 |
| Liquid Fuel Engine, Belfast and Northern Counties Railway | 317 |
| "Jubilee," Four-Wheels-Coupled Compound Locomotive, Belfast and Northern Conties Railway | 318 |
| "No. 73," Standard Passenger Engine, Great Northern Railway (Ireland) | 318 |
| Four-Coupled Bogie Express Engine, Great Southern and Western Railway | 319 |
| "Peake," a Locomotive of the Cork and Muskerry Light Railway | 319 |

# EVOLUTION

OF THE

# STEAM LOCOMOTIVE.

---

### CHAPTER I.

Trevithick's triumph; his first steam locomotives—Mistaken for the devil—The Coalbrookdale engine—A successful railway journey at Merthyr Tydvil—Description of the engine—"Catch-me-who-can"—The locomotive in London—Blenkinsopp's rack locomotive—Chapman's engine—Did Chapman build an eight-wheel locomotive?—Brunton's "steam horse"—Its tragic end.

TO Richard Trevithick, the Cornish mine captain and engineer, belongs the honour of producing the first locomotive—true, his original essay was a road locomotive. As long ago as 1796 he constructed a model locomotive which ran round a room; and on Christmas Eve, 1801, he made the initial trip with his first steam locomotive through the streets of Camborne. This machine carried several passengers at a speed in excess of the usual walking pace of a man. Trevithick was joined in the enterprise by his cousin Vivian, who provided the money to build the steam engines, and to patent them, their first patent being dated 24th March, 1802. It is described as "for improving the construction of steam engines, and the application thereof for drawing carriages on rails and turnpike roads and other purposes." It was claimed that their engine would produce "a more equable rotary motion on the several parts of the revolution of any axis which is moved by the steam engine, by causing the piston rods of two cylinders to work on the said axis by means of cranks, at a quarter turn asunder."

Among other improvements claimed in the specification, mention should be made of the return-tube boiler, bellows to urge the fire, and a second safety valve, not under the control of the driver.

A steam carriage with these improvements was constructed, and Vivian and Trevithick commenced a journey on it from Camborne to Plymouth, from which port it was shipped to London. On the road to Plymouth a closed toll-bar was met, and the steam carriage stopped for the gate to be opened. "What have us got to pay here?" demanded Vivian. The affrighted toll-keeper, shaking in every limb, and his teeth chattering, essayed to answer, and at last said, "No—na—na—na." "What have us got to pay, I say?" demanded Vivian. "Na—noth—nothing to pay, my de—dear Mr. Devil; do drive on as fast as you can. Nothing to pay.".

It must be remembered that to Cornishmen of a century ago the devil was a very real personage; and, seeing the horseless carriage proceeding with a fiery accompaniment, the poor toll-keeper thought he had at last seen his Satanic majesty. He also appears to have remembered that it is well "to be civil to everyone, the devil included; there is no knowing when you may require his good wishes." Hence the toll-keeper's reason for calling Vivian "my dear Mr. Devil."

As early as August, 1802, R. Trevithick (according to his life, as written by his son, F. Trevithick) appears to have constructed a railway locomotive at Coalbrookdale. This engine had a boiler of cast-iron $1\frac{1}{2}$in. thick, with an interior return wrought-iron tube. The length of the boiler was 6ft., and the diameter 4ft. The cylinder working this engine was 7in. in diameter, the stroke being 3ft. The next railway locomotive was that constructed for the Pen-y-darren Tramroad near Myrthyr Tydvil. Of this particular locomotive (Fig. 1) it is possible to obtain authentic particulars, although much that is legendary already clusters around this historic locomotive. For instance, we read that the locomotive in question had a brick chimney, and that it was demolished by colliding with an overhanging branch of a tree. Then the amount of the bet between Mr. Homfray, the owner of the tramroad, and his friend, as to whether the locomotive would successfully perform a journey from Pen-y-darren to the navigation at Plymouth, is a variable quantity. The amount staked has been stated to be £500 a side, and also £1,000 a side.

It is evident that some days prior to February 10th, 1804, the engine successfully performed the journey, and that overhanging trees and rocks considerably impeded the travelling, several stoppages having to be made whilst these obstacles were removed. Mr. Homfray, however, won the bet. On February 21st another trip was made by the locomotive. On this occasion the load consisted of 5 wagons, 10 tons

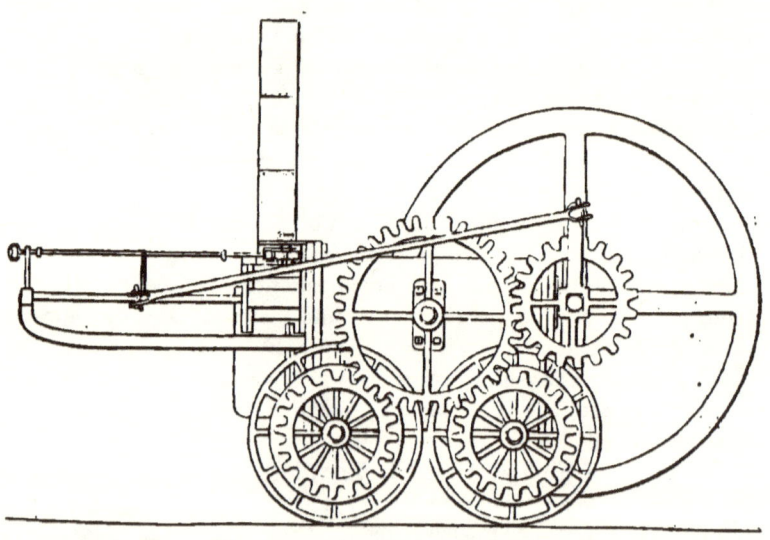

FIG. 1.—THE FIRST RAILWAY LOCOMOTIVE OF WHICH AUTHENTIC PARTICULARS ARE KNOWN

of bar iron, and 70 passengers, the weight of the engine, with water and fuel, being 5 tons; the journey of nine miles being performed in 4 hours 5 minutes, including several stoppages; the average speed when travelling being five miles an hour. On the return journey the engine hauled the empty wagons up an incline of 1 in 18 at the rate of five miles an hour. Several of the tramplates, which weighed only 28lb. per yard, were broken on the downward trip. Early in March the engine conveyed a load of 25 tons from the iron-works to the navigation.

It will be observed that this engine from the first decided the practicability of conveying loads by means of smooth wheels on smooth rails, simply by adhesion. Yet, strange to say, for several years after, it was the firmly-fixed belief of succeeding locomotive constructors that it was impossible to obtain sufficient adhesion between a smooth surface and a smooth rail to successfully work a locomotive. The result was the invention of many curious methods to overcome this apparent difficulty, which, as a fact, never existed, save in the minds of the designers of the early locomotives. These men do not seem to have been fully acquainted with the results of Trevithick's experiments on the Pen-y-darren tramroad in 1804.

A description of this locomotive prototype is of interest. The boiler was cylindrical, with a flat end. The fire-door and chimney were both at the same end, an extended heating surface being obtained by means of the return tube; above the fire-door was the single horizontal cylinder, the diameter of which was $8\frac{1}{4}$in.; a considerable portion of the cylinder was immersed in the boiler, the exposed portion being surrounded by a steam jacket. The stroke was 4ft. 6in.! The piston-rod worked on a motion frame extending in front of the engine. At the other end of the boiler was a fly-wheel some 9ft. 6in. in diameter, the motion being conveyed to it by connecting rods from the cross-head; a cog-wheel on the fly-wheel axle conveyed the motion by means of an intermediate wheel to the four driving-wheels, which are stated to have been 4ft. 6in. in diameter. The exhaust steam appears to have been turned into the chimney, not for the purpose of a blast, but only as an easy method of getting rid of the vapour. It will be remembered that Trevithick, in his patent specification, specially mentioned bellows for urging the fire, and was, therefore, not acquainted with the nature of the exhaust steam blast. It is important to bear this in mind, as the reader will find in a later chapter. This engine is stated to have blown up through not being provided with a safety valve, though Trevithick specially ordered one to be fixed to the boiler, but his instructions do not appear to have been carried out.

Trevithick made another locomotive, called "Catch-me-who-can." This ran on an ellipse-shaped railway specially laid down for it at Euston Square, London, and was visited by many people during the few days it was on view. Another locomotive was constructed from the drawings of Trevithick's Coalbrookdale locomotive of 1802, to the orders of Mr. Blackett, the owner of Wylam Collieries. This engine weighing 4½ tons, had a single cylinder 7in. diameter, 3ft. stroke, and, of course, a fly-wheel. For some reason or another this engine does not appear to have been used on the Wylam tramroad, but was used in a Newcastle foundry to blow a cupola. Mr. Armstrong, a former Locomotive-Superintendent of the Great Western Railway, was acquainted with this engine of Trevithick's at the time it was so employed at Newcastle.

Having given an outline of Trevithick's invention of the tramroad locomotive, and the other locomotive engines designed by him, we will deal with the locomotive built for J. Blenkinsopp (Fig. 2), of the Middleton Colliery, near Leeds, who, on April 10th, 1811, obtained a patent for a self-propelling steam engine, worked by means of a cog-wheel, engaging in a rack laid side by side with one of the rails forming the tramway.

The erroneous idea that the locomotive of itself had not sufficient adhesion between the smooth wheel and the surface of the rail to propel itself and draw a load was strongly entertained by Blenkinsopp, hence his patent rack and pinion system. Blenkinsopp having this opinion, which he published by means of his patent specification, caused succeeding inventors to fall into the same error regarding the adhesive properties of the locomotive, and consequently considerably retarded the development of the railway engine.

Although this engine is generally known as Blenkinsopp's, it was constructed by Matthew Murray, the Leeds engineer. The boiler was cylindrical, with slightly convex ends, a single flue ran through it, which was in front turned upwards, and so formed the chimney; the fire-grate was at the other end of the flue, as in the modern locomotive.

This engine was provided with two cylinders, and was, in this respect, an improvement on Trevithick's single-cylinder engines. The cylinders were 8in. in diameter, and placed vertically, the major portion of them being placed within the boiler. The stroke was 20in., and the motion was conveyed by means of cross-heads, working connecting-rods; these came down to two cranks on either side below the boiler. The cranks worked two shafts fixed across the frames, on which were toothed wheels, both working into a centre toothed wheel, which was provided with large teeth, these engaged on the rack rail previously described. The cranks were set at right angles, so that one piston was exerting power when the other was at its dead centre, and *vice versâ*. The engine was supported on the rails by four wheels 3ft. 6in. in diameter. The two cylinders were connected by a pipe which conveyed the exhaust steam and discharged it into the atmosphere through a vertical tube. The engine weighed 5 tons, burned 75lb. of coal per hour, and evaporated

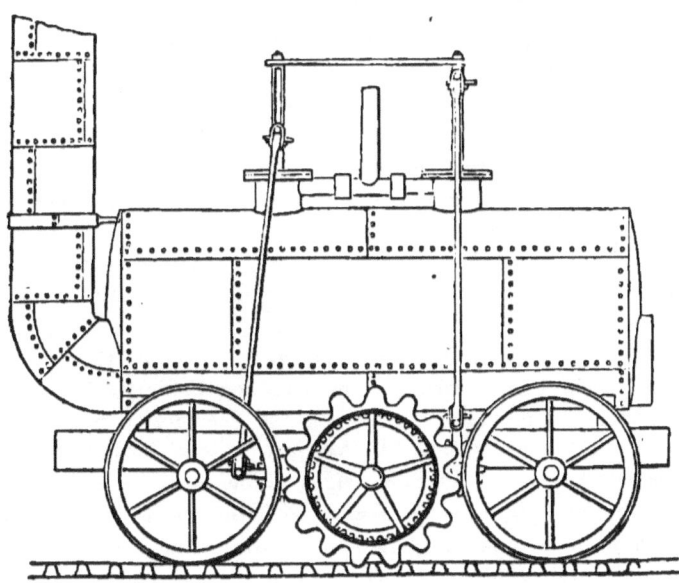

FIG. 2.—LOCOMOTIVE BUILT BY MURRAY FOR BLENKINSOPP'S RAILWAY

50 gallons of water in the same time. This locomotive could haul 94 tons on the level at 3½ miles an hour, or 15 tons up an incline of 1 in 15; its maximum speed was 10 miles an hour.. The engine cost £400 to construct, and worked from August, 1812, for a period of about 20 years, and in 1816 the Grand Duke Nicholas, afterwards Emperor of Russia, inspected the machine. The tramway on which it worked was about 3¾ miles long.

In September, 1813, Murray supplied two of Blenkinsopp's engines to the Kenton Colliery.

On December 30th, 1812, a patent was granted to William and Edward Chapman for a method of locomotion. A chain was stretched along the railway and fastened at each end; connected to the locomotive by spur gear was a barrel, around which the chain was passed. When the barrel rotated, the chain was wound over it, and since the chain was secured at either end, the engine was of necessity propelled. An engine on this principle was tried on the Heaton Colliery Tramroad, near Newcastle-on-Tyne. The machine was supported on wheels travelling on the rails. The boiler was of Trevithick's design, and fanners were used to excite the combustion of the fuel. The weight of Chapman's engine was 6 tons. After a few trials the scheme was abandoned, as it was found impracticable to successfully work such a system. Every eight or ten yards the chain was secured by means of vertical forks, which held it when disengaged from the drum of the locomotive.

By this method the pressure of one engine on the chain was limited to the fork on either side of the drum instead of being spread over the whole length of the chain, and it would, therefore, have been possible for several engines to have used the chain at one and the same time.

According to Luke Herbert and Lieut. Lecount, Chapman also built an 8-wheel locomotive for the Lambton Colliery. This engine, it was stated, had vertical cylinders, and the motion was conveyed by means of spur wheels. It weighed 6 tons loaded, and drew 18 loaded wagons, of a gross weight of 54 tons, from the colliery to the shipping place on the Wear; with the above load it attained a speed of four miles an hour up an incline of 1 in 115. The dimensions and capabilities accredited to this engine appear suspiciously similar to those related of the first Wylam locomotive.

On May 22nd, 1813, Mr. W. Brunton, of the Butterfly Ironworks, obtained a patent for a novel method of steam locomotion. This locomotive inventor was also suffering from the common belief that it was impossible to obtain sufficient adhesion between a smooth rail and smooth wheels, despite the successes that had already been obtained in this direction by Trevithick. He therefore built an engine supported on four flanged carrying wheels, but propelled from behind by means of two legs. Indeed, another inventor considered the idea of steam legs so natural that he constructed a steam road coach that was to be propelled by four legs, one pair partaking of the character and motion of the forelegs of a horse, and the other pair being fashioned on the model of the hind legs of the same quadruped.

In Brunton's leg-propelled steam locomotive (Fig. 3) we find that the boiler was cylindrical, with a single horizontal tube passing through it, and turned up in front in a vertical position, thus forming the chimney. The motion was obtained from a single horizontal cylinder, fixed near the top of the boiler, the piston rod projecting behind; the end of the piston rod was attached to a jointed rod, the bottom portion of which formed one of the legs. The upper portion of this rod was attached to a framework fixed above the boiler of the engine, which formed a fulcrum, and then by an ingenious arrangement of levers, an alternate motion was given to the second leg. Each leg had a foot formed of two prongs at the bottom; these stuck in the ground, and prevented the legs

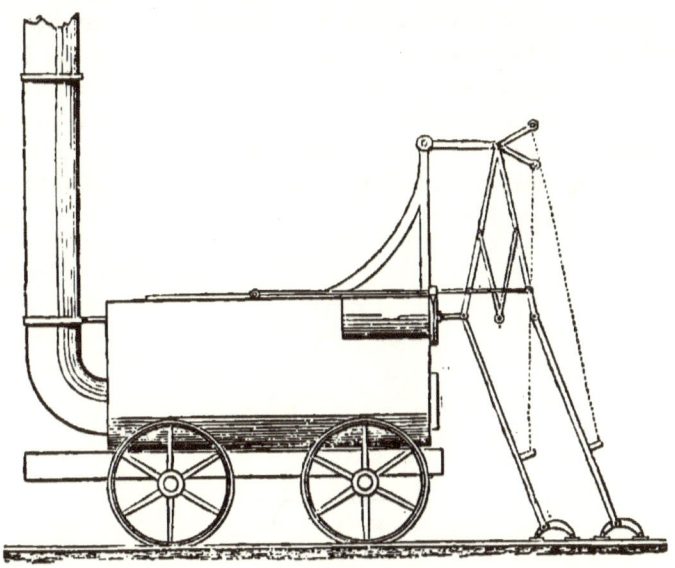

FIG. 3.—BRUNTON'S "MECHANICAL TRAVELLER" LOCOMOTIVE

from slipping. Upon steam being applied, the piston in the ordinary way would have travelled to the end of the cylinder, but the leg, having a firm hold of the ground, presented a greater resistance to the steam than did the weight of the engine, so the steam acting on the surface that presented the lesser resistance, caused the cylinder to recede, and with it the engine, to which it was, of course, firmly attached. By means of the reciprocating levers, a horizontal rod travelled on the top of the boiler and over a cog-wheel; then on the other side of this cog-wheel was another horizontal rod, which, actuated by the cog-wheel, travelled in a contrary direction, and being attached to the other leg of the engine, as the machine receded from the first leg, it drew the second leg close up to the back of the engine. The second leg was now ready to propel the engine, which it did upon the steam being applied to the other side of the piston, and the process was alternated with each admission of steam to the front or back of the piston.

Whilst the legs were returning towards the engine the feet were raised by means of straps or ropes fastened to the legs and passing over friction-wheels, movable in one direction only by a ratchet and catch, and worked by the motion of the engine.

Brunton called his locomotive a "mechanical traveller," and stated that the boiler was of wrought-iron, 5ft. 6in. long and 3ft. diameter, weighing $2\frac{1}{4}$ tons, stroke of piston, 2ft., and at $2\frac{1}{2}$ miles per hour, with a steam pressure of 45lb. per square inch, was equal in power to nearly six horses. This locomotive curiosity blew up at Newbottle in 1816, and about a dozen people were thereby either killed or seriously injured.

## CHAPTER II.

Who is entitled to the honour of constructing the Wylam locomotives?—The claims of Hackworth, Hedley and Foster—" Puffing Billy "—Rebuilt as an eight-wheel engine—Stewart's locomotive—Sharp practice causes Stewart to abandon locomotive building—George Stephenson as a locomotive builder—His hazy views as to his first engine—" Blucher "—The German General proves a failure—Stephenson and Dodd's engine—Stephenson's third engine, with (so-called) steam springs—Competent critics condemn Stephenson's engines—The " Royal William "—The " Locomotion "—Hackworth, General Manager of the Stockton and Darlington Railway—Horse haulage cheaper than Stephenson's locomotives—Hackworth to the rescue—The " Royal George," the first successful locomotive—The " exhaust " steam blast—Rival claimants and its invention—Locomotive versus stationary engine—" Twin Sisters "—" Lancashire Witch "—" Agenoria "—The " Maniac," " a Forth Street production."

WE have now arrived at a point in the evolution of the steam locomotive where the claims of several men are in competition. The facts as to the experiments and construction of the engines at Wylam are not disputed. The question at issue is as to whom the honour of the success should be given. Christopher Blackett, of the Wylam Colliery, as previously stated, ordered a locomotive of Trevithick, but never used it. He, however, determined to make a trial of steam haulage on his plate way, and in 1811 some kind of experiments were made, having in view the above-mentioned object. At this time Timothy Hackworth was foreman of the smiths (he would now be called an engineer), and William Hedley was coal-viewer at Wylam. The friends of both Hackworth and Hedley claim for their respective heroes the honour of these early essays in locomotive construction. But it is probable the honours should be shared by both, as well as by Jonathan Foster, who also assisted in the experiments and construction of the Wylam locomotives.

Hedley was colliery-viewer at Wylam, and therefore, most likely, Hackworth was, to an extent, under his orders, and probably had to defer to, and act under, the instructions of Hedley.

But Hackworth's position as foreman-smith did not preclude him from making suggestions and introducing improvements of his own into the locomotives under construction.

It is stated that Hedley was jealous because Hackworth obtained the praise for building the Wylam locomotives (or "Timothy's Dillies," as they were locally called), and to force Hackworth to leave Wylam, Hedley required him to do some repairs to the machinery on Sundays. Now, Timothy was a fervent Wesleyan, and spent his Sundays in local preaching, so he naturally refused to violate his conscience by working on that day. Consequently Hackworth sought employment elsewhere.

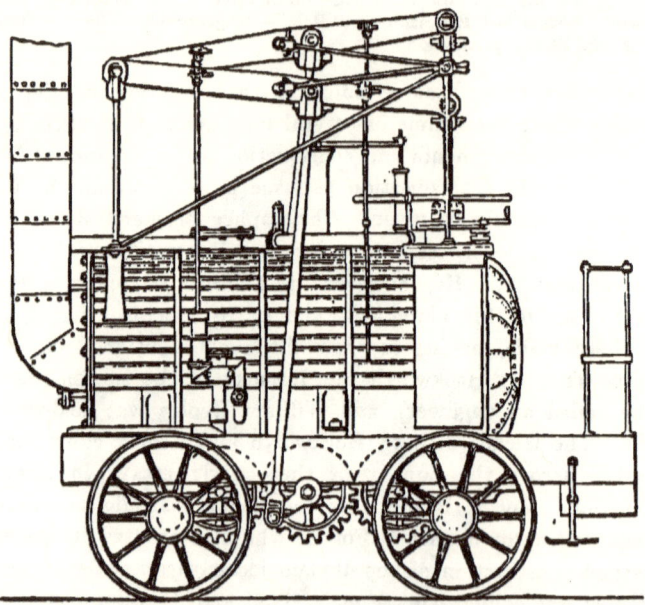

FIG. 4.—HACKWORTH'S "WYLAM DILLY," GENERALLY KNOWN AS HEDLEY'S "PUFFING BILLY"

On the other hand, it was a sore point with Hackworth that George Stephenson spent his Sundays at Wylam taking sketches and particulars of the locomotives at that time at work on the Wylam Railway, the result of which observations was apparent in the locomotive built by Stephenson at Killingworth in 1814.

The Wylam experimentalists in October, 1812, constructed a four-wheel vehicle driven by manual power working cranks connected with spur wheels. The carriage was loaded until sufficient weight had been placed upon it to cause the wheels to turn round without progressing.

The experiment, however, satisfied Mr. Blackett that locomotive engines with smooth wheels could be employed in drawing loads on his tramroad; and the construction of an engine was immediately proceeded with. This was completed and put to work early in 1813. It had a cast-iron boiler, and a single internal flue; the solitary cylinder was 6in. in diameter, and a fly-wheel was employed after the model of Trevithick's engine. The steam pressure was 50lb. This four-wheel engine drew six coal trucks at five miles an hour, and, therefore, did the work of three horses—not a very powerful example of a steam locomotive, it will be observed. This engine being somewhat of a failure, it was decided to build another, and one with a wrought-iron boiler and a return tube was constructed. In his engine (Fig. 4) it will be noticed the fire-box and chimney were both at the same end of the boiler. Two vertical cylinders were fixed over the trailing wheels of "Puffing Billy" (for it is this historical locomotive, now preserved in the South Kensington Museum, that is now being described). The piston rods were connected to beams of the ".Grasshopper" pattern, being both centred at the funnel end of the engine. The driving rods were connected with these beams at about their centres, and passed down to spur wheels, which, by means of toothed wheels on either side, communicated the motion to the four carrying wheels. The spent steam was conveyed from the cylinders to the chimney by means of two horizontal pipes laid along the top of the boiler. It was soon discovered that the cast-iron tram-plates, which were only of four square inch section, were unable to bear the weight of " Puffing Billy," and another change was decided upon.

The engine was therefore placed on two four-wheel trucks (Fig. 5), so that the weight was distributed on eight instead of four wheels, the same method of spur gearing was employed, and the whole of the wheels were actuated by means of intermediate cog-wheels. To prevent, as far as possible, the noise caused by the escaping steam, a vertical cylinder was fixed on the top of the boiler between the cylinders and the funnel. Into this chamber the spent steam was discharged, and from it the same was allowed to escape gradually into the chimney. In addition to the improvement of a return tube, with its extended heating surface, with which this class of engine was provided, the funnel was only 12in. in diameter, as compared with 22in. diameter as used by Stephenson in his early engines. As already stated, the

Wylam locomotives were locally called "Timothy's Dillies," after Timothy Hackworth, to whose inventive genius they were popularly ascribed. In 1830, the cast-iron plates on the road from Wylam to Leamington were removed, and the course was relaid with edge rails, so that the necessity for eight-wheel engines was at an end. "Timothy's Dillies" were then reconverted to four-wheel locomotives, and continued at work on the line till about 1862.

Not many locomotive writers are acquainted with the fact that

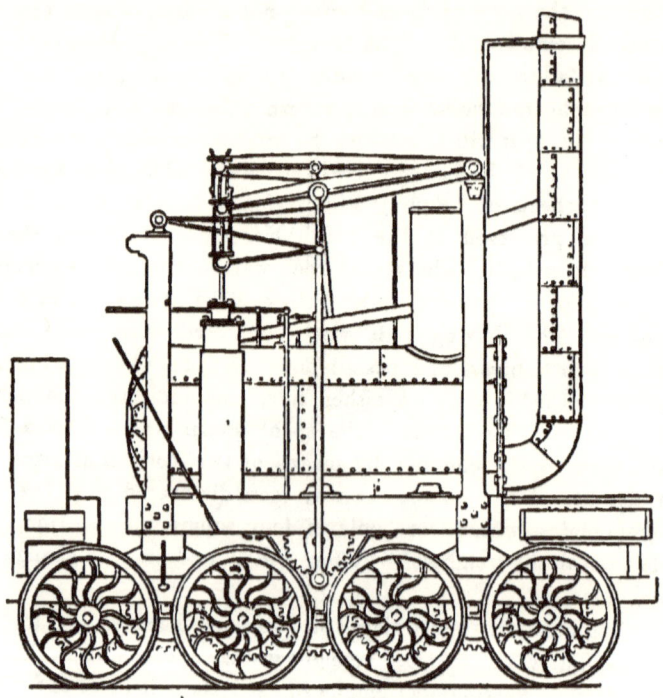

FIG. 5.—HACKWORTH'S OR HEDLEY'S SECOND DESIGN, AS USED ON THE WYLAM RAILWAY IN 1815

in 1814 William Stewart, of Newport, Mon., constructed a locomotive for the Park End Colliery Company, which was tried on the Lydney Railway, and found to work in a satisfactory manner. The Park End Colliery Company were paying about £3,000 a year to contractors for horse haulage of their coal to the Forest of Dean Canal, and Stewart undertook to do the same by locomotive power for half that sum. The Company accepted his terms, and he set about the construction of his engine. Whilst this was progressing the contractors who provided the horses were told at each monthly settlement that the Company were going to use a locomotive to haul the coal, as horse-power was too expensive. By means of these threats the contractors were induced each month to accept a less price than previously for "leading" the coal over the tramroad. Upon the specified date Stewart's locomotive was duly delivered on the line, and accepted by the Park End Colliery Company for doing the work required; but the engineer was informed that the horse-power contractors were then only receiving £2,000 a year for the work, and that as Stewart had agreed to provide locomotive power at one-half of the sum paid for horses, he would only receive £1,000 a year.

Stewart was so highly indignant at this piece of sharp practice that he refused to have anything further to do with the Park End Colliery Company, and at once removed his locomotive off their tramroad, and took it back to Newport.

The earliest attempts of George Stephenson in connection with the evolution of the steam locomotive now deserve attention. Stephenson himself is not very clear about his first engine, for, speaking at Newcastle at the opening of the Newcastle and Darlington Railway in 1844, he said that thirty-two years ago he constructed his first engine. "We called the engine 'My Lord,' after Lord Ravensworth, who provided the money for its construction." Both these statements are erroneous, for Stephenson did not build his first engine till 1814, and thirty-two years before 1844 would have been 1812. Then the engine could not have been called "My Lord," after Lord Ravensworth, for the title did not exist in 1814, the gentleman alluded to being only Sir Thomas Liddell till the coronation of King George IV. in 1821, when he was created Lord Ravensworth.

The "Blucher," as this engine was in fact usually called, was first tried on the Killingworth Railway on July 25th, 1814; she had a wrought-iron boiler, 8ft. long and 2ft. 10in. diameter, with a single flue 20in. diameter, turned up in front to form a chimney. The power was applied by means of two vertical cylinders located partly within the boiler, and projecting from its top, close together, and near the middle. The cylinders were 8in. diameter, the stroke 2ft. The motion was conveyed to the wheels by means of cross-heads and connecting-rods working on small spur wheels (Fig. 6), which engaged the four carrying wheels by means of cogged wheels fitted on the axles of the flanged rail-wheels; these were only 3ft. in diameter, and were 3ft. apart. The spur wheels engaged another cogged wheel, placed between them, for the purpose of keeping the cranks at right angles. No springs were provided for the engine, which was mounted on a wooden frame, but the water barrel was fixed to one end of a lever, and also weighted; the other end of this lever was fixed to the frame of the engine. This arrangement did duty for springs!

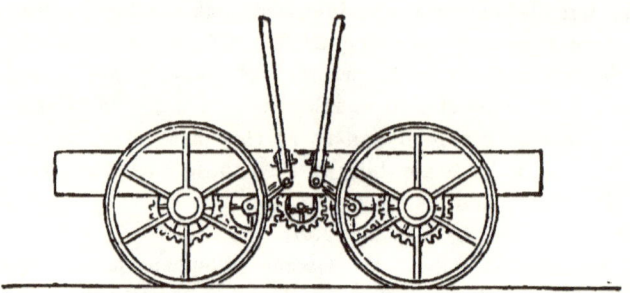

FIG. 6.—STEPHENSON'S INITIAL DRIVING GEAR FOR LOCOMOTIVES

The best work done by "Blucher" was the hauling of loaded coal-wagons, weighing 30 tons, up an incline of 1 in 450, at about four miles an hour. This first effort of Stephenson had no original points about it; the method of working was copied from the Wylam engines, whilst Trevithick's practice was followed with regard to the position of the cylinders—*i.e.*, their location, partly within the boiler. The average speed did not exceed three miles an hour, and after twelve months' working the machine was found to be more expensive than the horses it was designed to replace at a less cost. The absence of springs was specially manifested, for by this time the engine was so much shaken and injured by the vibration that the Killingworth Colliery owners were called upon a second time to find the money to enable Stephenson to construct another locomotive.

The second engine (Fig. 7) constructed by George Stephenson was built under the patent granted to Dodd and Stephenson on 28th February, 1815. In this engine vertical cylinders, partly encased in the boiler, were again employed; but their position was altered, one being placed at each extremity of the boiler over the wheels, the intermediate spur wheels formerly used for keeping the cranks at right angles were abandoned, and the axles were cranked. A connecting-rod was fitted on these cranks, thus coupling the two axles. To give greater adhesion, the wheels of the tender were connected with those of the engine by means of an endless chain passing over cogs on the one pair of engine wheels, and over the adjoining pair of tender wheels; by these methods six pairs of wheels were coupled. The mechanics engaged were not, however, capable of forging proper crank axles, and these had to be abandoned, and an endless chain coupling employed for the engine wheels, similar to the one connecting the tender and engine, as previously described.

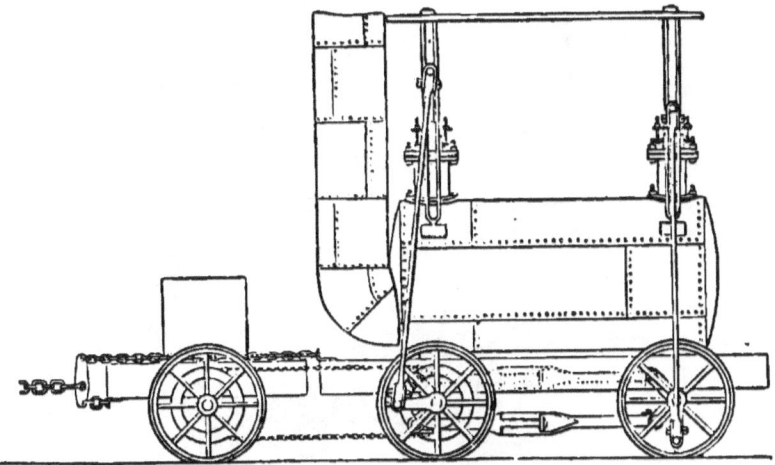

FIG. 7.—STEPHENSON AND DODD'S PATENT ENGINE, BUILT IN 1815

This engine had no springs, and, to avoid excessive friction arising from the bad state of the tramroad, Stephenson employed "ball and socket" joints between the ends of the cross-heads and the connecting-rods. In this way the necessary parallelism between the ends of the cross-heads and the axles was maintained. The spent steam in the engine was turned into the chimney, as in Trevithick's Pen-y-darren locomotive. This locomotive commenced to work on 6th March, 1815.

George Stephenson constructed a third engine (Fig. 8), under a patent granted to Lock and Stephenson on 30th September, 1816; this patent covered several matters, the most important in connection with the engine being malleable iron wheels, instead of cast-iron, and what has been described as "steam springs." The patentees called them "floating pistons"; of this description Colburn says emphatically "they are not," and the same authority continues, "and they (Lock and Stephenson) added, evidently without understanding the true action of the pistons, which were different in principle from the action of springs, that inasmuch as they acted upon an elastic fluid, they produced the desired effect, with much more accuracy than could

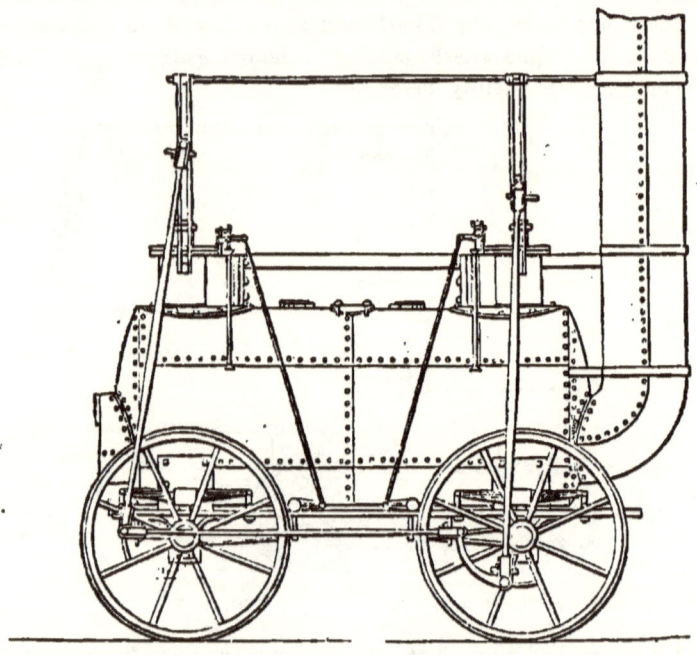

Fig. 8.—STEPHENSON'S IMPROVED ENGINE, AS ALTERED, FITTED WITH STEL SPRINGS (INVENTED BY NICHOLAS WOOD)

be obtained by employing the finest springs of steel to suspend the engine. The whole arrangement was, on the contrary, defective in principle and objectionable on the score of leakage, wear, etc.; and, as a matter of course, was ultimately abandoned."

In the drawings attached to the patent specification this engine is shown with six wheels, and the chain coupling is employed. Lecount says: "The six wheels were continued in use as long as the steam springs were applied, and when steel springs were adopted they were again reduced to four." So much praise has been given to Stephenson for the "great improvements" he is supposed to have introduced into the construction of the locomotive, that it will not be uninteresting if we here reproduce the extremely pertinent remarks of Galloway, the well-known authority on the steam engine, which go far to prove that it was only the great success obtained by George Stephenson from the construction of the Liverpool and Manchester and other railways, that caused historians and biographers to either magnify his locomotive successes, or to gloss over the evident faults in the design and construction of his engines. In his "History of the Steam Engine," published in 1827, Galloway says: These locomotive engines have been long in use at Killingworth Colliery, near Newcastle, and at Hilton Colliery on the Wear, so that their advantages and defects have been sufficiently submitted to the test of experiment; and it appears that, notwithstanding the great exertions on the part of the inventor, Mr. Stephenson, to bring them into use on the different railroads, now either constructing or in agitation, it has been the opinion of several able engineers that they do not possess those advantages which the inventor had anticipated; indeed, there cannot be a better proof of the doubt entertained regarding their utility than the fact that it has been determined that no locomotive engine shall be used on the projected railroad between Newcastle and Carlisle, since, had their advantages been very apparent, the persons living immediately on the spot in which they are used, namely, Newcastle, would be acquainted therewith.

"The principal objections seem to be the difficulty of surmounting even the slightest ascent, for it has been found that a rise of only one-eighth of an inch in a yard, or of eighteen feet in a mile, retards the speed of one of these engines in a very great degree; so much so, indeed, that it has been considered necessary, in some parts where used, to aid their ascent with their load, by fixed engines, which drag them forward by means of ropes coiling round a drum. The spring steam cylinders below the boiler were found very defective, for in the ascending stroke of the working piston they were forced inwards by the connecting-rod pulling at the wheel and turning it round, and in the descending stroke the same pistons were forced as much outwards. This motion or play rendered it necessary to increase the length of the working cylinder as much as there was play in the lower ones, to avoid the danger of breaking or seriously injuring the top and bottom of the former by the striking of the piston when it was forced too much up or down."

Stephenson must have felt himself to be a personage of some importance when he received an order from the Duke of Portland for a steam locomotive. The engine, which had six wheels, was duly built and delivered in 1817, when it was put to work on the tramroad connecting the Duke's Kilmarnock Collieries with the harbour at Troon; but, after a short trial, its use was abandoned, as the weight of the engine frequently broke the cast-iron tram-plates. It has been stated that "this engine afterwards worked on the Gloucester and Cheltenham Tramroad until 1839, when the Birmingham and Gloucester Railway bought the line, and took up the cast-iron tram-plates."

There is no doubt that a six-wheel engine with vertical cylinders partly encased in the top of the boiler, and called the "Royal William," was actually at work on this line—the fact having been commemorated by the striking of a bronze medal; but there is nothing to show that the "Royal William" and the engine built for the Kilmarnock and Troon Tramroad were one and the same locomotive; whilst it is certain that the Gloucester and Cheltenham Tramroad was not purchased by the Birmingham and Gloucester Railway, but jointly by the Cheltenham and Great Western Union Railway and the Birmingham and Gloucester Railway, the price paid being £35,000.

It would appear from a letter written by George Stephenson, and dated Killingworth Colliery, 28th June, 1821, that he had but little idea to what a great degree the development of the steam locomotive would be carried. The letter, which was addressed to Robert Stevenson, the celebrated Edinburgh engineer, proceeded as follows: "I have lately started a new locomotive engine with some improvements on the others which you saw. It has far surpassed my expectations. I am confident that a railway on which my engine can work is far superior to a canal. On a long and favourable railway I would start my engine to travel 60 miles a day, with from 40 to 60 tons of goods." Taking Stephenson's "day" to mean twelve working hours, his idea of maximum speed did not exceed five miles an hour at that time. Before this—in December, 1824—Charles MacLaren had published in the *Scotsman* his opinion that by the use of the steam locomotive "we shall be carried at the rate of 400 miles a day," or an average speed of 33 1-3 miles an hour.

Yet such is the irony of fate, that MacLaren, the true prophet, is forgotten, and George Stephenson is everywhere extolled.

The Hetton (Coal) Railway was opened on November 18th, 1822,

20   EVOLUTION OF THE STEAM LOCOMOTIVE

and five of Stephenson's "improved Killingworth" locomotives were placed upon the level portions. These engines were capable of hauling a train of about 64 tons, the maximum speed being four miles an hour.

FIG. 9.—"LOCOMOTION," THE FIRST ENGINE TO RUN ON A PUBLIC PUBLIC RAILWAY (THE STOCKTON AND DARLINGTON RAILWAY)

The Stockton and Darlington Railway, the first public railway, was opened on September 27th, 1825. The "Locomotion" (Fig. 9) was the first engine on the line. It was constructed at the Forth Street Works of R. Stephenson and Co., at Newcastle-on-Tyne. At this early period these now celebrated Forth Street Works were little better than a collection of smiths' forges.

Timothy Hackworth had been manager of these works, and he had a good deal to do with the construction of "Locomotion." His improvement of the coupling-rods in place of the endless chain previously used for the purpose by Stephenson is worthy of passing notice. George Stephenson expressed a very strong desire that Hackworth should remain in charge of the Forth Street Works, and went so far as to offer him one-half of his (Stephenson's) share in the business if he would remain. Hackworth agreed to do so if his name were added to that of the firm and he were publicly recognised as a partner; but this proposition was not accepted by Stephenson.

Hackworth then took premises in Newcastle, and intended to commence business as an engine-builder on his own account, he having already received several orders from the collieries, etc., where his skill was well known and appreciated. George Stephenson, having heard of Hackworth's plans for carrying on a rival engine factory at Newcastle, saw Hackworth, and persuaded him to relinquish the proposition and accept the office of general manager and engineer to the Stockton and Darlington Railway.

Hackworth commenced these duties in June, 1825, and removed to Darlington. The "Locomotion" had four coupled wheels, 4ft. in diameter; two vertical cylinders, 10in. in diameter, placed partly within the boiler; the stroke was 24in.; steam pressure, 25lb. per square inch; weight in working order, 6½ tons. The tender was of wood, with a coal capacity of three-quarter ton, and a sheet-iron tank holding 240 gallons; weight loaded, 2¼ tons. The tender was supported on four wheels, each of 30in. diameter. This engine worked on the Stockton and Darlington Railway till 1850. In September, 1835, "Locomotion" engaged in a race with the mail coach for a distance of four miles, and only beat the horses by one hundred yards! She was used to open the Middlesbrough and Redcar Railway on June 4th, 1846, being under the charge of Messrs. Plews and Hopkins on this occasion, when she hauled one carriage and two trucks, and took thirty-five minutes to cover the eight miles. From 1850 to 1857 she was used as a pumping engine by Pease at his West Collieries, South Durham, after which she was mounted on a pedestal at North Road Station, Darlington. This engine was in steam upon the Darlington line during the celebration of the Stockton and Darlington Railway jubilee in September, 1875.

She has been exhibited as follows:—1876, at Philadelphia; 1881, Stephenson Centenary; 1886, Liverpool; and 1889, Paris. In April, 1892, she was removed from North Road to Bank Top, Darlington.

The Forth Street Works in 1826 supplied three more engines to the Stockton and Darlington Railway, named "Hope," "Black Diamond," and "Diligence." These locomotives possessed many faults; indeed, they were frequently stopped by a strong wind, and the horse-drawn trains behind the locomotive-propelled ones were delayed because the engines could not proceed. "Jemmy" Stephenson (brother to George) was the principal engine-driver, and he was known far and near as most prolific in the use of oaths of a far from Parliamentary style.

"Jemmy" would be cursing his engine and the horsemen cursing "Jemmy" for the delay; and, indeed, the usual result was a general skirmish. We have already stated that Hackworth was a deeply religious man, and these scenes of lawlessness made a deep impression on his mind, so that he sought for some means to improve the locomotives, the radical fault of which was the shortness of steam —Hackworth knowing that if things progressed smoothly "Jemmy" would have fewer occasions to display his oratorical gift. After eighteen months' working of the Stockton and Darlington Railway it was found that locomotive haulage was much more expensive than horse power; indeed, for every pound spent on horse power about three pounds were paid for locomotive power for doing an exactly similar amount of work.

The £100 stock of the Stockton and Darlington Railway quickly fell to £50, and the shareholders began to get alarmed.

There were two opposite interests at stake—that of the general body of shareholders and that of the locomotive builders (Messrs. Pease and Richardson), who were also large shareholders in, and directors of the Stockton and Darlington Railway, as well as partners in the firm of R. Stephenson and Co. The question as to retaining the use of locomotive engines was fully discussed at a meeting of the principal proprietors, and Hackworth, as manager and engineer of the railway, was asked to give his opinion on the point. He replied: "Gentlemen, if you will allow me to make you an engine in my own way, I will engage that it shall answer your purpose." To have refused him permission would have shown clearly to the other proprietors that Pease and Richardson did not care for the principles of steam locomotion, but that it was the locomotives constructed at the Forth Street Works they wished to retain. Therefore, after some discussion, it was agreed that "as a last experiment Timothy shall be allowed to carry out his plan."

Hackworth's opportunity had now arrived, and the result was the production of the first really successful locomotive steam engine.

But although the shareholders "as a last experiment" gave Hackworth leave to build a locomotive on his own plan, they do not appear to have had much belief in the success of the venture, for the boiler of an old locomotive was given him to use in the construction of the new engine.

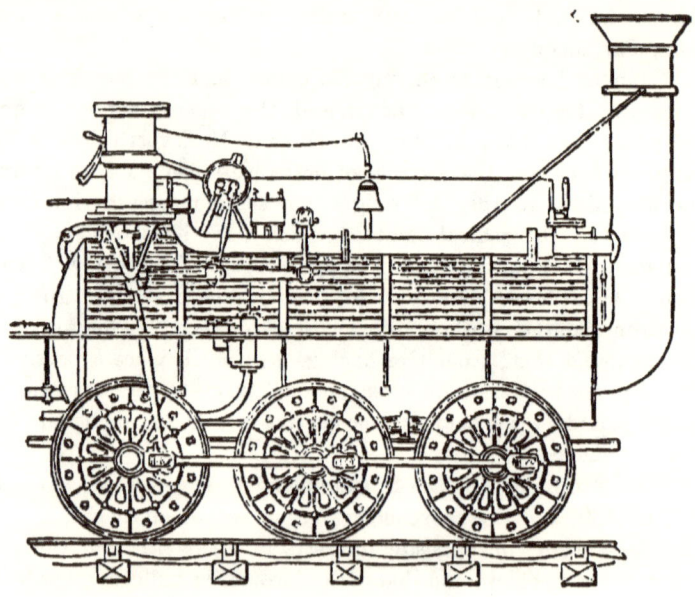

FIG. 10.—THE FIRST SUCCESSFUL LOCOMOTIVE, HACKWORTH'S "ROYAL GEORGE," STOCKTON AND DARLINGTON RAILWAY, 1827

The engine was originally a four-wheel engine, provided with four cylinders, two to each pair of wheels, and it is stated to have been the first built in which a single pair of wheels was worked by two pistons actuating cranks placed at right angles to each other. She was built by Wilson, of Newcastle, and was the fifth engine supplied to the Stockton and Darlington Railway.

This boiler was a plain cylinder, 4ft. 4in. in diameter, and 13ft. long. A wrought-iron tube of ⊂⊃ shape provided the heating surface, the vapour from the furnace travelling from the fire-grate up and down the tube to the chimney, which was at the same end of the boiler as the grate; indeed, the chimney was an elongation of the tube continued through the end of the boiler and turned up vertically.

This return tube gave the new engine twice the heating surface of the ordinary engines, which were only provided with one straight tube. The locomotive was called the "Royal George" (Fig. 10), and was supported on six coupled wheels, each of 4ft. diameter.

The cylinders were placed in a vertical position over the pair of wheels farthest from the chimney. They were 11in. diameter, the stroke being 20in. Four of the wheels were provided with springs, but the pair connected to the pistons were not so fitted, the position of the cylinders rendering it impossible for springs to be used. The other improvements to be noted in the construction of the "Royal George" are :—

(1.) Springs instead of weights for the safety-valves.
(2.) The short-stroked pumps.
(3.) Self-lubricating bearings fitted with oil reservoirs.
(4.) The cylinders placed central with the crank journals and the centre of its orbit.
(5.) The first example of six coupled wheels.
(6.) The first really spring-mounted locomotive, the springs performing the double functions of "bearing springs" and "balance beams."

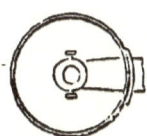

FIG. 11.—HACKWORTH'S BLAST PIPE IN THE "ROYAL GEORGE".

(7) A portion of the exhaust steam used as a jet beneath the fire-grate and part also for heating the feed water; and last and most important—so important, indeed, that it has been described as the "life-breath of the high-pressure locomotive"—*the Steam Blast*. (Fig. 11.)

Trevithick, Nicholson, Stephenson, Gurney, and others have been credited with the production of this valuable contrivance, but an inquiry into the facts conclusively proves that before Hackworth built the "Royal George," the real nature of the exhaust steam blast was not understood by any of those who have since been credited with the invention.

Doubtless several locomotive experimentalists, after various endeavours to get rid of the spent steam, at last turned the escape pipe into the chimney, as the most practical way of discharging the exhaust steam. Trevithick did so, and George Stephenson and others simply followed Trevithick's example, but knew nothing of the value of the exhaust steam as a means of increasing the heating powers of the locomotive.

The claims of both Stephenson and Trevithick appear to be founded on the use of the words "steam blast" by N. Wood in his "Treatise," when describing the exhaust steam arrangement. This he probably did, not understanding the true nature of the blast, or contracted orifice, as invented by Hackworth.

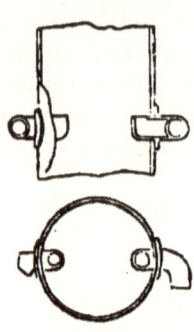

FIG. 12.
WASTE STEAM PIPE
IN STEPHENSON'S
"ROCKET"

It is abundantly evident that Trevithick was absolutely ignorant of the effect of the blast on the fire, for in his patent (No. 2,599) no mention is made of it, although the specification is most minutely drawn; indeed, thirteen years later Trevithick actually patented "fanners, etc., for creating an artificial draught in the chimney." Nicholson, in his patent (No. 2,990) dated November 22nd, 1806, also says, "The steam must be high pressure; *the steam draught cannot be produced by exhaust steam.* This clearly shows he was not aware of the exhaust steam blast; indeed, he expressly states that *exhaust steam cannot be used.* With regard to George Stephenson, the fact that as late as July 25th, 1828, he wrote to Timothy Hackworth, "We have tried the new locomotive engine ('Lancashire Witch') at Bolton; we have also tried the blast to it for burning coke, and I believe it will answer. *There are two bellows, worked by eccentrics underneath the tender.*" It will, therefore, be observed that Stephenson's "blast" was produced by bellows. This letter was written ten months after Hackworth had successfully used the steam blast in the "Royal George."

It will be shown later that it was only at the Rainhill trials, in October, 1829, that Stephenson learned Hackworth's secret of the blast pipe. Although Gurney, in 1822, used a coned pipe, he expressly states that the steam must be continuously ejected at a high velocity from a high-pressure boiler, which distinctly shows he did not use exhaust steam as Hackworth did.

Walker and Rastrick were the engineers engaged by the directors of the Liverpool and Manchester Railway to report on the advantages to be gained from the adoption of stationary or locomotive engines on the Liverpool and Manchester Railway. They decided in favour of the former, but they stated in their report, "Hackworth's engine ('Royal George') is undoubtedly the most powerful that has yet been made, as the amount of tons conveyed by it, compared with the other engines, proves." The first year's work of the "Royal George" consisted of conveying 22,442 tons of goods 20 miles, at a cost of only £466, whilst the same amount of work performed by horses cost £998, showing a saving by the use of the "Royal George" of £532 in one year. The "Royal George" was numbered, 1½, in the books of the S. and D. R.

This was the first time that a locomotive engine had worked for a whole year at a cheaper rate than horses. Upon this result being known to the Stockton and Darlington Railway directors, one of them exclaimed, "All we want is plenty of Timothy's locomotives." The "Royal George" worked night and day upon the Stockton and Darlington Railway until December, 1840, when she was sold to the Wingate Grange Colliery for £125 *more than her original cost.*

R. Stephenson and Co. in 1828 supplied a six-wheel coupled engine, "Experiment," to the Stockton and Darlington Railway. This locomotive had inclined outside cylinders, 9in. diameter, with a stroke of 24in.; the wheels were 4ft. diameter. This engine did not give nearly so satisfactory results as Hackworth's "Royal George."

Reference must here be made to Stevenson's locomotive, "Twin Sisters," used in the construction of the Lancashire and Manchester Railway. She had two fire-boxes and boilers, and two chimneys; she was supported on six coupled wheels of 4ft. diameter; the cylinders were outside in an inclined position. The "Lancashire Witch" (previously mentioned) was built by Stephenson and Co. in 1828 and sold to the Bolton and Leigh Railway. She was supported on four coupled wheels, 4ft. diameter; the cylinders were outside, 9in. diameter, fixed in an inclined position, projecting over the top and at the rear of the boiler. The engine is only mentioned for the purpose of noticing the fact that the fire was urged by means of bellows, worked by eccentrics fixed on the leading axle of the four-wheeled tender, which was specially built with outside frames for the purpose of allowing sufficient room to locate the bellows, etc. Yet some people have assurance enough to state that at the time Stephenson built this engine, and provided it with bellows for the purpose of urging the fire, he was fully acquainted with the nature and advantages of the steam blast!!

In the South Kensington Museum there is preserved the "Agenoria," a locomotive built for the Shutt End Railway by Foster, Rastrick and Co. in 1829, the engine being put to work on June 2nd in that year. It is a four-wheel engine, with vertical cylinders, 7½in. diameter, placed at the fire-box end; the stroke is 3ft., and the motion is taken from two beams fixed over the top of the boiler, which is 10ft. long and 4ft. diameter. The slide valve eccentrics are loose upon the axle, and to enable the engine to work both ways a clutch is provided, as also is hand gear to the valves, to enable the axle to make a half turn, and so bring either the forward or backward clutch into action. The chimney was of abnormal height. The "Agenoria" worked for some thirty years.

In 1829 R. Stephenson and Co. supplied an engine named "Rocket," No. 7, to the Stockton and Darlington Railway, similar in general design to "Experiment," No. 6 (already referred to). This engine was delivered at the time Hackworth was attending the Rainhill locomotive contest, and a director of the Stockton and Darlington Railway wrote to Hackworth, describing the shortcomings of this engine as follows:—"The new one last sent was at work scarcely a week before it was completely condemned and not fit to be used in its present state. The hand gear and valves have no control in working it. When standing without the wagons at Tully's a few days ago it started by itself when the steam was shut off, and all that Jem Stephenson could do he could not stop it; it ran down the branch with such speed that old Jem was crying out for help, everyone expecting to see them dashed to atoms; the depôts being quite clear of wagons, this would have been the case had not the teamers and others thrown blocks in the way and fortunately threw it off. A similar occurrence took place on the following day in going down to Stockton. As soon as the wagons were unhooked at the top of the run, away goes 'Maniac,' defying all the power and skill of her jockey, old Jem; nor could it be stopped until it arrived near the staiths. Had a coach been on the road coming up, its passengers would have been in a most dangerous position. The force-pump is nearly useless, having had, every day it was at work, to fill the boiler with pails at each of the watering-places. No fewer than three times the lead plug has melted out. This 'Maniac' was a Forth Street production, and at last was obliged to be towed up to the 'hospital' by a real 'Timothy' in front, on six wheels, and actually had twenty-four wagons in the rear as guard. It is now at head-quarters at Shildon."

Such was the opinion expressed by a director of the Stockton and Darlington concerning a Stephenson locomotive!

## CHAPTER III.

The Liverpool and Manchester Railway Locomotive Competition—The conditions of the contest—The competitors—The " Novelty "—The " Sanspareil "—The secret of the steam-blast stolen—Mr. Hick's history of the " Sanspareil "—The " Rocket "—Colburn's comparison of the " Rocket " and " Sanspareil "—Booth's tubular boiler fitted to the " Rocket "—The prize divided—History of the " Rocket "—The " Perseverance " wihtdrawn from competition—The " Cycloped " horse-propelled locomotive—Winan's manumotive vehicles for the Liverpool and Manchester Railway—The directors purchase a dozen.

ALTHOUGH Walker and Rastrick had reported to the directors of the Liverpool and Manchester Railway in favour of stationary engines, there were some of them who were enlightened enough to be desirous of giving steam locomotives a fair trial. The Stephensons being locomotive engine builders, naturally were not behindhand in fully and frequently describing the superiority of locomotive traction. Finally, at the suggestion of Mr. Harrison, the directors offered a prize of £500, to be awarded to the locomotive that at the trial appeared to be the best machine competing. The following is a copy of the notice detailing the conditions of the competition:—

" Railway Office, Liverpool, 25th April, 1829.

" Stipulations and Conditions on which the Directors of the Liverpool and Manchester Railway offer a premium of £500 for the most improved Locomotive Engine:—

" 1st. The said engine must effectually consume its own smoke, according to the provisions of the Railway Act, 7, George IV.

" 2nd. The engine, if it weighs six tons, must be capable of drawing after it, day by day, on a well-constructed railway, on a level plane, a train of carriages of the gross weight of twenty tons, including the tender and water tank, at a rate of ten miles per hour, with a pressure of steam on the boiler not exceeding fifty pounds per square inch.

" 3rd. There must be two safety-valves, one of which must be completely out of the control of the engineman, and neither of which must be fastened down while the engine is working.

" 4th. The engine and boiler must be supported on springs, and rest on six wheels, and the height from the ground to the top of the chimney must not exceed fifteen feet.

"5th. The weight of the machine, with its complement of water in the boiler, must at most not exceed six tons; and a machine of less weight will be preferred if it draw after it a proportionate weight; and, if the weight of the engine, etc., does not exceed five tons, then the gross weight to be drawn need not exceed fifteen tons, and in that proportion for machines of still smaller weight; provided that the engine, etc., shall still be on six wheels, unless the weight (as above) be reduced to four tons and a half or under, in which case the boiler, etc., may be placed on four wheels. And the Company shall be at liberty to put the boiler, fire-tube, cylinders, etc., to a test of pressure of water not exceeding 150 pounds per square inch, without being answerable for any damage the machine may receive in consequence.

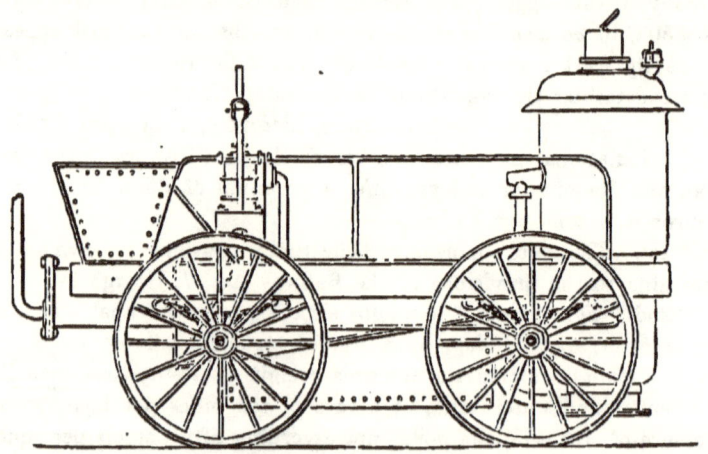

FIG. 13.—THE "NOVELTY," ENTERED BY BRAITHWAITE AND ERICSSON FOR THE RAINHILL PRIZE

"6th. There must be a mercurial gauge affixed to the machine with index rod showing the steam pressure above forty-five pounds per square inch.

"7th. The engine to be delivered complete for trial at the Liverpool end of the railway not later than the 1st of October next.

"8th. The price of the engine which may be accepted not to exceed £550, delivered on the railway, and any engine not approved to be taken back by owner.

"N.B.—The Railway Company will provide the engine tender with a supply of water and fuel for the experiment. The distance within the rails is four feet eight inches and a half."

At this period there were but few men who understood even the outlines of locomotive construction, and unfortunately all of these did not take part in the competition. The fifth condition, limiting the weight of the loaded locomotive to six tons, probably deterred some makers from competing. Others had commenced constructing locomotives for the competition, but were unable to finish them by the date mentioned in the conditions.

The actual entries were as follows: 1. Braithwaite and Ericsson's "Novelty"; 2. Timothy Hackworth's "Sanspareil"; 3. R. Stephenson's "Rocket"; 4. Burstall's "Perseverance"; and 5. Brandreth's "Cycloped."

The "Novelty" (Fig. 13) was far and away the favourite engine at Rainhill, its neat appearance and smartness attracting universal attention. It was a "tank" engine, and probably the first locomotive constructed to carry its supply of water and coal on the engine, being thus complete without a tender. This raised a difficulty in apportioning the load, as in the conditions it was arranged that the tender was to be counted as part of the load hauled. The machine with water and coal weighed 3 tons 17 cwt. 14 lb.; the allowance made for the tender and fuel reduced the theoretical weight of the "Novelty," as an engine only, to 2 tons 13 cwt. 2 qr. 3½ lb.; the gross weight hauled, including the locomotive, being 10 tons 14 cwt. 14 lb.

The "Novelty" was first tried upon October 10th, 1829—she had not previously been upon a railway—and it was found necessary to make some alterations to her wheels. Timothy Hackworth, although he had an engine running in competition with the "Novelty," generously offered to repair the defect, and he personally took out the broken portion, welded it, and replaced it in position with his own hands.

The trials were conducted upon a level portion of line at Rainhill, on a course only one and a half miles in length, and at either end an additional eighth of a mile was allowed for the purpose of getting up the speed and stopping after the run of a mile and a half. The engines had to make forty runs over the course, or a distance of sixty miles in all, which was computed to be equal to a return journey between Liverpool and Manchester.

After running two trips of one and a half miles each, the pipe from the pump to the boiler burst, in consequence of the cock between the boiler and pump having, by accident, been closed. The "Novelty" and train covered the first trip in five minutes thirty-six seconds, and the return in six minutes forty seconds; being at the rate of 16.07 and 13½ miles an hour respectively. After being repaired, the engine, with its train, made an unofficial trip, and developed a speed of 21 1-6 miles an hour. Without a load the "Novelty" attained a speed of nearly thirty miles an hour.

The "Novelty" was again tried on October 14th, but upon its third trip part of the boiler gave way, and it was decided to withdraw the locomotive from competition.

The boiler of the "Novelty" was partly vertical and partly horizontal; the latter portion was about 12ft. long and 15in. in diameter. In the former was the fire-box, surrounded by water, coke being supplied through what at first might be mistaken for the funnel of a steam fire-engine. This was, however, kept air-tight, the fuel being introduced by means of a descending hopper. The area of the fire-grate was 1.8 sq. ft., the fire-box heating surface 9½ sq. ft., and the heating surface of the tubes, 33 sq. ft.

The air entered below the fire-bars by a pipe traversing the length of the engine, and connected with bellows fixed above the frame at the other extreme of the engine. The bellows were worked by the engine, so that the "Novelty" was provided with a forced draught. The heated air was forced through a tube, which made three journeys through the horizontal portion of the boiler, and was consequently 36ft. in length. It was 4in. in diameter at the grate end, and 3in. at the other extreme, where it was turned up as a chimney. The cylinders were located over the pair of wheels at the bellows end of the machine. They were fixed vertically, the diameter being 6in., and length of stroke 12in. The piston rods worked through the top covers, and by means of cross-heads, side-rods, and bell-cranks the motion was conveyed to the crank axle beneath the vertical portion of the boiler, although, as previously mentioned, the cylinders were over the other pair of wheels. The wheels were 4ft. 2.1in. in diameter, and chains were provided for coupling the wheels together; but these were not used at Rainhill.

The water was carried in a tank located between the axles below the frame. The construction of the "Novelty" was only decided upon on August 1st, 1829, but so expeditiously was the work carried out that she was constructed in London and delivered in Liverpool—a lengthy journey at that time—by September 29th, 1829. Her distinguishing colours at Rainhill were copper and blue.

After the conclusion of the Rainhill Competition several alterations were made in the design of this engine, the position of the cylinders being altered from vertical to horizontal by Watson and Daglish, and in 1833 she was working on the St. Helens and Runcorn Gap Railway.

Although, through an accident, the "Novelty" had to be withdrawn from competing for the prize at Rainhill, the directors of the

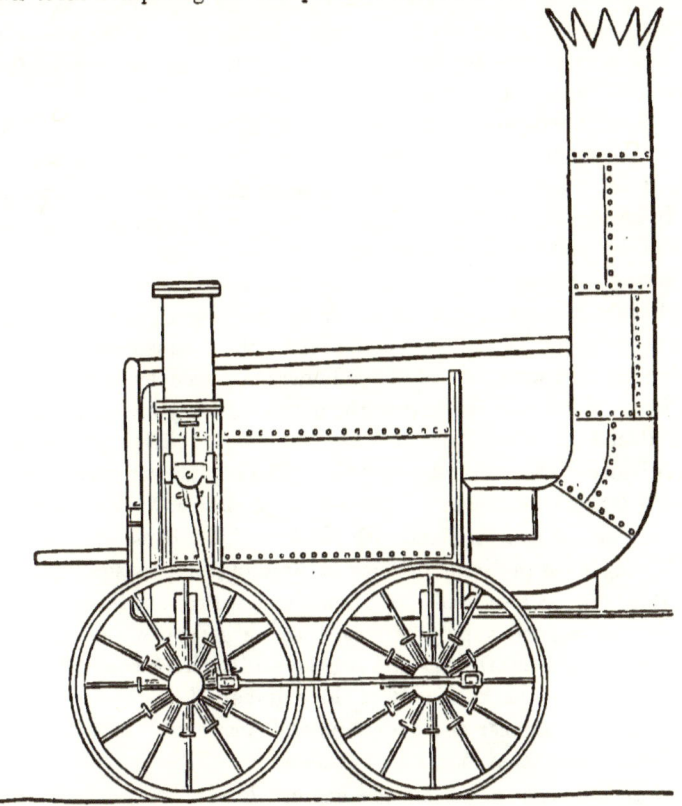

Fig. 14.—HACKWORTH'S "SANSPAREIL," ONE OF THE COMPETITORS AT RAINHILL

Liverpool and Manchester Railway were so well satisfied with her performances that they gave Braithwaite and Ericsson an order for some locomotives of the same design. A description of these will be found in Chapter IV.

The engine next in order was Timothy Hackworth's "Sanspareil," (Fig. 14), now preserved in the South Kensington Museum.

## EVOLUTION OF THE STEAM LOCOMOTIVE

The engine-shops at Shildon were not in a position to construct the whole of this locomotive; consequently Hackworth was forced to obtain the boiler and cylinders from other makers. The former was constructed at Bedlington Ironworks, and was of cylindrical form, 6ft. long, 4ft. 2in. diameter, with one end flat and the other hemispherical. The heating surface was provided by means of a double return tube, the fire-grate and chimney being both at the same end.

The area of the fire-grate was 10 sq. ft., the heating surface of same 15.7 sq. ft., the remaining heating surface $74\frac{1}{2}$ sq. ft.

The cylinders were constructed by R. Stephenson and Co., and six had to be made before two perfect ones were obtained, the sixth one, indeed, only being fitted at Liverpool when the contest was in progress.

It has been stated that these cylinders were purposely constructed in a faulty manner to prevent the "Sanspareil" beating the "Rocket." This may or may not be true, but it is very evident that, save for Stephenson's imperfect workmanship in this respect, the "Sanspareil" would have won the £500 prize. When the "Sanspareil" was competing for the prize, one of the cylinders supplied by Stephenson and Co. burst, and it was found that the metal was only one-sixteenth of an inch thick! A nice state of things certainly! The cylinders were 7in. diameter, the stroke being 18in. The engine was carried on four wheels, 4ft. 6in. diameter. Total weight of engine, 4 tons 15 cwt. 2 qr.

She was, of course, fitted with Hackworth's exhaust steam blast.

During some preliminary trips at Rainhill, Stephenson was greatly surprised to see how well the "Sanspareil" ran, and he noticed she always had a good supply of steam, so he got upon the engine and had a ride on her. During this trip he said to Hackworth, "Timothy, what makes the sparks fly out of the chimney?" Hackworth touched the exhaust pipe near the cylinders, and answered, "It is the end of this little fellow that does the business."

After Stephenson got off the engine, John Thompson, the driver (he was Hackworth's foreman at Shildon), said to Hackworth, "Why did you tell him how you did it, sir? He will be trying to fit up the 'Rocket' in the same way." Hackworth said he did not think so, but Thompson determined to watch the "Sanspareil" all night. He therefore locked himself in the shed containing the engine that night, but towards daybreak sleep overcame him, and when he awoke he saw two men getting out of the window of the shed, and he found the chimney door of the "Sanspareil" open, and some materials inside the chimney. The secret of the exhaust steam blast was stolen! The next evening the "Rocket" again appeared; this time she was fitted with a similar contrivance. The above is Hackworth's foreman's version of the theft, but the "Practical Mechanic's Journal" for June, 1850, gives the tale as told by the man who committed the theft.

When in repair, the "Sanspareil" ran faster, took a heavier load, and consumed less coke than the "Rocket," and whilst the latter was remodelled within twelve months of the Rainhill contest, the former worked with practically no alteration until 1844. In 1864 she was presented to the South Kensington Museum by Mr. John Hick, M.P., Bolton.

The following is an extract from her history, as supplied by Mr. Hick to the Museum authorities:—

"After the Rainhill trial the engine was purchased by the Liverpool and Manchester Railway Company, and used by them for various purposes. In 1831, the engine was purchased by Mr. John Hargreaves, of Bolton, and was employed by him in the conveyance of passengers and general traffic on the Bolton and Leigh Railway for several years. In 1837, Mr. Hargreaves had the engine thoroughly repaired, and put on a pair of new cylinders of larger dimensions than the old ones, so as to increase the power. The original wood-spoked wheels were also removed at this time, and replaced with cast-iron hollow-spoked wheels.

"One pair of these are under the engine at the present time. The engine continued regularly at work in conveying coals, general goods, and passengers until 1844, when, being found much too small and short of power for the rapidly increasing traffic, Mr. Hargreaves took her to his colliery at Coppull, near Chorley, Lancashire, where the engine was fixed near a coal-pit. One axle and one pair of wheels were removed, and upon the other toothed gear was fitted, in order to give motion to winding and pumping apparatus, and the engine commenced its work as a regular fixed colliery engine, pumping and winding in the most satisfactory manner until the end of the year 1863; having raised many thousand tons of coal and many million gallons of water, and even at the time above named was in fair working order, and only removed because the coal in the pit was exhausted.

"I hope the old engine will now find a permanent resting place in the Kensington Museum, where her end will be peace, if not pieces. Mr. Hargreaves has kindly given me the old engine, in consequence of my having told him of my intention with regard to her. And having restored her as far as possible by collecting and putting together the available materials, I have pleasure in presenting this interesting relic to the Museum."

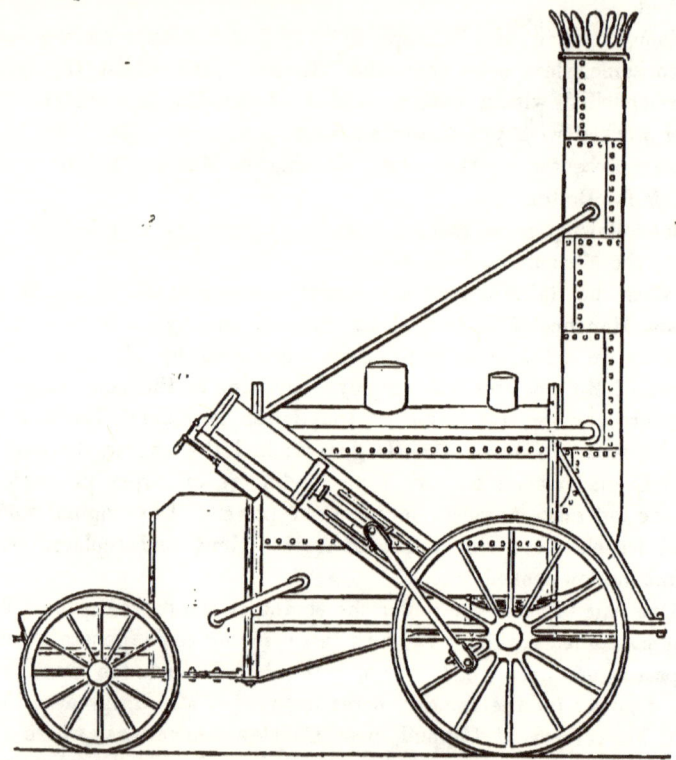

FIG. 15.—STEPHENSON'S "ROCKET," THE WINNER OF THE RAINHILL PRIZE OF £500.

The "Rocket" (Fig. 15), was entered in the name of Robert Stephenson, and was constructed at the Forth Street Works, Newcastle-on-Tyne, in 1829. Her distinguishing colours were yellow and black, with a white chimney. She was the first engine to be tried at Rainhill. Her weight was 4 tons 5 cwt.; load (including tender), 12¾ tons; total, 17 tons. During the first twenty trips she attained a maximum speed of 24.43 miles an hour, the average being 13.42 miles an hour; during the second twenty trips an average speed of 14.2 miles an hour was the result, with a maximum speed of 24 miles an hour. These short trips of one and a half miles each just suited the design of the "Rocket," as the steam raised before starting on each trip was sufficient to work her the one and a half miles; had the trips been longer, she—not then being furnished with a proper blast, but with that illustrated on Page 25—would probably have stopped for want of steam. Of this Z. Colburn significantly states: "The 'Rocket,' on the first day of her trial, derived but little benefit from the discharge of the exhaust steam up the chimney; and, indeed, made steam nearly as freely when standing as when running." Without a load, or tender even, she attained a speed of 29½ miles an hour. The authority just quoted says: "The real power of the 'Sanspareil' is to be estimated by its rate of evaporation, which was one-third greater than that of the 'Rocket,' and thus the 'Sanspareil,' after allowing for its greater weight, was the most powerful engine brought forward for trial. . . As far as it had gone, the mean rate of speed (of the 'Sanspareil') was greater than that of the 'Rocket' up to the same stage of the experiment."

The boiler of the "Rocket" was cylindrical, with flat ends, 6ft. long, 3ft. 4in. diameter; the fire-box was 3ft. long, 2ft. broad, and about 3ft. deep; between the box and the outer casing was a space of 3in. filled with water. The cylinders were placed at an angle of 45 degrees at the fire-box end, the connecting-rod being attached to a pin on the leading wheels, which were 4ft. 8½in. diameter, that of the cylinders being 8in.; the stroke was 16½in.

The "Rocket" had a great advantage over other engines because she was supplied with a tubular boiler, containing 25 tubes of 3in. diameter. The idea of the tubular boiler did not originate with the Stephensons. Mr. Booth, the Secretary of the Liverpool and Manchester Railway, suggested their use in the "Rocket"; but before this the tubular locomotive boiler had been patented by a Frenchman (M. Sequin), on February 22nd, 1828. Mr. Booth, however, states that he was unaware of the French patent, and, so far as he was concerned, the tubular boiler was an original discovery. The use of these tubes increased the evaporating power of the boiler three-fold, and at the same time reduced the consumption of coke 40 per cent.; yet the "Rocket," with this great advantage, was not equal to the "Sanspareil," until the former was fitted with Hackworth's blast. When this had been done, the "Rocket" was capable of hauling 20 tons (engine included) up an incline of 1 in 96, at 16 miles an hour, for a distance of one and a half miles. The prize of £500 was divided between Robert Stephenson, the constructor of the "Rocket," and Mr. Booth, the suggester of the tubular boiler, which enabled that locomotive to be entitled to the prize. Tubular boilers had been successfully used in steam road coaches as early as 1821.

After running a year or so, the "Rocket" was re-built, the cylinders being placed in a slightly inclined position over the trailing wheels, but still working the leading wheels; a smoke-box was added, and other improvements introduced.

The "Rocket" was bought in the year 1837, from the Liverpool and Manchester Railway, by Mr. J. Thompson, of Kirkhouse, the lessee of the Earl of Carlisle's coal and lime works.

Here the engine was worked for five or six years on the Midgeholme line—a local line belonging to Mr. Thompson—for forwarding his coals from the pits towards Carlisle.

Soon after the engine was placed on this line the great contest for East Cumberland took place, when Sir J. Graham was superseded by Mayor Aglionby, and she was used for conveying the Alston express with the state of the poll from Midgeholme to Kirkhouse. Upon that occasion the "Rocket" was driven by Mr. Mark Thompson, and accomplished her share of the work, a distance of upwards of four miles, in 4½ minutes, thus reaching a speed nearly equal to 60 miles an hour. On the introduction of heavier and more powerful engines, the "Rocket" was "laid up in ordinary" in the yard at Kirkhouse. This historic steam locomotive is now preserved in the South Kensington Museum. It must not be forgotten, however, that the "Rocket" has been rebuilt, and its design considerably altered, since the Rainhill competition of 1829.

The last of the steam locomotives entered for trial at Rainhill remains to be described. The "Perseverance" was constructed by Mr. Burstall, of Edinburgh. He was already known as a maker of steam road coaches. Unfortunately for the success, or rather want of success, of the "Perseverance," Mr. Burstall designed his railway locomotive on much the same lines as his steam coaches.

The "Perseverance" had the misfortune to have some damage done to its wheels, etc., when being unloaded at Rainhill off the wagon on which it had been conveyed from Liverpool. A preliminary trial was made, and Mr. Burstall, finding the engine was unable to attain a higher speed than about six miles an hour, withdrew his locomotive from competition.

The boiler was horizontal, and the water was admitted to shallow trays placed over the fire, and in this way was immediately converted into steam. The cylinders were vertical, and worked horizontal beams placed above them; the wheels were worked by cranks fixed on the beams about half-way between the cylinders and the centre pivots of the beams. The second pair of wheels was driven by means of an axle with bevel wheels at each end, which conveyed the motion from the one axle to the other.

This engine was distinguished by having the wheels painted red.

Although not "steam" locomotives, we think it right to give a few details of the "Cycloped," (Fig. 16), and also of Winans' manumotive carriage, both of which were exhibited at Rainhill. The former was

FIG. 16.—WINANS' "CYCLOPED" HORSE LOCOMOTIVE

worked by a horse or horses fastened on a frame supported by four wheels; the horses walked at a speed of one and a quarter miles an hour, on an endless platform formed of planks of wood. The horses being firmly attached to the frame could not go forward when they essayed to walk, and the consequence of their using their legs was the revolving of the floor, which worked round drums geared to the driving wheels. This motion caused the vehicle to move forward on the rails at a speed of about three miles an hour, with a load of fifty passengers. Had the horses moved at a quicker rate, the speed of the "Cycloped" would have been increased in a proportionate ratio.

Winans' carriage was worked by two men, who turned a windlass, which actuated the wheels. It accommodated six passengers, and it was facetiously proposed that those passengers who worked at the windlasses should be conveyed by such vehicles at reduced rates. Although we now smile at the simplicity of such vehicles ever having been suggested for working on a railway, the Directors of the Liverpool and Manchester Railway were considerably taken with the idea of Winans' man-propelled carriages, and they engaged two well-known engineers to report on their adaptability for passenger traffic on the railway. As might be expected, the experts reported against the proposed use of Winans' machines; but, despite this adverse report, the Directors of the Liverpool and Manchester Railway actually bought twelve of these "manumotive" carriages of Winans. The purchase was made prior to March, 1830, and as we do not read of their being used after the railway was opened in September, 1830, we may conclude that during the six months that elapsed between the purchase and the opening of the line the Directors had come to the same conclusion regarding the machines as did the engineers who reported against their use on the railway.

## CHAPTER IV.

An important improvement in the locomotive—Bury's original "Liverpool," the first inside cylinder engine—Bury's own account of his invention—Other authorities agree with Bury—Extract, supplied by the Secretary of the L. & N.W.Rly., from the minute books of the Liverpool and Manchester Rly.—An early authentic list of Bury's locomotives—Description of Bury's "Liverpool"—Last heard of on the Bolton and Kenyon Railway—The "Invicta" for the first Kentish railway—Still preserved by the S.E.R.—First official trip on the Liverpool and Manchester Railway—Formal opening of the L. & M.R.—The locomotives that took part in the ceremony—The "William the Fourth" and "Queen Adelaide" for the L. & M.Rly.—Hackworth's "Globe" for the Stockton and Darlington Railway—The romance of her construction, life, and end—Stephenson's "Planet"—Some of her feats on the L. & M.Rly.—Heavier locomotives for the L. & M.Rly.—Dodd's engine for the Monkland and Kirkintilloch Rly.—Historical locomotive sold by auction for 20 guineas—Bury's "Liver" for the L. & M.Rly.—More Hackworth "iron horses" for the Stockton and Darlington Rly.—Despite their peculiarities, they prove most successful—The "Caledonian."

We have now to deal with one of the most important improvements in the locomotive—viz., that introduced by Mr. Edward Bury, of the Clarence Foundry, Liverpool, in the design of his celebrated "Liverpool," (Fig. 17). Of late years many extraordinary statements concerning various types and designs of locomotives have been made, and the "romancing" relative to the original "Liverpool" is perhaps the most conspicuous, whilst at the same time its utter incorrectness is easily proved.

One of these statements is that "the first engine built by Bury at Clarence Foundry was an outside cylinder engine, the 'Dreadnought,' which was completed March 30th, 1830, but proved a failure. However, he lost no time, but, with the assistance of his foreman, Mr. Kennedy, got out working drawings for a new engine, to be named the 'Liverpool.' This engine, No. 2 in the locomotive order book, and class A in the description book, was commenced early in January, 1831; it was completed in March of that year, and in May, 1831, it was put to work on the Petersburg Railroad of America. It had four coupled wheels of 4ft. 6in. diameter."

Now, as to the facts, Bury's books were sold by auction by his creditors on August 15th and 16th, 1851; and, even if the books are now in existence (which is extremely unlikely), it is obviously impossible for them to contain the particulars quoted above, for the very simple and conclusive reason that the facts relative to the original "Liverpool" are quite different to the statement just quoted.

There are three improvements with which Bury is justly credited in the locomotive now under review—viz., the adoption of (1) horizontal inside cylinders below the smoke-box, (2) cranked driving axle, and (3) coupled driving wheels of the (then) great diameter of six feet.

In describing this historical "Liverpool" locomotive we cannot do better than quote Bury, the maker and designer of it, and Kennedy, his foreman, who constructed it. The former, at a meeting of the Institute of Civil Engineers, held on March 17th, 1840, read a paper on the locomotive, and, speaking of the inside cylinder engine, said: "This form of engine was adopted by the author as early as 1829, when he constructed the 'Liverpool,' which was the original model engine, with horizontal cylinders and cranked axles. It was set to work on the Liverpool and Manchester Railway in July, 1830."

FIG. 17.—BURY'S ORIGINAL "LIVERPOOL," THE FIRST ENGINE WITH INSIDE CYLINDERS AND CRANKED DRIVING AXLE. COUPLED WHEELS, 6ft. DIAMETER.

About 1843 there was considerable discussion amongst engine builders and locomotive engineers as to the relative safety of inside and outside cylinder engines, and also regarding the superiority of the four-wheel or six-wheel locomotive. Bury and Co. thereupon issued a circular giving a history of the locomotive practice of their firm, and the various advantages claimed for their locomotive designs.

From this circular we extract the following remarks, as bearing upon the point now under discussion:—"It was the good fortune of the conductor of this foundry to originate the construction of four-wheel engines, with inside framing, crank axles, and cylinders placed in the smoke-box. . . . The first engine on this principle was manufactured in this foundry in 1829, prior to the opening of the Liverpool and Manchester Railway." Such are Mr. Bury's statements concerning the original "Liverpool."

We will now see what his partner, Mr. James Kennedy, the then President of the Institution of Mechanical Engineers, had to say regarding the "Liverpool."

At a meeting of the Institute of Civil Engineers, held on November 11th, 1856, a communication was read from Mr. Kennedy, in which he stated that "the late George Stephenson had told both Bury and Kennedy, after having seen the 'Liverpool' engine on the Liverpool and Manchester Railway, that his son, the present Robert Stephenson, had taken a fancy to the plan of the 'Liverpool,' and intended to make immediately a small engine on the same principle." This he afterwards did, Stephenson's "Planet" being the said engine "on the same principle." Kennedy went on to state that "the letter-book of the firm (Bury and Co.) for the year 1830 contained the whole of the correspondence on the subject between the Directors of the Liverpool and Manchester Railway and Bury."

The reader can readily judge as to which statement is likely to be correct—those of such well-known men as Bury and Kennedy, which are concise, straightforward statements of known and accepted facts, or the recently published remarks concerning the "books, etc."

Fortunately, students of locomotive history are not even obliged to decide either one way or the other on the statements pro and con already quoted concerning the original "Liverpool," but are able to gain independent and conclusive evidence on this important point in locomotive history. For the purpose of finally clearing up the point, the writer communicated with the Secretary of the London and North Western Railway, asking him to examine the Directors' Minute Books of the Liverpool and Manchester Railway for the year 1830, to see if these authentic documents contained any reference to Bury's "Liverpool." Mr. Houghton most generously had the search we required made, and the result was as might have been expected. But let the letter tell its own tale.

"London and North Western Railway,
"Secretary's Office, Euston Station, N.W.
"June, 3rd, 1896.

"Dear Sir,—With further reference to your request for information relative to Bury's locomotives, I have had the Minute Books of the Liverpool and Manchester Railway searched for the years 1829-30.

"Towards the end of 1830 the Board sanctioned the [further] trial of the 'Liverpool,' and it was consequently allowed to work on the railway in competition with one of Mr. Stephenson's engines. The engineer was dissatisfied with the size of the wheels, which were 6ft. instead of his maximum 5ft.; and there was a long controversy as to the respective merits of circular and square fire-boxes, which was ultimately referred to arbitration, when the square boxes recommended by Mr. Stephenson were given the preference.—Yours truly, "(Signed) T. HOUGHTON."

The above letter conclusively settles the points in dispute—viz., that the "Liverpool" was tried on the Liverpool and Manchester Railway in, 1830, and that the diameter of the wheels was 6ft.

We have thus pricked the specious bubble that stated the "Liverpool" was duly commenced to be built in 1831, and that the diameter of the wheels was but 4ft. 6in.!

Readers may wonder why such obviously inaccurate statements should be published. One can only conjecture. Many lists of early locomotives have during the past few years been published. These should, however, be accepted with the very greatest caution. The following table of dimensions of Bury's early engines appeared as long ago as September 18th, 1857, in the *Engineer*. As this was nearly forty years before "locomotive lists." had any marketable value, there can be no reason to call in question its accuracy:—

| No. of Engines | Diameter of Cylinder. | Length Stroke | Diameter of Wheels. | No. of Tubes. | Diameter of Tubes. | No. of Tubes. | Diameter of Tubes. | Length of Tubes. | Inside Diameter of Fire-box. | Area of Tubes. | Total Surface. |
|---|---|---|---|---|---|---|---|---|---|---|---|
| No. | Ins. | Ins. | Ft. Ins. | No. | Ins. | No. | Ins. | Ft. Ins. | Ft. Ins. | Inches. | Sq. Feet |
| 2 | 12 | 18 | 6 0 | 79 | 1⅜ | 52 | 1¼ | 7 11½ | 3 9 | 19 | 450·25 |
| 3 & 4 | 11 | 16 | 5 0 | 73 | ⅞ | 24 | 1¼ | 7 1 | 3 0 | 16⅜ | 303·58 |
| 5 | 8 | 16 | 4 6 | 40 | 1¼ | 13 | 1¼ | 6 5 | 2 7½ | (6) 12¼(7) | 160·28 |
| 6 & 7 | 12 | 18 | 5 6 | 76 | 2 | 13 | 1¼ | 8 6 | 3 7 | 18¼ 181/16 | 390 |
| 8 & | 9 | 16 | 5 0 | 40 | 1¼ | 13 | 1¼ | 6 11 | 2 10¼ | 12⅞ | 162 |
| 10 | 9 | 16 | 4 4 | 43 | 2 | 0 | 1¼ | 7 1 | 3 0 | 14¼ | 169·1 |
| 11 | 8 | 16 | 4 6 | 40 | 1¼ | 13 | 1¼ | 6 7¼ | 2 9¼ | 12¼ | 155·2 |
| 12 & 13 | 10 | 16 | 4 6 | 51 | 2 | 0 | 1¼ | 7 2¼ | 3 0¼ | 15¼ | 222·74 |
| 14 | 8 | 16 | 4 6 | 43 | 1¼ | 9 | 1¼ | 6 8 | 2 10¼ | 12¼ | 155 |
| 15, 16, & 17 | 12 | 18 | 4 6 | 58 | 2¼ | 2 | 2 | 8 2 | 3 7 | 19¼ 19 1/19 | 318 40 |
| 18 & 19 | 10 | 16 | 4 6 | 76 | 2 | — | 1⅜ | 7 2½ | 3 2 | 15¼ | 251 |

At the present time there exists a market for early locomotive details; as with other marketable commodities, given a demand, a supply (of some kind) will be forthcoming.

We have a copy of the report prepared by the arbitrators, appointed by the Directors of the Liverpool and Manchester Railway, to inquire into the question of the round or square fire-box, as mentioned in Mr. Houghton's letter. The report was made by John Farey and Joshua Field, two celebrated engineers of that period, and was in favour of the square fire-box.

It will now be of interest to give a description of Bury's original "Liverpool," which was designed and her construction commenced in 1829. She contained many unusual features. Instead of a tubular boiler a number of convoluted flues were used. The fire was urged by bellows fixed under the tender; the driver stood at one end of the engine in front of the smoke-box, and the fireman at the other end, behind the fire-box; the cylinders were horizontal, placed inside the frames beneath the smoke-box; their diameter was 12in., the stroke being 18in.; the four wheels were 6ft. in diameter, and were coupled, and the driving axle was, of course, cranked.

The "Liverpool," in this her original state, was used as a ballast engine in the construction of the Liverpool and Manchester Railway, but not being very successful, was withdrawn. After some alterations, she was again put to work on July 22nd, 1830. Then the crank axle appears to have broken, and she was again removed for repairs, and again put to work on the Liverpool and Manchester Railway on October 26th, 1830. After the report previously mentioned, the Directors refused to purchase the "Liverpool," and Bury removed her to the Bolton and Kenyon Railway. Here she attained a speed of 58 miles an hour with twelve loaded wagons. On this line one of her wheels broke, and the driver was killed. As a result of this accident, she was then rebuilt and sold to Hargreaves, the contractor, for locomotive power on the Bolton and Kenyon Railway, and continued to work on that line for some years.

The Canterbury and Whitstable Railway was opened on May 3rd, 1830, and was the first locomotive line in the South of England. The original engine, the "Invicta" (Fig. 18), is still preserved by the South Eastern Railway at Ashford, but it is a mere chance that this engine did not disappear nearly sixty years ago. The Canterbury and Whistable Railway Company, after a short time, let the working of the line to contractors, and they preferred to work it by horse-power, and we find that in October, 1839, the contractors were advertising the "Invicta" for sale, describing her as of "12 horse-power, 18in. stroke, cylinders 24in. long, 9½in. diameter, wheels 4ft. in diameter." Fortunately for students of early locomotives, there was no demand for the engine anywhere in the neighbourhood of Whitstable, there then being no other locomotive line nearer than Greenwich, on which she could have been used; so no buyer was forthcoming, and the "Invicta" was thereupon laid up. The dimensions of the "Invicta," as supplied to us by Mr. J. Stirling, are as follows: Cylinders, 10½in. diameter, fixed in inclined position over leading wheels, and working the trailing wheels; stroke, 18in.; four-coupled wheels, 4ft. diameter; wheel base, 5ft.; boiler, 10ft. 5in. long, 3ft. 4in. diameter, containing a

FIG. 18.—THE "INVICTA," CANTERBURY AND WHITSTABLE RWY., 1830

single flue 20in. diameter; distance from top of boiler to rails, 6ft.; from top of chimney to rails, 11ft. 1in.; chimney, 15in. diameter; total length over all, 13ft. 6in. At the bottom of the chimney is a kind of smoke-box, measuring about 2ft. 4in. high, 1ft. 8in. long, and 2ft. 4in. wide. The South Eastern Railway exhibited the "Invicta," at the jubilee of the Stockton and Darlington Railway in 1875, and at the Newcastle Stephenson Centenary in 1881. The "Invicta," when originally built, is said to have had a tubular boiler.

The Directors of the Liverpool and Manchester Railway in 1829 ordered of Stephenson and Co. seven engines of somewhat similar design to the "Rocket." The Directors made their first trip by railway from Liverpool to Manchester and back on Monday, June 14th, 1830. The train was drawn by the "Arrow," and consisted of two carriages and seven wagons; the total weight, including the engine, was 39 tons, the journey to Manchester being made in two hours one minute, whilst the return trip to Liverpool only took one and a half hours, a speed of 27 miles being attained for some distance.

The Liverpool and Manchester Railway was formally opened on September 15th, 1830, when the "Northumbrian" (Fig. 19), driven by George Stephenson, hauled the train consisting of the Duke of Wellington's carriage, the band, etc., on one line, whilst the "Phœnix," "North Star," "Rocket," "Dart," "Comet," "Arrow," and "Meteor," each hauled a train upon the other line. Starting from Liverpool, the eight trains proceeded towards Manchester. At Parkside Mr. Huskisson was run over by the "Rocket," and he was placed on the "Northumbrian" and conveyed to Eccles in 25 minutes, or at the rate of 36 miles an hour.

FIG. 19.—THE "NORTHUMBRIAN," THE ENGINE THAT OPENED THE LIVERPOOL AND MANCHESTER RAILWAY

The Duke of Wellington's carriage was now left without an engine, and a curious sight was witnessed; a long chain was obtained, and the trains which had been up to this point hauled by the "Phœnix" and "North Star," consisting of ten carriages, were joined together. The chain was then fixed to the Duke of Wellington's train on the other line, and so the rest of the journey was performed by the two engines and ten carriages on one line hauling another train upon a parallel set of rails. It may be of interest to observe that the carriage built for the Duke of Wellington was provided with eight wheels, so it will be noticed that eight-wheeled passenger stock is not at all a modern introduction, but, on the contrary, has been in use ever since the opening of the first railway built for the conveyance of passengers. The vehicle in question was 32ft. long and 8ft. wide.

The two engines ordered by the Directors of the Liverpool and Manchester Railway of Braithwaite and Ericsson after the style of the "Novelty," were named "William the Fourth," (by special permission of that monarch) and "Queen Adelaide." They were delivered to the Liverpool and Manchester Railway immediately the railway was opened, and on September 22nd, 1830, the "William the Fourth" ran off the rails on the Sankey Viaduct. A very considerable number of trials were made with these locomotives on the Liverpool and Manchester Railway; but, as was the case with Bury's "Liverpool," Stephenson strongly objected to any other maker's engines being used on the line, and he was, therefore, always ready to find out some fault in the engines not of his construction tendered to the company. Braithwaite and Ericsson claimed four great advantages for their class of engines—viz., (1) the total absence of all smoke; (2) the dispensing with a chimney; (3) a saving of 120 per cent. in the cost of the fuel, and of 30 per cent. in the space required to store it; (4) a saving of 400 per cent. in the space occupied by the boilers.

Several improvements were introduced into the "William the Fourth," and "Queen Adelaide," so that they differed somewhat from the "Novelty." They were provided with four-wheeled tenders, which were placed in front of the engines. The four wheels of the engines were 5ft. in diameter, the wheel base being 6ft. 9in. The horizontal portion of the boiler was 8ft. long, the vertical portion, containing the fire, etc., being 6ft. 6in. high and 4ft. diameter. The cylinders were vertical, but worked downwards; they were located one on each side of the vertical portion of the boiler, and a little to the rear of the leading wheels, to which the motion was conveyed by means of bell-cranks and connecting-rods—the latter joined the axle within the wheels, so that the driving axle was cranked.

The next engine that requires our attention is the celebrated "Globe" (Fig. 20), designed for the Stockton and Darlington Railway by Timothy Hackworth, and built by R. Stephenson and Co. The "Globe" was built for passenger traffic; she was provided with a steam dome, and was the first locomotive built with this advantageous appendage for obtaining dry steam. The valve motion was reversible by a single lever. The heating surface was provided for by means of a single fire-tube, whilst behind the fire-bridge, and extending to the chimney, were seven small radiating tubes crossing the main flue.

This idea of Hackworth's was afterwards introduced by Galloway in his stationary engine boilers, and patented by him. The engine "Globe" had a cranked axle and inside cylinders.

Hackworth explained the construction of the "Globe" to the Directors of the Stockton and Darlington Railway, and he was instructed to go to Newcastle and arrange for the building of the "Globe" by Stephenson and Co. He saw the officials at the Forth

Street Works on March 3rd, 1830, and after the examination of the plans there, it is stated that one of the officials objected to the crank axle, saying "it would certainly involve a loss of power, as the efficient length of lever could only be calculated from the inside of the journal to the axle's centre." It is well known that Geo. Stephenson had previously seen Bury's "Liverpool," and said of it, "My son has taken a fancy to the plan of the 'Liverpool' engine, and intends to make

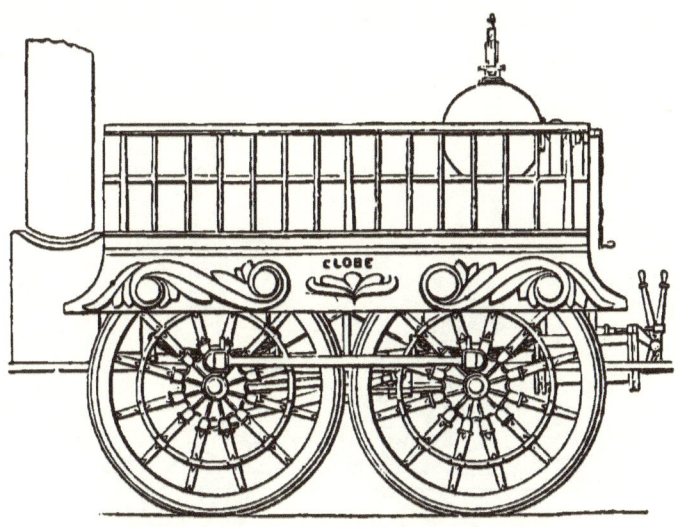

FIG. 20.—HACKWORTH'S "GLOBE" FOR THE STOCKTON AND DARLINGTON RAILWAY. THE FIRST LOCOMOTIVE WITH A STEAM DOME

immediately a small engine on the same principle." Hackworth's reply to the objection to the crank axle was "that he held Stephenson responsible only for supplying good workmanship, and not for any failure of the design, should such occur."

On March 3rd, 1830, Hackworth, in company with Harris Dickinson, one of R. Stephenson and Co.'s foremen, drove over to Bedlington Iron Works to order the boiler plates required for the construction of the "Globe."

Hackworth remained at Newcastle till March 6th, and being satisfied that the construction of the "Globe" would be immediately proceeded with, he returned to Darlington, having obtained a promise of quick delivery. The boiler plates were delivered at the Forth Street Works, April 14th, 1830.

The completion of the engine was, however, delayed until after R. Stephenson and Co. had delivered the "Planet," inside cylinder locomotive, to the Liverpool and Manchester Railway. The "Globe" opened the Stockton and Middlesbrough Branch of the Stockton and Darlington Railway on December 27th, 1830. Her speed frequently exceed 50 miles an hour with passenger trains.

In consequence of a deficiency of water, she blew up in 1839. The engine was provided with a copper globe for the purpose of obtaining dry steam—hence her name "Globe." She had four wheels, each of 5ft. diameter.

Fig. 21.—STEPHENSON'S "PLANET," LIVERPOOL AND MANCHESTER RAILWAY

Stephenson soon put into practice the borrowed idea of inside cylinder locomotives, to his own advantage, and on October 4th, 1830, he delivered the first engine of his construction containing inside cylinders, placed in the smoke-box, as suggested to R. Stephenson by Trevithick. This locomotive was named the "Planet," and was constructed for the Liverpool and Manchester Railway. The cylinders were 11in. diameter, stroke 16in. The boiler was 6ft. 6in. long, 3ft. diameter, and contained 129 tubes. She weighed eight tons; the driving wheels were 5ft. diameter, and were placed just in front of the fire-box. The leading wheels were 3ft. diameter, and projected beyond the front of the smoke-box.

E

The frames were outside the wheels, and were of oak lined with iron plates. As the "Planet" embodied several improvements not before used in the engines constructed by Stephenson for the Liverpool and Manchester Railway, it is natural that the locomotive should be able to perform better service than the earlier ones. On November 23rd, 1830, she conveyed a train from Manchester to Liverpool in one hour, including a stop of two minutes for water.

On December 4th, 1830, the "Planet" (Fig. 21) hauled a mixed train, weighing 76 tons without the engine and convoy (tender) from Liverpool to Manchester in two hours thirty-nine minutes' running time.

Stephenson continued to supply various locomotives to the Liverpool and Manchester Railway with different minor improvements; thus the "Mercury," built in December, 1830, had the outside frame placed above the driving axle, an improvement on the "Planet," which had the frames below the driving axle. But all these early engines of Stephenson were of a very unsatisfactory character. Pambour, writing in 1834, says of them: "When an engine requires any repair, unless it be for some trifling accident, it is taken to pieces and a new one is constructed, which receives the same name as the first, and in the construction of which are made to serve all such parts of the old engine as are still capable of being used with advantage. The consequence of this is that a reconstructed or repaired engine is literally a new one. The repairs amount thus to considerable sums, but they include also the renewal of the engines."

The directors of the Liverpool and Manchester Railway soon found the method of working their heavy trains with four or five locomotives was far from economical, and Stephenson was required to supply more powerful engines for the merchandise traffic. He, therefore, built the "Samson" and "Goliath." These were only four-wheel engines, but all the wheels were made of one size and coupled together. The former was delivered in January, 1831, and on February 25th she conveyed a train weighing 164 tons (without reckoning the weight of engine or tender) from Liverpool to Manchester in two and a half hours. The dimensions of the engines were: Cylinders 14in. diameter, stroke 16in., wheels 4ft. 6in. diameter, heating surface 457.10 sq. ft.

In 1831, the Directors of the Monkland and Kirkintilloch Railway decided to work their line by locomotives, and instructed Mr. Dodd, their engineer, to design engines for the purpose. He, however, merely adopted the plan used in the construction of the "Locomotion" (Stockton and Darlington Railway), with the cylinders placed partly within the boiler over the wheels, working by means of cross-heads and connecting-rods. He also adopted the tubular boiler, which was, of course, wanting in the "Locomotion." The engines were constructed by Murdoch and Aitken, of Glasgow, and were the first locomotives built in that city. The first was put to work on May 10th, 1831, and the second on September 10th, 1831. The boilers of these two locomotives were lagged with wood, and metallic packing was for the first time employed in connection with the pistons. The cylinders were 10½in. diameter, stroke 24in., steam pressure 50lb. The locomotives were supported on four coupled wheels, the coupling-rods having ball-and-socket joints at each end. A speed of six miles an hour was attained with Dodd's engines, and, although of rough design, they were much more economical in fuel and repairs than the engines supplied about the same time by Stephenson to the neighbouring Glasgow and Garnkirk Railway. These latter two engines were named the "St. Rollox" and "George Stephenson." Their dimensions were as follows:—

|  | Diameter of cylinders. | Stroke. | Wheels. | | Weight in working order. |
| --- | --- | --- | --- | --- | --- |
|  |  |  | Driving. | Leading. |  |
| St. Rollox | 11in. | 14in. | 4ft. 6in. | 36¼in. | 6 tons. |
| George Stephenson | 11in. | 16in. | 4ft. 6in. | 4ft. 6in. | 8 tons. |

The gauge of this line was only 4ft. 6in. The "St. Rollox" cost the G. and G. Railway about £750; that company sold it to the Paisley and Renfrew Railway for £350, and the latter, in December, 1848, when the gauge of their line was altered, disposed of the locomotive by auction for £13. It had wooden wheels. At the same auction the other two locomotives of the Paisley and Renfrew Railway were also sold, and realised only 20 guineas each, although about ten years previously the Paisley and Renfrew Railway had paid Murdoch, Aitken, and Co. £1,100 for each of them. They were six-wheel tank engines. The Scotch engines we have just been describing, all burnt coal in place of coke, and as they caused a good deal of smoke they were much objected to on that account.

We have previously stated that upon the advice of two engineers the Directors of the Liverpool and Manchester Railway had refrained from purchasing more locomotives from E. Bury, but other people soon saw the good points of his engines, and in 1832 the Liverpool and Manchester Railway considered it policy to purchase another locomotive from the Clarence Foundry. This engine was called the

"Liver." She had cylinders 11in. diameter, 16in. stroke, and driving wheels 5ft. diameter. The "Liver" worked very successfully, and in 1836 her fire-box was altered to burn coal, but this experiment turned out somewhat of a failure.

Towards the end of 1831, and during 1832, the increasing traffic on the Stockton and Darlington Railway made a considerable increase in the number of locomotives necessary. Hackworth designed two new classes of engines to work the trains. One type was known as the "Majestic" class, and six engines of this description were soon at work.

The "Majestic" locomotives had each six coupled wheels. The heating surface was obtained from a tube 9ft. long, 2ft. 6in. diameter, one end of which communicated with the fire-grate; the other was divided from the boiler by a partition plate, inserted in which were 104 copper tubes 4ft. long, and reaching to the smoke-box. It should be observed that the boiler was 13ft. long. The cylinders were fixed in a vertical position in front of the smoke-box, the connecting-rods working on a straight shaft or axle parallel with the wheel axles: this driving shaft was coupled by outside rods to the six wheels. The slide valves had "lap," and were worked by two eccentrics, which also worked the force pumps. The engine was reversed by means of a single lever. This class of engines included:—

"Majestic," built by Hackworth.
"Coronation," built by Hawthorn.
"William the Fourth," built by Hackworth.
"Northumbrian," built by Hackworth.
"Director," built by Stephenson.
"Lord Brougham," built by Hackworth.

All of them were built from Hackworth's designs, the leading dimensions being: Cylinders, 14½in. diameter, stroke, 16in.; boiler, 13ft. long, 3ft. 10in. diameter; weight of engine—empty, 10¼ tons; full, 11¾ tons. The other class of engines designed by Hackworth at this time included:—

"Darlington," built by Hawthorn.
"Shildon," built by Hackworth.
"Earl Grey," built by Hawthorn.
"Lord Durham," built by Stephenson.
"Adelaide," built by Stephenson.
"Wilberforce," built by Hawthorn.

"Wilberforce," an illustration of which is given (Fig. 22), was built by Hawthorn, of Newcastle, and commenced to work in 1832; it had six coupled wheels 4ft. in diameter; the cylinders were 14¾in., with 16in. stroke. Like many of the locomotives of that period, the "Wilberforce," as will be observed, had two tenders, one at each end of the engine. On the tender at the front end, which only carried coals (the fire-door being at the chimney end of the engines), the fireman stood; whilst the other tender, at the footplate end, carried water in a barrel, and also the tool boxes. The engine wheels were made of two separate castings or rings, and the axles were all straight, the crank shaft being carried in separate bearings beneath the footplate. There were no tail lamps in those early days; to make up for this deficiency a cresset containing burning coal was used. In some cases, when it was necessary to indicate the destination of the engine, or the section to which it belonged, as many as three of these cressets of glowing coals were employed on the same locomotive.

FIG. 22.—"WILBERFORCE," A STOCKTON & DARLINGTON RAILWAY LOCOMOTIVE

On certain favourable gradients the "Wilberforce" was capable of taking 36 loaded chaldron wagons, equal to about 171 tons, and its coal consumption is given as 68lb. per mile. During the year ending June, 1839, this engine ran 16,688 miles, conveyed 635,522 tons over one mile, and cost £318 10s. 8d., or 4.5d. per mile run, for repairs. The wages of the driver and fireman during the same period amounted to £353 12s. 8d.

The engines of this class, in their time, performed a greater amount of work than any others then existing. As late as 1846 one of the principal officials of the Stockton and Darlington Railway said of them: "Take them, weight for weight, they surpass any engine on the line."

The cylinders were 14½in. diameter, 16in. stroke; the valve gearing, wheels, etc., were similar to the "Majestic" class, but the cylinders were fixed on a framing extending 6ft. beyond the boiler over the driving shaft, which was coupled to the six wheels, each of 4ft. diameter.

The heating surface of the engines was on a different system, a "return multitubular fire-tube" being employed. This comprised a principal tube 8ft. long and 28in. diameter at the fire-grate end, and 24in. at the other. Here was fixed a D-shaped box; from this, 89 copper tubes conveyed the heated air back through the boiler to the semi-circular box fixed at the fire-grate end; the chimney came out of this smoke-box extension. These flues proved most economical, many lasting as long as six years, and, when necessary, duplicate ones could be fixed, and the engine again at work in three days. The boiler was 10ft. long and 4ft. 4in. diameter, weight of engine 10¼ tons empty, 11¾ tons loaded.

The "Magnet," built by Hackworth, at Shildon, in 1832, was an improvement on the above. The cylinders were 15in. diameter, 16in. stroke. The fire-tube at the furnace end was 2ft. diameter, and was divided in the middle by a 4in. fire-brick partition. The number of return tubes was 110. These were 7ft. 6in. long. Hackworth was at this time hauling all the trains on the Stockton and Darlington Railway by contract, at the rate of 2·5d. per ton of goods per mile; afterwards reduced to a still lower price. He paid the Stockton and Darlington Railway interest at 5 per cent. on the cost of locomotives employed on the line, which were the property of the Stockton and Darlington Railway Company, but leased to him.

FIG. 23.—GALLOWAY'S "CALEDONIAN," BUILT FOR THE LIVERPOOL AND MANCHESTER RAILWAY IN 1832

An engine named "Caledonian" (Fig. 23) was supplied to the Liverpool and Manchester Railway in 1832, by Galloway, Borman and Co. She had inside frames, four coupled wheels 5ft. diameter, and a domed fire-box. The curious point about the locomotive was the location of the cylinders, which were placed on the frame in front of the smoke-box, and were fixed vertically, with the piston-rods working through the upper cover, connecting-rods working downwards to the leading wheels, the axle of which was below the frames, in front of the smoke-box.

As might be expected, the "Caledonian" was far from being an easy-running locomotive, and, after several times running off the rails, she was rebuilt with inside cylinders and a crank axle.

# CHAPTER V.

A Stephenson "bogie" engine for America—The genesis of a world-famous locomotive fiim—Its initial effort in locomotive construction, the "Experiment"—Her cylinder valves—Two early Scotch locomotives—Stephenson favours 6-wheel engines, and constructs the "Patentee"—Forrester's "Swiftsure"—Opening of the Newcastle and Carlisle Rwy.—The "Comet"—R. Stephenson's early "ultimatum," the "Harvey Combe"—Hackworth to the front with a locomotive novelty—The first locomotive in Russia—The "Goliath"—The "Tyne" and her steam organ—Other early Newcastle and Carlisle Rwy. engines—An engine driver's reminiscences—No eight hours day then—The "Michael Longridge"—Opening of the Grand Junction Rwy.—Its first locomotives.

R. Stephenson and Co., in 1833, constructed a locomotive for the Saratoga and Schenectady Rail Road of America, which deserves mention from the fact that it had a leading bogie, rendered necessary because of the sharp curves on the Saratoga and Schenectady Rail Road. R. Stephenson named this locomotive the "Bogie," because the low wagons used on the quarries at Newcastle were locally called "bogies," and it was from these vehicles that he developed the idea of providing a small truck to carry the leading end of the locomotive in question. Ever since 1833 the swivelling truck used for supporting locomotives and other railway rolling stock has, in England, been designated the "bogie."

Richard Roberts, of the firm of Sharp, Roberts and Co. (the predecessors of Sharp, Stewart and Co., Limited), in the year 1833, turned his attention to locomotive construction. His initial effort was of a somewhat novel kind. Four locomotives of his first design were constructed, one—"Experiment"—for the Liverpool and Manchester Railway, and the others for the Dublin and Kingstown Railway. The cylinders, which were 11in. diameter, were placed in a vertical position on the frames, just at the point were the boiler entered the smoke-box. By means of cross-heads and side-links the motion was conveyed to a bell crank, and so transmitted by a connecting-rod to the driving wheels. There was, of course, a similar arrangement of cylinder, crank, etc., on both sides of the engine. The stroke was 16in. The driving wheels, 5ft. in diameter, were placed in front of the fire-box, and had inside bearings; the leading wheels were located below the vertical cylinders, and had outside bearings. The pump was placed in a horizontal position above the frame over the driving wheels, and was worked by a rod actuated by the vertical member of the bell-crank.

The "Experiment" (Fig. 24) was unsuccessful, and was rebuilt, when a third pair of wheels was added, and the position of the cylinders, bell-crank, etc., altered. The valves were also of a novel kind, patented by Mr. Roberts in 1832. Colburn thus describes them: "The valve, of wrought-iron, was formed of two concentric tubes or pipes, the larger pipe having holes perforated to admit steam from the steam-pipe into the annular space. This annular space was closed steam-tight at each end of the valve, and steam could only escape from it alternately to each end of the cylinder through the slots. The exhaust steam passed from one end of the cylinder directly into the waste pipe, and from the other end it traversed the interior of the

FIG. 24.—ROBERTS'S "EXPERIMENT," WITH VERTICAL CYLINDERS, BELL-CRANK, CONNECTING-ROD, AND CYLINDER VALVES

pipe of the cylindrical valve. These valves did not work well, as they did not expand equally with their cast-iron casings when heated by steam. For this reason the cylinder valves were soon abandoned. It should be mentioned that, in Mr. Roberts' first engines, the valve for each cylinder was worked with a motion derived from the opposite side of the engine. No eccentrics were employed, the requisite motion being taken from a pin near the fulcrum of each bell-crank, and transmitted thence through suitable gearing to the valve attached to the cylinder on the opposite side of the engine."

The engines used on the Dundee and Newtyle Railway, constructed in 1833, partook somewhat of the character of Roberts's "Experiment," inasmuch that right-angled cranks and vertical cylinders were employed, the diameter of the latter being 11in., and stroke 18in. These engines were named "Earl of Airlie" and "Lord Wharncliffe," and were constructed by J. and C. Carmichael, of Dundee. Both these engines were delivered at the end of September, 1833. The "single" driving wheels were placed in the leading position, the axle being just behind the smoke-box. The cylinders were placed on the side frames, about midway between the two ends.

The piston-rods worked upwards, and the motion was conveyed by means of rods from the piston cross-heads. These connecting-rods passed down outside the pistons, and were connected to one end of the bell-cranks, which were fixed beyond the cylinders, with the pivots over the centre of the second pair of wheels. From the lower ends of the bell-cranks the driving-rods were pivoted, the other ends being connected to the outside cranks of the driving wheels. The fire-box end of the engines was supported on a four-wheel truck or bogie. These engines weighed $9\frac{1}{2}$ tons each, and cost £700 each. An ordinary four-wheel wagon, fitted with a water-butt, was used for a tender.

An engine of similar design was ordered from Stirling and Co., of the East Foundry, Dundee, and delivered on March 3rd, 1834. Mr. A. Sturrock, the first manager of Swindon Works, and afterwards locomotive superintendent of the Great Northern Railway, helped to construct this engine, which was named "Trotter." Mr. Sturrock was at the time an apprentice at the East Foundry.

The gauge of the Dundee and Newtyle Railway was only 4ft. 6in., but when the line was taken over by the Dundee and Perth Railway the gauge was altered to the normal gauge of Great Britain. The original engine, "Earl of Airlie," after some alteration, of course, could not run on the railway, but for some years after the change the "Earl of Airlie" was employed as a stationary pumping engine.

Stephenson's four-wheel passenger engines with a short wheel base were found to be very unsteady at the very moderate speeds then attained, and he, therefore, added a pair of trailing wheels, thus constructing a six-wheel "single" passenger engine. Stephenson considered that the moderate wheel base of these small engines with six wheels would, on the easy curves of the Liverpool and Manchester Railway, offer considerable resistance, so he took out a patent, in which he provided that the middle or driving pair of wheels should be without flanges (or flanchès, as they were then called). He claimed that by this modification the six-wheel passenger engine would pass round curves with much less strain and greater safety. The first engine so constructed by Stephenson he designated the "Patentee," and she was delivered to the Liverpool and Manchester Railway in January, 1834. She had outside frames, inside cylinders, 18in. stroke, 12in. diameter; the driving wheels were 5ft. diameter.

George Forrester and Co., Vauxhall Foundry, Liverpool, in 1834 constructed a six-wheel engine named "Swiftsure." This locomotive possessed many novel features. It had outside horizontal cylinders; the frames were also outside, thus making the cylinders a considerable distance apart. The connecting-rods were keyed on cranks, at some distance outside the frames, whilst the fact that the driving wheels were not counterbalanced caused the engines of this class to be most unsteady at even moderate speeds,

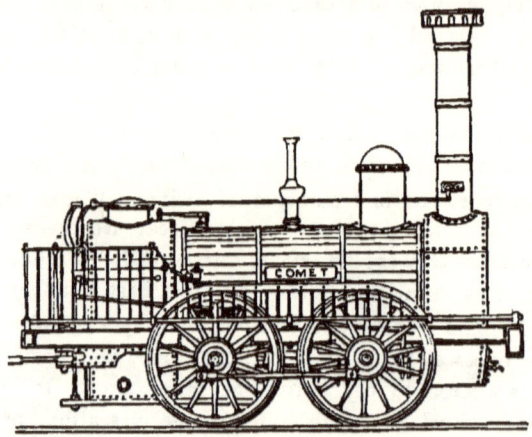

FIG. 25.—HAWTHORN'S "COMET," THE FIRST ENGINE OF THE NEWCASTLE AND CARLISLE RAILWAY, 1835

and they were soon known by the sobriquet of "Boxers." Colburn says: "A few pounds of iron properly disposed in the rims of the driving wheels would have redeemed the reputation of these engines." The arrangement of cylinders and frames allowed the leading wheels to be placed well forward, the total length of the frames of the "Swiftsure" being 17ft. The driving wheels were 5ft. diameter, and the cylinders 11in.; the stroke was 18in.

In the "Boxer" Forrester employed his patent valve gear, with vertical gab ends and four eccentrics.

A portion of the Newcastle and Carlisle Railway was opened March 9th, 1835, and R. and W. Hawthorn constructed the first engines for that railway. No. 1 was the "Comet," (Fig. 25), a four-wheel (coupled) locomotive; the cylinders (12in. diameter, 16in. stroke) were placed below the smoke-box, the connecting-rods passing under the leading axle. The wheels were 4ft. diameter. Hawthorn's valve gear was used in the engines of this class, which was actuated by four fixed eccentrics. The "Comet" continued to work on the Newcastle and Carlisle Railway for a number of years, and was afterwards used as a stationary engine for driving the steam saws at the Forth Bank Engine Works, Newcastle. She was so engaged up to and subsequently to 1863.

About 1836 short-stroke locomotives came into favour, and Tayleur and Co. built ten for the Liverpool and Manchester Railway. Although the cylinders were 14in. diameter, the stroke was only 12in. We need scarcely add the experiment was not successful, although some of the original broad-gauge engines were built with short strokes. These will, however, be dealt with fully later on.

In 1836 R. Stephenson and Co. constructed the "Harvey Combe" locomotive. She was a ballast engine, and was engaged in the construction of the London and Birmingham Railway. R. Stephenson had a minute description of this engine written by W. P. Marshall, and the work in question is stated to be "the most perspicuous and the illustrations of the most elaborate kind of any work describing a locomotive."

The fact that at once strikes the intelligent reader as peculiar is that, although the "Harvey Combe" was designed "for conveying the earth excavated in the construction of a line of railway," as Marshall "perspicuously" puts it (but which we should shortly describe as a "ballast" engine), she is a "single" engine! and, therefore, is not much like a modern six-coupled ballast engine. She cost £1,400, and was of 50 horse-power.

The principal dimensions of the "Harvey Combe" were: Cylinders, 12in. by 18in.; driving wheels, 5ft., and leading and trailing 3ft. 6in. diameter; 102 tubes, 1⅜in. internal diameter; total heating surface, 480ft.; weight, empty, 10 tons; with fuel and water, 11 tons 18 cwt. No flanges were provided to the driving wheels. Although the "Harvey Combe" was built for, and had rough usage as, a ballast engine, yet, when at the end of 1837 Nicholas Wood was making experiments for the purposes of his report to the

Great Western Railway as to the broad-gauge, the "Harvey Combe" was the principal narrow-gauge engine with which he experimented. With a gross load (including engine, etc.) of 81 tons, she attained a speed of 25 to 53 miles an hour, and consumed 0.47lb. of coke per ton per mile. With a gross load of 50 tons the speed reached was only 32.88 miles an hour, with the above coal consumption.

In 1836, Hackworth built a locomotive of novel construction—viz., with double-acting ram or trunk engines, by means of which piston-rods were dispensed with, the connecting rods being pivoted directly on to the piston and oscillated within the trunk.

This was the first locomotive engine ever seen in Russia. She commenced work on the Zarskoe-Selo Railway on November 18th, 1836, a religious service being held and the locomotive consecrated before the first train was run. Of this engine the Russian Emperor remarked in English, "It is the finest I ever saw." An old officer of the Stockton and Darlington Railway, informs the writer that a locomotive on the double-acting trunk principle was also built by Hackworth for that line, and so far as his memory serves him, he believes it was the "Arrow" passenger engine. The "Arrow" had leading and trailing wheels 3ft. 6in. diameter; driving wheels, 5ft. 6in. diameter; 135 tubes in the boiler of 1¾in. diameter; cylinders, 20in. in diameter, and with a stroke of only 9in.!

We have already mentioned the first engine (the "Comet") supplied to the Newcastle and Carlisle Railway, but several of the other early locomotives used on that line were powerful ones, and their design in advance of the generality of locomotives then in use. Thus, the "Goliath," one of the first engines supplied to the line by Hawthorn, in March, 1837, hauled a train consisting of 63 wagons of coal, weighing 267 tons, 12 miles in less than 40 minutes.

The "Goliath" had six-coupled wheels 4ft. diameter, cylinders 14in. diameter, 18in. stroke. Total heating surface 550.91 sq. ft. Weight, empty, 11¾ tons; in working order, 13 tons. The "Atlas,' built by R. Stephenson and Co. in 1836, drew a train of 100 wagons, loaded with coal, coke, and lime, and weighing 450 tons, 10¾ miles in 45 minutes, but this was on a falling gradient, varying from 1 in 215 to 1 in 106. This locomotive was also six-coupled, the wheels being 4ft. diameter; cylinders, 14in. by 18in. stroke; heating surface, 553.77 sq. ft., weighing 10 tons 6 cwt. empty, and 11 tons 6¾ cwt. in working trim. Another small locomotive on the Newcastle and Carlisle Railway, named "Tyne," built by Hawthorn, is worthy of notice, for the reason that the first steam organ was fitted to the engine. This was the invention of the Rev. James Birket, of Ovingham. It was fixed on the top of the fire-box, and was thus described: "The organ consists of eight pipes, tuned to compass an octave, but without any intervening tones or semi-tones. This is the first attempt to adapt a musical instrument to the steam engine capable of producing a tune, and though not so perfect as to admit of all the pleasing variety and combination of sound capable of being produced by the instrument to which we have compared it, there is no doubt but very considerable improvements will be made in this steam musical instrument by the inventor, who is a skilful musician as well as an ingenious mechanic."

The "Tyne" had cylinders $13\frac{1}{2}$in. by 16in. stroke, and four wheels, 4ft. 6in. diameter; she weighed only $9\frac{1}{2}$ tons. After working for many years, a pair of trailing wheels 3ft. 6in. diameter was added, thus making her a six-wheel engine, with the leading and driving wheels coupled. She continued to work on the Newcastle and Carlisle Railway till the end of 1857, when she was sold, but even at that time the "Tyne" was in good working order. Three other old locomotives were sold at the time—viz., "Eden," "Meteor," and "Lightning."

The "Eden" was built by R. Stephenson and Co. in 1836, and had four-coupled wheels of 4ft. 6in. diameter, and a third pair 3ft. 6in. diameter; cylinders, 14in. by 15in. stroke, afterwards increased to 16in. stroke. Weight, empty, 10 tons, 6 cwt.

The "Meteor" was built by Bury and Co., of Liverpool, and had only four wheels of 4ft. diameter; cylinders, 12in. diameter. The stroke at first was 15in., but afterwards was made 16in. Steam pressure, 55lb. She was provided with hand gear, the slide valves working into the front of the steam chest by means of weight bars located between the front buffer beam and the smoke-box end. The piston connecting-rods, of course, actuated the rear axle, but the eccentric sheaves were upon the leading axle, so that if the crank pins upon which the side rods worked went a bit loose, the side rods had to be disconnected, and the valves worked by the gear handles. This was rather hard work for the driver and fireman, who, upon such occasions, took it in turns to thus work the valve gear. This Bury locomotive opened the line from Blaydon to Newcastle on Sunday, October 31st, 1839. The man who was fireman on this engine at that time thus relates his experiences:—"The 'Meteor' engine was sent to Redheugh Station to work the passenger trains between that station and Blaydon, also coal trains and other things, with this tiny engine of about eleven tons all told. We formed the connection at Blaydon with all trains to and from the west. For this new arrangement of running I was to be called out of bed by a watchman close after two o'clock each morning, to gather up my fire bars, put them into the box, and get a fire as best I could as usual, and have steam ready by 5 a.m. to take our first train from Gateshead to Blaydon at 5.20 a.m. I had also to clean most of the little engine, the driver doing part. I had to clean up the shed, take all ashes out, coke the tender, etc. To turn the engine the tender had to be taken off, and pushed on one side to get past it, and reunited as often as we made a short trip. There is nothing like it in the divorce court. For this work my pay was 2s. 8d. per day, commencing at 5 a.m., when my driver made his appearance, little overtime being allowed, and we did well to finish by 8.45 p.m. I worked about 18½ hours daily, with one exception, weekly, and on this particular time we had our boiler to clean out, and had to fill by hand buckets—this after our train work was finished. Water being a little scarce in the shed, it was frequently necessary to haul out of the river Tyne and carry to the shed, and pour into the boiler by the safety valve or man-hole by the driver, the fireman having the honour of carrying it from the river quay.

"This work took so much labour and time that our only rest on that particular night and morning was upon the soft side of a plank while the steam was rising in the engine boiler, to leave for Blaydon at 5.20 a.m. with our usual first train. Then we were again at work until 8.45 p.m. There was not a guard for our passenger train, so I had the closing of the carriage doors, etc., to attend to, to fill up my spare time, and to keep myself awake. We had to load coals during part of the day from Wylam, etc., to Dunston, so that there was not much fear of falling asleep. I was coupler and guard for this work. When not otherwise engaged I had my cleaning to attend to, and tubes to keep clean daily, so I was really never committed for going to sleep during working hours. I was at this work over the winter almost the whole of 1839-40, when early one morning I had a fall from the boiler top in the shed, and came down the wrong end first. I injured one shoulder very much, which laid me off work one month.

I kept at work all the day after falling, but only one arm was of any use to me, and I was compelled to give up.

"A bone-setter in North Shields had to do the needful for me, as they have often had to do for others before and afterwards."

The "Lightning" was an engine with dimensions similar to those of the "Eden," previously described.

Longridge and Co., of Bedlington, supplied the Stanhope and Tyne Railway in 1837 with a very powerful locomotive named the "Michael Longridge." She had six coupled wheels, 4ft. diameter; cylinders, 14in. diameter; and a stroke of 18in.

The Grand Junction Railway was opened in July, 1837, and R. Stephenson and Co. (together with other builders), supplied the original locomotives. Stephenson's engines at this time had become a little more dependable, for we find it chronicled that three of them which had run uninterruptedly since they were first employed had, between July 8th and September 30th, 1837, accomplished the following distances—viz: the "Wildfire," 11,865 miles; "Shark," 10,018 miles; and "Scorpion," 11,137 miles; and, moreover, they were then

FIG. 26.—"SUNBEAM," BUILT BY HAWTHORN FOR THE STOCKTON AND DARLINGTON RAILWAY

still running in perfect working condition. They were six-wheel locomotives, with leading and trailing wheels 3ft. 6in. diameter, driving, 5ft. diameter; cylinders, 12½in. by 18in. stroke; weighing in working order 9 tons 12 cwt.

In 1837, No. 43, of the Stockton and Darlington Railway, the "Sunbeam" (Fig. 26) was turned out by Hawthorn. It was a "single" engine, having driving wheels 5ft. in diameter, and cylinders 12in. in diameter, with 18in. stroke. The "Sunbeam" worked well for 19 years, and in 1863 was reported as being "still in good working order, but too small for the present heavy traffic. The boiler of the "Sunbeam" was 8ft. long by 3ft. 2in. in diameter, and contained 104 copper tubes. The "Dart," No. 41, was built by Hackworth in 1840, at Shildon, and was a four-wheeled engine, the wheels being 4ft. 6in. in diameter. The boiler, containing 122 tubes, was 8ft. 2in. long and 3ft 3in. in diameter. The fire-box was 4ft. high, 3ft. 10in. long, and 3ft. wide. The boiler pressure was 100lb., and the heating surface of the engine 602 square feet; the cylinders were 14in. in diameter, and the stroke 16in. The extreme length of the engine and tender was 35ft. 3in., and the regular speed attained is said to have been thirty miles an hour.

## CHAPTER VI.

An important epoch in locomotive history.—The first broad-gauge engines.—Absurd incorrect statements regarding these locomotives.—The facts concerning same; extracts from directors' report.—Brunel and the engine bui'ders.—The delivery of the first engines to the Great Western Railway.—Further extract from the directors' report.—Daniel Gooch appears on the scene.—Trial of the broad-gauge engines.—Table of the original Great Western engines.—The "Vulcan" — "Æolus" — "Bacchus" — "Venus" — "Apollo" — "Mars" and "Ajax," 10ft.-wheel engines.—The builders' account of one of these giants. —"Ajax," a sister engine.—10ft. disc wheels.—Dr. Lardner.—The "boat" engines.—T. R. Crampton and the "Ajax."—The "Ariel."—"Atlas."— "Hurricane," a locomotive monstrosity with 10ft. driving wheels.—The "Thunderer," a geared engine on Harrison's system.—Gooch's opinion of these two curious locomotives.—The Haigh Foundry geared engines, described by an eye-witness.—Table showing results of trials with the original broad-gauge engines.—The last of "Lion," "Planet," and "Apollo."

WE have now come to an important era in the evolution of the steam locomotive—viz., the first appearance in the arena of broad, or 7ft., gauge locomotives. Readers are probably aware that very much has been written on the subject of the early Great Western Railway locomotives during the past few years, and a surprising lack of knowledge of the subject has been exhibited by people taking part in discussions that have arisen. The facts are clearly established, so that it would be waste of time to recapitulate the many inaccurate statements that have been made relative to the original broad-gauge locomotives. Thus we read that "the first portion of the Great Western Railway was opened in 1837," also that "Mr. Brunel designed the 'Hurricane.'" These statements are, of course, utterly at variance with the facts, but they prepare one for yet more extraordinary statements on the same subject, such as "the directors of the Great Western Railway having appointed Mr. (afterwards Sir Daniel) Gooch as locomotive superintendent, the duty devolved upon him to design and provide the necessary engines. Mr. Gooch, having inspected all the locomotives on other railways, considered that 5ft. 6in. wheels were far too small; he therefore designed the engines for the Great Western with driving wheels of 6ft., 7ft., and 8ft. diameter, and placed orders for their construction with the leading builders of that time."

To commence with, therefore, it will be as well to give the exact particulars as to the ordering and delivery of the original broad-gauge locomotives, for the opening of the first portion of the Great Western Railway.

The facts as given in the directors' reports to the shareholders, stated at the meetings of the shareholders, or mentioned in the various reports of Brunel, Wood, and Hawkshaw, are as follows:— The first locomotive engines were ordered prior to August, 1836. The directors in their report of that date thus mentioned them:—" Difficulties and objections were at first supposed by some persons to exist in the construction of engines for this increased width of rails, but the directors have pleasure in stating that several of the most experienced locomotive engine manufacturers in the North have undertaken to construct these, and several are now contracted for, adapted to the peculiar track and dimensions of this railway, calculated for a minimum velocity of thirty miles an hour."

Instead of the builders having personal interviews to obtain orders for engines, as has been recently stated, it appears from Brunel's report of August, 1838, that he "left the form of construction and the proportions entirely to the manufacturers, stipulating merely that they should submit detailed drawings to me for my approval. This was the substance of my circular, which, with your sanction, was sent to several of the most experienced manufacturers. Most of these manufacturers, of their own accord, and without previous communication with me, adopted the large wheels as a necessary consequence of the speed required. As it has been supposed that the manufacturers may have been compelled or induced by me to adopt certain modes of construction, or certain dimensions in other parts, by a specification—a practice which has been adopted on some lines—and that these restrictions may have embarrassed them, I should wish to take this opportunity to state distinctly that such is not the case."

Then, as to the delivery of the engines, from the directors' report it is clear that on August 12th, 1838, eleven locomotives were actually on the line. According to a statement drawn up by Mr. C. A. Saunders, the superintendent of the Great Western Railway, for the purposes of Mr. N. Wood's report, the following engines were then in use on the railway:—" North Star," " Æolus," " Venus," " Neptune," " Apollo," " Premier," and " Lion." This leaves four engines to be accounted for. Sir Daniel Gooch states that the six engines built by the Vulcan Foundry Company could be depended upon. We ca.., therefore, take it for granted that the "Vulcan" and "Bacchus" were two of the four, whilst the geared "Thunderer" was delivered before April 26th, 1838, and the "Ariel" before June 1st, 1838.

The directors stated that the railway company had only accepted eight of these engines, and the three others required alterations before the engineer would accept them.

This report continues with the following significant paragraph:— "The directors are under the necessity of declining to receive two engines made for them, in consequence of a material variation in the plan of them since it was submitted to and approved by their engineer." These two engines may be the "Ajax" and her sister 10ft. wheel engine, the "Mars," constructed by Mather, Dixon and Co., or the two geared engines built by the Haigh Foundry Company; although it is probable that the two latter engines had not been delivered at this date. Besides the eleven engines already on the line, and the two refused by the engineer, the directors stated that nineteen others were then in course of construction, making a total of thirty engines. Of the seven engines mentioned as being in use on the line, according to Mr. Brunel, only four were really used for the passenger service, the fifth being kept with steam up to take the place of one of the other four in case of a breakdown, and the other two were used for conveying ballast, etc., for the construction of the line. According to Hawkshaw's report, dated October 4th, 1838, fourteen engines had at that time been delivered to the Great Western Railway, and seven more were approaching completion, the nine remaining to complete the thirty not having then been put in hand. Mr. Daniel Gooch commenced his duties as locomotive superintendent of the Great Western Railway on August 18th, 1837. At this period the following engines had been ordered for the Great Western Railway:—Six from the Vulcan Foundry, where Gooch had served under Stephenson; four from Mather, Dixon and Co., Liverpool; two from Hawthorn and Co., Newcastle; two from the Haigh Foundry Company, and, curiously, two from R. Stephenson and Co.

Mr. Gooch states in his "diaries" that these two engines were constructed for a Russian railway with a 6ft. gauge, and that he himself prepared the working drawings from which they were constructed. There, however, appears to be some doubt as to whether it was a Russian or American railway for which the two locomotives in question were originally built. When ready for delivery the purchase money was not forthcoming, so the careful firm of R. Stephenson and Co. did not part with the "North Star" and her sister engine. They afterwards widened the frames, fitted longer axles to the two locomotives, and then sold them to the Great Western Railway as 7ft. gauge engines.

The "Vulcan," built by the Vulcan Foundry Company, was the first engine delivered to the Great Western Railway. One of Mather and Dixon's 10ft. wheel engines arrived a few days after, having been sent by sea from Liverpool to Bristol in December, 1837, and forwarded by canal from Bristol to West Drayton. A preliminary trial of these two engines was made on Wednesday, January 18th, 1838, and the following extract details the working of the two locomotives on this occasion:—"A full trial was made during the whole of Wednesday in running the engines on two or three miles of the line near West Drayton, between London and Maidenhead. The object of the trial was to prove the rails, and most satisfactory was the result, both as to the increased width of gauge and the use of continuous bearers of kyanised wood confined by piles, on which plan the line is constructed. An engine with 8ft. drawing wheels, made by Messrs. Tayleur and Co., Warrington, weight 23 tons, with the tender, water, coke, etc., and another engine made by Messrs. Mather, Dixon and Co., weight about 19 tons, with the tender, etc., ran the whole day without producing the slightest vibration either in the rails or the wood under them. The rails are, in fact, so beautifully firm, smooth, and true, that the engines glided over them more like a shuttle through a loom or an arrow out of a bow than like the effect on any previous railway. There is literally no noise—no apparent effort—nor can there ever be discovered any difference between the centre and the joint in the rails. A maximum speed was not attempted, as on so short a piece the momentum would be no sooner attained than it would require to be lowered, in preparation for stopping the engine. A speed of forty-five to fifty miles an hour was attained, and when the engines are run, as they will be, either next or the following week, on an eight or ten-mile length, there is no doubt they will as easily run at a very much greater speed."

The following table gives particulars of the original locomotives as supplied to the Great Western Railway. These engines were ordered by Brunel before Sir D. Gooch was appointed Locomotive Superintendent; the first duty of the latter was to inspect these locomotives, then in course of construction, and he was not at all pleased with their dimensions:—

| Builder. | Name of Engine. | Diameter of Driving Wheels. | Diameter of Cylinder. | Stroke. | Grate area. | Heating Surface. | | Total |
|---|---|---|---|---|---|---|---|---|
| | | | | | | Tubes | Fire-box. | |
| | | feet. | inches. | inches. | feet. | sq. ft. | sq. ft. | sq. ft |
| Vulcan Foundry Co. | Vulcan | 8 | 14 | 16 | 9·58 | 534 | 35·0 | 589 A |
| " | Æolus | 8 | 14 | 16 | | 530 | 57·15 | 587 A |
| " | Bacchus | 8 | 14 | 16 | | 530 | 57·15 | 587 A |
| " | Venus | 8 | 12 | 16 | | 458 | 52·35 | 510 B |
| " | Neptune | 8 | 12 | 16 | | 458 | 52·35 | 510 B |
| " | Apollo | 8 | 12 | 16 | | 458 | 52·35 | 510 B |
| Mather, Dixon & Co. | Mars | 10 H | 14 | } 10 or { 11 | 10·22 10·2 J | 417 417 | 57·3 57·3 | 474 C 474 C |
| " | Ajax | 10 | 14 | | | | | |
| " | Premier | 7 | 14 | 14 | | 326 | 51·71 | 377 D |
| " | Ariel | 7 | 14 | 14 | | 326 | 51·71 | 377 D |
| R. Stephenson & Co. | North Star | 7 | 16 | 16 | 13·0 | 654 | 70·10 | 724 E |
| " | Morning Star | 6½ | 16 | 16 | 13·0 | 65½ | 70·10 | 705 |
| Sharp, Roberts, & Co. | Lion | 6 | 14 | 18 | | 427 | 51·17 | 478 |
| " | Atlas | 6 in. | 14 | 18 | | 427 | 51·17 | 478 |
| Haigh Foundry Co. | Viper | 6 | 4 | | | | | F |
| " | Snake | 6 | 4 | | | | | F |
| Hawthorn & Co | Thunderer | 6 | 16 | 20 | 17·12 | 515 | 108·26 | 623 G |
| " | Hurricane | 10 | 16 | 20 | 17·12 | 515 | 108·26 | 623 G |

A. These engines had the driving axles above the frames.
B. Gooch, N. Wood, Whishaw, C. A. Saunders, Z. Colburn, and other reliable authorities all state that these three engines had cylinders 12 inches in diameter.*
C. J. Locke in his evidence before the Gauge Commissioners in 1846 stated that the stroke of these engines was only 10 or 11 inches.
D. It will be noted that these two engines, also built by Mather, Dixon and Co., had very short strokes.
E. Stroke was afterwards increased to 18 inches.
F. These engines were geared, so that the driving wheels were equal to 12 feet diameter.
G. The engines and boilers were on separate carriages. The "Thunderer" was geared up 3 to 1.
H. Although the "Mars" was built with 10ft. driving wheels, it is probable that the size was reduced after her trial trips on the G.W.R. After running 10,000 miles the G.W.R. Co. sold the "Mars."
J. It is not certain that the "Viper" and "Snake" were the geared engines built by the Haigh Foundry Co., but they are generally accepted as such.

The "Vulcan," it would seem, was a conspicuous failure. The Great Western Railway officials did not consider her good enough to be used in the experiments made during the autumn of 1838 for the purpose of Nicholas Wood's report to the Great Western Railway in connection with the gauge controversy. Whishaw only gives an account of one trip to West Drayton and back with the "Vulcan." This was made on the 12th of August, 1839, when, with a load of 18 tons, she attained a speed of 50 miles an hour on a falling gradient,

* There is some question as to this, as at one time the diameter of the cylinders was 14 inches.

the average speed for the trip of 13 miles being 28.32 miles an hour. On the return trip, with a load of only 14½ tons, the average speed was only 21 miles an hour. The "Vulcan" was afterwards converted into a tank engine, and worked the traffic on a branch line for a few years.

The "Æolus" appears to have been a somewhat better engine than her sister (although, by the way, Sir D. Gooch states that, excluding the "North Star," the engines from the Vulcan Foundry were the only once he coudld depend upon). N. Wood, in his tables, states that ".Æolus" was capable of hauling 32 tons at fifty miles an hour, with a consumption of 0.76lb. of coke per ton per mile, the water evaporated in an hour being 115.3 cubic feet. The greatest load drawn by ".Æolus" during N. Wood's experiments was 104 tons, the speed attained being 23 miles an hour, and the consumption of coke .30lb. per ton per mile. Whishaw details four experiments with this engine, the most successful being on November 6th, 1838, when with a load of about 20 tons she attained an average speed of 31.39 miles an hour; the maximum on this occasion being 48 miles an hour. Whishaw's remarks concerning another journey are worth repeating. It was on July 21st, 1838, when "Æolus" took a train consisting of three first-class carriages, two open and one closed second-class carriages, and two stage coaches on trucks, or a load of 96,164lb., or about 43 tons, and essayed a trip to Maidenhead; but "after about two and a half miles the train was suddenly stopped, and remained *in statu quo* for 21¾ minutes. In the meantime, '.Æolus' moved slowly away to recover her strength, and having sufficiently exercised herself, returned after a lapse of 21¾ minutes to lead the train forward"; but the engine did not appear to have quite recovered her strength by this exercise (!) for she had to stop at Slough, where she took water. This journey took 150 minutes to complete; but, deducting the 34 minutes spent in four stoppages, the average travelling rate was 11.71 miles per hour.

On January 11th, 1840, the "Æolus" is stated to have made a remarkable trip. At this time certain Chartists were being tried at Monmouth, and the *Dispatch*, a Democratic Sunday paper, published detailed reports of the trial. Special messengers were despatched by road from Monmouth to Maidenhead, where an engine (the "Æolus") was engaged to carry the messengers to London. She is said to have covered the first ten miles in seven minutes, or at the rate of 85 miles an hour. Here the preceding train was overtaken, and the whole journey of 31 miles was completed in about twenty-five minutes.

Whishaw records a trip with "Bacchus" on December 13th, 1839, when, with a train of two second and one first-class (four-wheel) carriages, she covered 13 miles at an average speed of 29 miles an hour, the highest speed attained on the trip being 44.11 miles an hour. On January 9th, 1840, Whishaw made a trip to West Drayton and back with the "Bacchus." On the down journey, with a load of three coaches, 50 miles an hour was attained. On the up trip a similar maximum speed was attained three times, twice for a distance of a quarter-mile, and once for a half-mile.

With "Venus" Whishaw records one experiment with a load of 25½ tons, made up of one open second-class, one first-class carriage, and two stage coaches on trucks. The average speed was 21 miles an hour, the highest being 48 miles an hour. The "Venus" was not much used during the first four months following the first opening of the Great Western Railway, her total mileage during that period being only 240 miles. Mr. Gooch found this engine was so extremely unsteady that he did not make use of her, save when no other engine was conveniently available—hence her small mileage. The "Venus" was afterwards rebuilt as a tank engine, and her driving wheels reduced to 6ft. in diameter. When so rebuilt she worked the Tiverton branch traffic for some years.

The "Apollo" drew the first up-train on the Great Western Railway, leaving Maidenhead for Paddington at 8 a.m. on June 4th, 1838; whilst the next day, when leaving Maidenhead with the afternoon train of 13 carriages, she broke down, in consequence of a tube bursting, the train being delayed for some hours, and great excitement being caused in London consequent upon the exaggerated reports of the mishap.

It will be noticed that in the table of the original Great Western Railway locomotives we have given the diameter of the cylinders of "Venus," "Neptune," and "Apollo" as 12in., and we have also given the names of several men (whose probity is unimpeachable) as our authorities on the point. Nor is that all the weight of evidence in favour of 12in. being the original diameter of the cylinders. N. Wood, in his report to the Great Western Railway directors, specially refers to the point, thus: " . . . . The performance of engines, such as 'Venus,' 'Neptune,' and 'Apollo,' with 12in. cylinders." This is in addition to the statement contained in Wood's Table, No. 3, where also he gives the dimensions as 12in. It is now, however, stated that the cylinders of these engines were 14in. in diameter.

We now have to deal with the two locomotives with 10ft. driving wheels, constructed by Mather, Dixon and Co. for the Great Western Railway.

Fortunately, one of the people who assisted in the construction of these engines is still living, and in the *Engineer* for January 3rd, 1896, he gave a detailed account of the building of the locomotive, and also a drawing of the " Grasshopper " (a nickname for the " Ajax " or " Mars "), which is here reproduced :—

FIG. 27.—THE "GRASSHOPPER," ONE OF THE TWO BROAD-GAUGE ENGINES ("AJAX" AND "MARS"), WITH 10FT. DRIVING WHEELS, DISC PATTERN, BUILT FOR THE GREAT WESTERN RAILWAY COMPANY BY MATHER, DIXON AND CO.

The gentleman in question has favoured the writer with the following particulars concerning this engine :—"The engine was designed by John Grantham, draughtsman at Mather, Dixon, and Co., North Foundry, Liverpool. The outside view resembled a steamer, the driving-wheel splashers like a paddle-box, and the handrail plates, brought to the buffer planks, shaped like the stem of a vessel, and intended to take the wind pressure off the front end of the engine. The great diameter of the driving wheel shows that Brunel had something to say about it—perhaps ordered it to be made twice the size of any other then made. The staff employed in the works then were: John Grantham, principal of drawing office, afterwards partner; Robert Hughes, manager of the marine department, afterwards of the Royal Arsenal, Woolwich, and inspector of steamships; Mr. Banks, locomotive foreman, well known at Derby on the Midland

Railway; Mr. Buddicomb, first locomotive superintendent of the Grand Junction Railway, and of the locomotive works at Rouen, France; Josiah Kirtley, first locomotive superintendent of the Midland Counties; George Harrison, first locomotive superintendent Scottish Central, and manager at Brassey's, Birkenhead; Mr. Potts, afterwards of the firm of Jones and Potts, Newton-in-the-Willows, locomotive builders, where the first solid locomotive wheel was made by the wheelsmith Frost.

"All the above-named were apprentices and journeymen with me in my time.

"William Tait, of the firm of Tait and Mirlees, Scotland Street, Glasgow, was the erector of the 10ft. wheel locomotive; I worked as mate with him on the same engine. Tait was manager of Neilson's Hyde Park Locomotive Works, Glasgow, in 1845, and his mate— John Wilson—was manager from 1864 to 1884 under Mr. James Reid, sole owner of Neilson's Works. James Smith Scarf welded the 10ft. tyres. The crank axles were forged at the Mersey Forge, when Mr. Norris was manager, and turned by Charles Ackers. Ned Bursing turned the rims and tyres on a large lathe, driven by the gearing of the boring mill. I remember, having worked on the same lathe, that they had to cut a curved piece out of the shop wall for clearance."

The "Ajax" and "Mars" (Fig. 27), the 10ft. wheel engines supplied by Mather, Dixon, and Co., had the driving wheels of peculiar construction. Instead of the usual spokes, the circumference and the centres were connected by means of iron plates, bolted together in segments, and slightly convex in form.

These disc wheels were constructed under a patent granted to Mr. B. Hicks, of Bolton, in October, 1834. The primary object of Mr. Hicks's patent was not, however, the disc wheels, but a threecylinder engine, with the cylinders placed vertically above the crank axle. Steam was only to be admitted at the top of the piston, so that the force of the steam was always pressing downwards; by this method Mr. Hicks expected to considerably augment the adhesive properties of the engine. We cannot discover that an engine with three such cylinders was ever constructed, although the disc wheels were used in the "Mars," "Ajax," and other locomotives.

As will be seen from the illustration of the "Grasshopper," these two 10ft. wheel engines had a projecting front, and the splashers covering the wheels above the frames were made to represent paddleboxes of a steamboat. For these reasons, Dr. Lardner says, they were generally known as the "boat engines," and he goes on to remark that they were found incapable of working the passenger trains (probably in consequence of the time lost in starting and stopping the monsters), and were used to haul the ballast trains during the construction of the Great Western Railway. Mr. Brunel gave the following evidence relative to these 10ft. wheel engines before the Gauge Commissioners in 1845 :—" Three engines were made for 10ft. The idea did not originate with me, but it was proposed by certain manufacturers, and although I expressed some fear of the feasibility of constructing 10ft. wheels, I thought it worth the trial. They were made, and it so happened that the three engines to which they were applied totally failed in other respects, and the whole engine was cast aside. . . . . . The engines to which I refer were a pair made in Liverpool by a maker there, who was also making other engines for us. I take the whole responsibility, of course, of having allowed the 10ft. wheel to be made; but the engines, from other circumstances, were not successful, and the construction of the wheels was one which we should certainly never again adopt. It was an entire plate, and that with such a diameter is heavy, and offers such an enormous surface to the side wind that it certainly would not do to adopt it. In the other engine ('Hurricane'), which was tried with a 10ft. wheel, the wheel worked very well, but accidental circumstances threw the engine out of use; the wheels got broken by an accident which would have broken any wheels, and no further attempt was made to use it."

Mr. T. R. Crampton, the designer and patentee of the famous Crampton engines, gives the following particulars of the "Ajax":—" Area of fire-grate, 10.22ft.; total heating surface, 474.0ft.; diameter of driving wheels, 10ft.; diameter of cylinders, 14in.; length of stroke, 20in.; surface in fire-box, 57.3ft.; cubic contents of both cylinders, 7.09ft.; proportion of capacities to the wheel, 1 : 1.41."

The "Ariel" appears to have come into collision with the "Hurricane" at Bull's Bridge (Hayes) on November 6th, 1838, whilst the "Lion" broke down near the same spot at five o'clock on July 30th, and was unfortunate enough to run over and kill a man at Ealing on November 6th, 1838.

About midnight on March 3rd, 1839, the "Atlas" was hauling a ballast train of 25 wagons towards Paddington, and instead of stopping at the usual place, the train continued on into the engine-house, colliding with the "North Star," and doing considerable damage to that renowned locomotive; then, proceeding on its victorious career, it next charged the wall of the engine-house, and, finally, came to a stop. Upon inquiring into the cause of the accident it was discovered that both the driver and stoker were asleep on the engine, and that the train had been running for some miles with no one in charge. Although there were fifty men on the wagons, none of them were seriously injured.

Great excitement was caused in London on the evening of October 26th, 1838, by the report that Mr. Field (a partner in the firm of Maudslay and Field, the well-known engineers) had been run over and killed by the "Hurricane" (Fig. 28), but this was not quite correct. The true facts were as follows:—Dr. Lardner and his assistant, a youth of 19, named Field, were making experiments at Acton on the deflection of the rails, for the purpose of Wood's report to the directors, and were using the up line. The "Hurricane" was the engine employed, and this engine came down from Paddington on the up line for their use. Young Field was stooping down to measure the amount of deflection as the engine passed, and just at the moment overbalanced himself in front of the "Hurricane," and, although it was only travelling at the rate of five miles an hour, it could not be pulled up in the short space, and he was, unfortunately, run over and killed.

In December, 1836, T. E. Harrison patented an arrangement for carrying

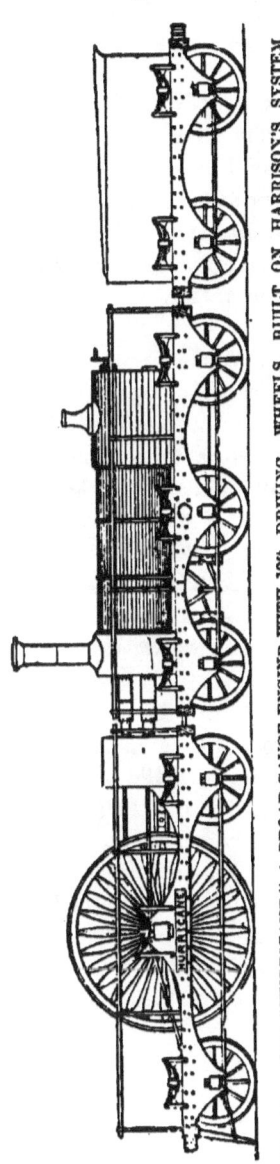

Fig. 28.—THE "HURRICANE," A BROAD-GAUGE ENGINE WITH 10ft. DRIVING WHEELS, BUILT ON HARRISON'S SYSTEM

the boiler of the locomotive on one carriage and the machinery on another, the idea being that when repairs were necessary to the boiler portion it could be disconnected from the machinery, and another boiler carriage substituted, and *vice versâ*. Considering the amount of repairs necessary to locomotives at this early period of their evolution, great economy was expected from the adoption of the arrangement.

The "Thunderer" (Fig. 29) was constructed in 1837 by Hawthorn's of Newcastle. The boiler portion of the machine was carried on six wheels, and viewed from its exterior, it appeared to be similar to an ordinary locomotive. In front, at the chimney end, was the machinery carriage, carried on four-coupled wheels of 6ft. diameter. The gearing being 3 to 1, therefore, one revolution of the prime driving wheels caused the travelling wheels to turn three times, thus making them equal to driving wheels 18ft. in diameter.

The cylinders were horizontal, and the connecting-rods were attached to a double-cranked axle, on which was the cogged wheel; this worked a pinion on the axle of the driving wheels. The axle of the driving wheels had a motion up and down, to allow for imperfections in the road; and the cogged wheel and pinion were kept at the requisite distance in gear by the supports of the cranked axle being fixed over and connected with those of the driving wheels, and thus moving in conjunction with them. Two eccentrics on the cogged wheel axle worked the slides with the usual levers and hand-gear, and the exhaust steam from the cylinders was discharged into the chimney.

The two carriages were connected by a bar, and the steam pipes had a ball-and-socket joint for lateral motion, with a metallic ring packing; they also were composed of two parts which slid one within the other, allowing by this means a motion in the direction of their length. The tank was under the boiler, and the engine-wheels were coupled, in order to have the whole weight for the purpose of obtaining adhesion. To keep the teeth at the right pitch, and prevent backlash on reversing the motion, the pinion was in two parts, one of which was movable round the axle, and by means of keys these might be set so as to place the two halves of the teeth a little out of the right line, and thus tighten their action.

The diameter of the boiler was 44in., that of the 135 tubes, 1⅝in. (internal); the tubes were 8ft. 7in. long. The fire-box was provided with a mid-feather.

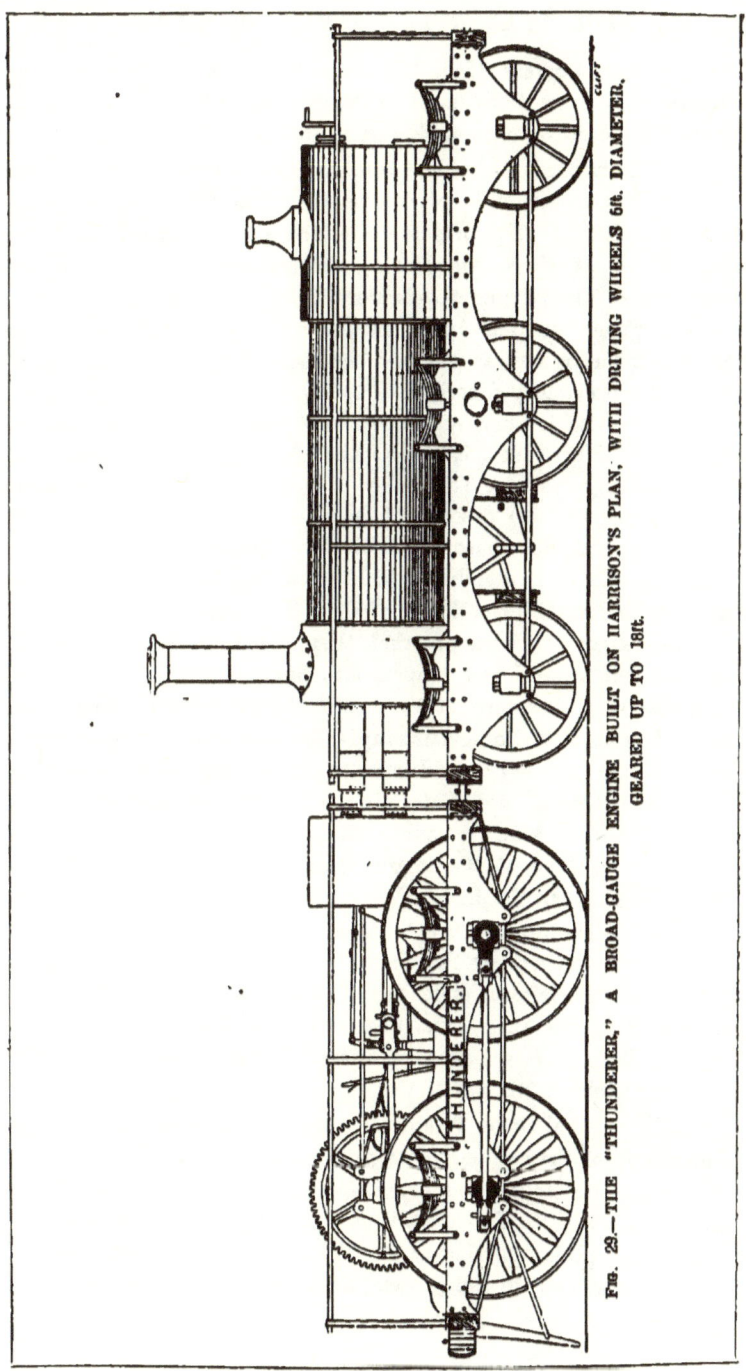

FIG. 29.—THE "THUNDERER," A BROAD-GAUGE ENGINE BUILT ON HARRISON'S PLAN, WITH DRIVING WHEELS 6ft. DIAMETER, GEARED UP TO 18ft.

On Friday, January 18th, 1839, the "Thunderer" drew a heavy ordinary train from Maidenhead to Paddington in 32 minutes, including the time occupied in stopping at Slough.

The "Hurricane" was of the same general design and dimensions as the "Thunderer," with, of course, the wide difference as to the mode of working. The machinery vehicle of the "Hurricane" was supported on six wheels, the leading and trailing being 4ft. 6in. diameter, whilst the driving wheels were 10ft. in diameter, the piston-rod connections working direct on the crank axle. The axle-boxes were above the frames, as was also the case with the two 10ft. wheel locomotives previously described.

In a so-called locomotive history what purports to be an illustration of the "Hurricane" is given; the wheels are there shown with direct radiating spokes. The spokes of both these curious locomotives were, however, of the V description, as shown in the illustrations (Figs. 28 and 29), and in Colburn's "Locomotive Engineering."

At the end of September, 1839, when the 31 miles of the line was open to Twyford, the driver of the "Hurricane," having obtained a promise from the directors that they would provide for his wife and family if an accident happened to him, undertook to drive the "Hurricane" to Twyford at the speed of 100 miles an hour; and, allowing three miles for getting up speed and stopping, it is stated that he successfully covered 28 miles at the rate of 100 miles an hour.

In 1846, Grissell and Peto, the well-known railway contractors, undertook the task of removing the mammoth bronze equestrian statue of the Duke of Wellington from Mr. Wyatt's studio in the Harrow Road, near the Great Western Railway locomotive shops, to Hyde Park. The car weighed 20 tons, and was borne by four wheels 10ft. in diameter, lent by the Great Western Railway, one pair being open-spoked wheels from under the "Hurricane," the other pair being constructed of disc sheet iron, and were from under the "Mars" or "Ajax." Both pairs are clearly illustrated in the *Illustrated London News* for October 10th, 1846.

Of the original Great Western locomotives there now only remain to be described the two geared engines supplied to the Great Western Railway by the Haigh Foundry Company. Unfortunately, little is known of these. Sir D. Gooch thus writes of them (after describing the spur and pinion gearing of the "Thunderer") :—" The same plan of gearing was used in the two engines built by the Haigh Foundry;

their wheels were 6ft.* diameter, and the gearing 2 to 1, but the cylinders were small. I felt very uneasy about the working of these machines, feeling sure they would have enough to do to drive themselves along the road." In the face of this emphatic and distinct statement of Sir D. Gooch respecting the two geared engines built by the Haigh Foundry Company, it has been stated that Sir D. Gooch was referring to the Haigh Foundry valve gear! Fancy reading "the same (spur and pinion) plan of gearing was used in the two engines built by the Haigh Foundry . . . the gearing being 2 to 1," and then being told that it was the Haigh valve gear that was meant!

In addition to Sir D. Gooch's statement, we are fortunate to have the evidence of an independent person. This eye-witness, who saw one of the Haigh geared-up engines at Paddington in August, 1838, gives a very interesting and lucid account of this engine and its trial trips. He writes:—"I have just returned from witnessing the performance of an engine on the Great Western Railway, built by the Haigh Company, upon somewhat of a new principle, which combines what the writer deems to be essential to the perfectibility of the locomotive engine—namely, slower motion of piston with increased speed of engine. The experiment was completely successful, and, although Mr. Harrison has abandoned his plan, the principle of giving increased speed by the application of tooth and pinion gear is fully established by this experiment.

"The engine started from Paddington with five carriages to Maidenhead, and returned with five carriages and two wagons loaded with iron, and frequently travelled at the rate of 40 miles an hour.

"The engine then took the five o'clock train with passengers to Maidenhead, and performed the journey at the rate of 36 miles an hour with from 120 to 150 passengers."

It will be noticed in the above statement that Harrison had already discontinued the 3 to 1 gearing of the "Thunderer." Sir D. Gooch says that he had to rebuild one-half of the original engines to make them of any service. It is more than probable that the two Haigh geared engines were thus rebuilt. Indeed, the fact that the books of the Great Western Railway show that the "Snake" and "Viper" had driving wheels 6ft. 4in. in diameter is evidence that

* The records at Swindon Locomotive Works show that the "Snake" and "Viper" had wheels 6ft. 4in. in diameter.

such was the case, as the geared engines when delivered had wheels 6ft. in diameter, and allowing that the small spur wheels were in a certain position, it would only be necessary to remove the spur wheels, slightly alter the length of the connecting-rods, and place wheels of 6ft. 4in. diameter on the crank axle to make ordinary locomotives of the engines in question.

It is also possible that the discs of the "Ajax" wheels were cut down to 8ft., and new tyres provided, which would account for the fact that in 1842 Whishaw gives the diameter of "Ajax's" driving wheels as only 8ft.

The following interesting table gives the result of the working of some of the original Great Western Railway locomotives:—

| Names of Engines. | LOAD. | | | RATE OF TRAVELLING. | | | COKE CONSUMED. | | | |
|---|---|---|---|---|---|---|---|---|---|---|
| | Carriages, etc. | Engine and Tender. | Gross Load | Distance. | Average time in 22½ miles. | Mean Rate. | Total Quantity. | Lbs. per mile. | Lbs. per ton per mile. | |
| | Tons. | Tons. | Tons. | Miles. | Min. | Miles per hour. | In lbs. | | Goods. | Gross load. |
| North Star | 40·5 | 28·5 | 69 | 8,848 | ·884 | 25·45 | 420,784 | 47·5 | 1·17 | ·69 |
| Æolus | 40·5 | 28·4 | 68·9 | 7,292 | ·287 | 23·81 | 353,360 | 48·4 | 1·19 | ·71 |
| Venus | 40·5 | 26·5 | 67 | 240 | ·100 | 22·5 | 12,656 | 52·7 | 1·3 | 78 |
| Neptune | 40·5 | 26·5 | 67 | 4,728 | ·949 | 23·83 | 188,384 | 39·8 | ·95 | ·59 |
| Apollo | 40·5 | 26·5 | 67 | 4,392 | ·942 | 23·81 | 193,080 | 43·9 | 1·08 | ·65 |
| Premier | 40·5 | 25 | 65·5 | 3,024 | ·99 | 22·73 | 159,936 | 52·8 | 1·3 | ·87 |
| Lion | 40·5 | 24 | 64·5 | 3,973 | ·96 | 23·43 | 226,576 | 57 | 1·4 | ·89 |

In consequence of the deficiency in the heating surface of many of the original broad-gauge engines, they had but a short career; among the first discarded were the "Ajax," "Planet," "Lion," "Apollo," "Hurricane," and "Thunderer."

Although their lives as locomotives were ended, they were made to perform the functions of stationary engines; thus, during repairs to the beam engine in the fitting shops at Swindon Works in 1846 or 1847, the "Lion" and "Planet" supplied the motive power to actuate the machinery, while the "Apollo" supplied steam to work the first Nasmyth's steam-hammer erected at Swindon.

## CHAPTER VII.

Opening cf the London and Birmingham Railway—" Wallace," with feed-water heating apparatus.—Dr. Church's tank engine, "Eclipse."—Balanced locomotives.—Smoke-consuming locomotives.—Opening of the London and Southampton Railway.—" Soho," a locomotive without eccentrics.—A double flanged wheel engine.—Hancock's attempts to supply railway locomotives.—American engines for England.—Particulars of the engines and their working. —Gooch commences to design engines for the Great Western Railway.—His patent steeled tyres—Gray introduces expansive working.—Trial of his valve gear.—The "long boiler" fallacy.—Stephenson's design for the York and North Midland Railway.—Rennies build a powerful locomotive.—Inventor of the link motion: Howe, Williams, or Stephenson?—America claims the credit for the improvement.—Beyer's single-plate frame engines.—Early Crewe engines.—Robertson fits a steam brake to a locomotive.—Engines for working the Cowlairs incline.—Bodmer's reciprocating or "compensating" engines.— Tried on the Sheffield and Manchester, South Eastern, and London and Brighton Railways.--They prove failures.—McConnell's "Great Britain."—Dewrance's coal-burning "Condor."

EDWARD BURY, the celebrated locomotive engineer, of Liverpool, contracted to supply the London and Birmingham Railway with locomotives. The first portion of the line was opened on June 20th, 1837, and four-wheel Bury engines of his well-known types hauled the trains. Fig. 30 shows one of his standard passenger engines for the London and Birmingham Railway.

In 1838 Kimmond, Hutton, and Steele, of Dundee, built a locomotive, named "Wallace," for the Dundee and Arbroath Railway, at a cost of £1,012, including the tender. This engine had inside frames and inclined horizontal outside cylinders, 13in. diameter, 18in. stroke; the driving wheels were 5ft. 6in. diameter, the leading and trailing being 3ft. 6in. diameter; the valve chests were on top of the cylinders. The exhaust steam was turned into the tender for the purpose of heating the feed-water. The "Wallace" was described as being, "without exception, one of the most splendid and beautifully finished pieces of mechanism; indeed, all present who had seen the 'Scorpion,' 'Spitfire,' and other celebrated English engines, gave the preference to the 'Wallace.'" The gauge of the Dundee and Arbroath Railway was 5ft. 6in.

Dr. Church, a celebrated scientific experimentalist of Birmingham, constructed a four-wheel tank engine in 1838, named the "Eclipse."

EVOLUTION OF THE STEAM LOCOMOTIVE 83

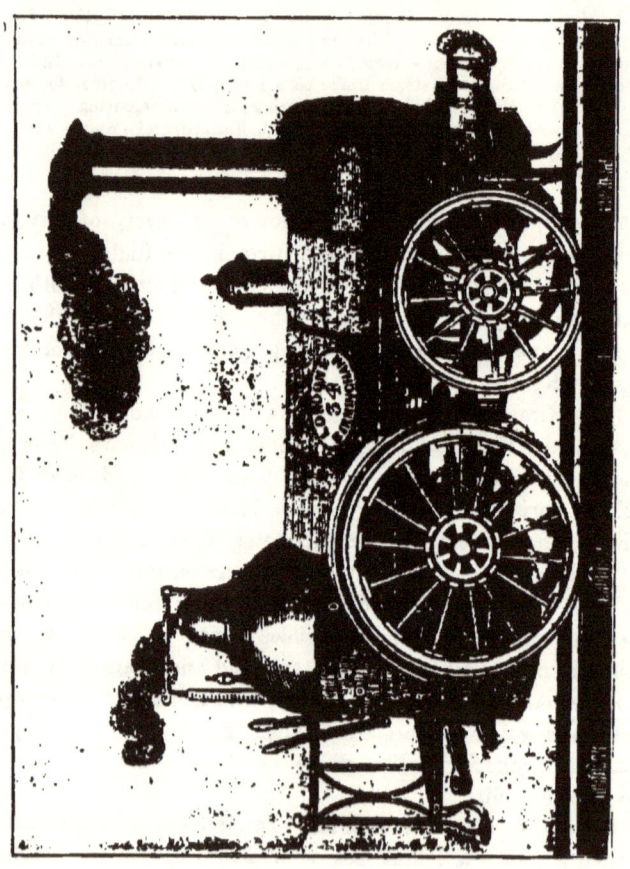

FIG. 30.—BURY'S STANDARD PASSENGER ENGINE FOR THE LONDON AND BIRMINGHAM RAILWAY

This locomotive was used in the construction of the London and Birmingham Railway. The cylinders were placed outside in a horizontal position, and were 11½in. diameter, the stroke being 24in. The leading or driving wheels were 6ft. 2½in. diameter, and are said to have been the largest used up to that time on the narrow-gauge, being 2½in. larger than the 6ft. wheels of the original "Liverpool." The trailing wheels were 3ft. diameter. The water tanks were placed beneath the boiler, and when loaded the driving wheels sustained a weight of 9 tons, and the trailing 5 tons. The "Eclipse" hauled a load of 100 tons, and when running "light" attained a speed of 60 miles an hour. It will be observed that for the size of the driving wheels, weight of engine, design, and speed, the "Eclipse" was a considerable advance on the narrow-gauge practice then obtaining. The "Eclipse," after being rebuilt, was at work at Swansea in 1861.

In 1838, two important improvements were introduced in locomotive construction—viz., the balancing of the reciprocating parts of the engine, and the partially successful use of coal in place of coke as fuel. Heaton, an engineer of Birmingham, introduced the balancing of locomotive wheels. This was in August, 1838, when he made a model engine on the suggestion of a director of the London and Birmingham Railway. The "Brockhall," one of the engines of the Company, was repaired at the Vulcan Works, Birmingham, early in 1839, and was then fitted with Heaton's improvement. Sharp, Roberts and Co. had, in the previous December, supplied an engine to the London and Southampton Railway fitted with balancing weights just within the wheel rim ; while Heaton's weights took the form of an extension of the crank-throws on the opposite side of the axle, a method still employed in modern engines. The first locomotive that ever burned coal in a satisfactory manner, without the smoke causing a nuisance, was the "Prince George," a six-wheel engine belonging to the Grand Junction Railway. In 1838 it was fitted with Chanter's patent furnace, the fire-bars of which sloped from the fire-box door to the tube-plate at an angle of 45 degrees; over the fire-bars was a deflector. The motion of the engine caused all the fuel to fall to the lower end. Early in 1839 another six-wheel engine belonging to the Grand Junction Railway, the "Duke of Sussex," with cylinders 13in. by 18in., was fitted with a Chanter furnace. This time the fire-bars did not slope so much, and on a trip from Crewe to Liverpool the engines covered several consecutive miles at the speed of 60 miles an hour, the officials of the company at the same time declaring that the engine emitted no more smoke than the engines burning coke.

The first portion of the London and Southampton Railway (now the London and South Western) was opened on May 12th, 1838, from London to Woking. The original locomotives were, with four exceptions, six-wheel "single" engines, with driving wheels 5ft 6in. diameter. Fig. 31, "Garnet," is an illustration of one of these locomo-

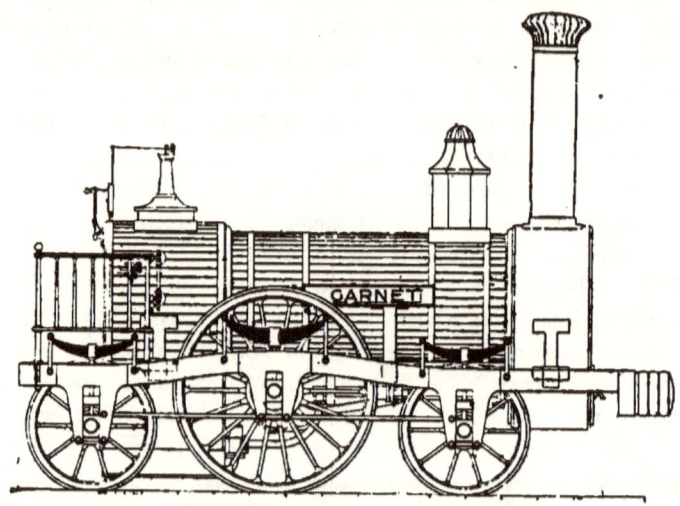

FIG 31.—"GARNET," ONE OF THE FIRST ENGINES SUPPLIED TO THE LONDON AND SOUTHAMPTON RAILWAY

tives; the cylinders were 13in. diameter, and the stroke 18in. The leading and trailing wheels were 3ft. 6in. diameter. The "Garnet" weighed 13 tons empty.

In 1839, Peel, Williams, and Peel, of Soho Works, Ancoats, sent the first locomotive constructed by them to the Liverpool and Manchester Railway. This engine was named "Soho," and took a train of 25 loaded wagons, weighing 133 tons 18 cwt. 2 qrs., from Liverpool to Manchester; whilst for a fortnight before this she was running with the ordinary passenger trains, and "no failure had taken place, and the trains having usually been brought in before their time." The improvement introduced into this engine consisted of a new method of working the valves. The "Soho" had no eccentrics, but in place of them were two spur wheels, staked on to the crank axle, driving two other wheels of equal diameter placed immediately over them, so as to preserve the distance between the centres constantly the same, and unaffected by the motion of the engine on its springs. The wheels last mentioned were attached to a short axle, carrying at each end a small crank arm, which drove a connecting-rod attached to the valve spindle.

Fenton, Murray, and Jackson, of Leeds, in 1839, supplied a six-wheel engine named "Agilis" to the Sheffield and Rotherham Railway. We have only very meagre details relating to this locomotive, but she is said to have had flanges an each side of the wheels, and also "that if either one or all the eccentrics which move the valves were broken, disarranged, lost off, or taken away, she is still under the control of the engineer, who can safely conduct her along the railway nearly as well as if those parts had remained entire." No explanation is given as to "how it was done"!

In 1840, Walter Hancock, of Stratford, Essex, who was well known as a steam road-coach builder, constructed a locomotive on somewhat the same system as his steam coaches. This engine was tried on the Eastern Counties Railway. The boiler was of peculiar design, containing a number of separate chambers, each enclosing several tubes. Each chamber or set of tubes connected with two general reservoirs, one at the bottom for the supply of water, the top one being a reservoir for the steam. The connection from each chamber to the water, steam pipes, and reservoirs had self-acting valves, so that should an accident happen to any one chamber the self-acting valves were closed by the pressure of the steam above, or the water beneath, so that the remainder of the boiler retained its efficiency, the only result of the accident being a reduction of the heating surface. An accident of this kind was not so serious as a burst tube, as the damaged portion was automatically thrown out of use. Another advantage of this locomotive was the great heating surface contained in a comparatively small space; a further improvement was a reciprocating set of fire-bars. The cylinders were vertical, and actuated an independent crank shaft; the progressive motion was conveyed to the wheel axle by means of endless chains working over pulleys fixed on the driving wheel axles, the diameter of the pulleys being graduated, so that the engine could be geared up or down, as either speed or power was required.

As the machinery did not directly drive the wheels, it was possible to put that portion out of gear when it became necessary to work the feed pumps, etc. This was a considerable improvement on the usual locomotive, which upon such occasions either had to make a few trips for the purpose of supplying the boiler with water, or else perform over a "race."

In 1839, Norris, the locomotive builder of Philadelphia, U.S.A., made an offer to the directors of the Birmingham and Gloucester Railway to provide engines for working the severe gradient known as the Lickey Incline, 2 miles 3.35 chains in length. The agreement stipulated that the "locomotive engines were to be of a higher power, greater durability, and less weight than could be obtained in' this country. They were to be subjected to 15 trials within 30 days, and prove their capability by drawing up a gradient of 1 in 330 a load of 100 tons gross weight, at a speed of 20 miles an hour, and up a gradient of 1 in 180 a load of 100 tons at the speed of 14 miles an hour." If the American locomotives fulfilled these conditions the Birmingham and Gloucester Railway were under a contract to accept ten of the engines, at a price not exceeding £1,600 each, including the 20 per cent. import duty. Captain Moorsom, the engineer of the railway, stated that the "engines had not strictly complied with the stipulated conditions, yet he considered them good, serviceable engines." It will be observed that no guarantee was given as to what work these engines would accomplish on the Lickey Incline.

The first three engines to arrive were the "England," "Columbia," and "Atlantic," and, according to the arrangements between the builder and the Birmingham and Gloucester Railway, they underwent a series of trials on the Grand Junction Railway before the directors of the Birmingham and Gloucester Railway accepted the engines. These trials took place during April and May, 1839, between Birmingham and Liverpool, a double journey of 156 miles being frequently made in one day. The requisite load could not always be obtained, and it then became necessary to add empty wagons to the train to make up the right weight. The trains on some of the occasions exceeded 220 yards or 1·8 mile in length. With a steam-working pressure of 62lb. per square inch, the results tabulated were as follows:—On a rising gradient of 1 in 330, with a load ranging between 100 and 120 tons, the speed ranged from 13 4-5 miles to 22½ miles an hour; on an incline of 1 in 177, with a load of 100 tons, the variation in speed ranged between 9 4-5 miles and 13 4-5 miles an hour. Twenty-one trial trips were made, and in only five were the stipulated performances carried out, in five others doubt existed as to the work performed, but in eleven the engines failed to do the required amount of work.

These experiments showed a curious result with regard to the fuel consumed. The aggregate rise of the gradients from Liverpool to Birmingham is about 620ft.; that from Birmingham to Liverpool is about 380ft. (exclusive in both cases of the Liverpool and Manchester Railway); the difference, therefore, up to Birmingham is about 240ft.

In seven journeys of 596 miles up to Birmingham, the engine conveyed 682 tons gross, evaporated 12,705 gallons of water, and consumed 177 sacks of coke (1½ cwt. each). In seven journeys of 596 miles down from Birmingham, the same engine conveyed 629 tons gross, evaporated 12,379 gallons of water, and consumed 177 sacks of coke. It would thus appear that the consumption of fuel was the same in both cases, and the only difference was the evaporation of 326 gallons of water more in the journey up than in the journey down, conveying nearly the same load both ways. The construction of these engines was very simple, and the work plain. The boiler was horizontal, and contained 78 copper tubes 2in. diameter and 8ft. long, with an iron fire-box. The cylinders, 10½in. diameter, were inclined slightly downwards, and so placed that the piston-rods worked outside the wheels, thus avoiding the necessity of cranked axles.

The framing of these American engines was supported by six wheels; the two driving wheels of 4ft. diameter were placed close before the fire-box; the other four wheels, of 30in. diameter, were attached to a truck, which carried the front end of the boiler, and was connected with the frame by a centre-pin, on which it turned freely, allowing the truck to accommodate itself to the exterior rail of the curve, and, with the assistance of the cone of the wheels, to pass round with very little stress upon the rails.

|  | Tons | cwt. |
|---|---|---|
| The weight of the engine with the boiler and firebox full was... | 9 | 11¼ |
| That of the tender with 21 cwt. of coke and 520 gallons of water was... | 6 | 4¾ |
| Total weight | 15 | 15¼ |

These engines, when empty, weighed only eight tons each.

Another of the American bogie engines supplied to the Birmingham and Gloucester Railway was named the "Philadelphia." She was a more powerful locomotive than the three mentioned above, and Captain Moorsom, the engineer of the railway, in a letter dated from Worcester on June 22nd, 1840, gives an interesting account of her trial on the Lickey Bank. "Seventy-six chains in the incline of 1 in $37\frac{1}{2}$ were made ready with a single way, and three chains nearly level were laid temporarily to rest upon before starting. The road was quite new, and consequently not firm or well gauged, and the works going on close at hand occasionally covered the rails with dirt. The wagons used were of a large class, like those on the Manchester and Leeds line, and weighed when empty rather more than $2\frac{1}{2}$ tons, and at first worked very stiffly. They were loaded with 4 tons, and generally weighed, including persons upon them, about $6\frac{3}{4}$ tons. The 'Philadelphia' weighed (as she worked) about 12 tons 3 cwt., and her tender weighed nearly 7 tons, being in all 19 tons. She had $12\frac{1}{2}$in. cylinders, 20in. stroke, 4ft. driving wheel not coupled. The weight on her driving wheels was 6 1-3 tons (as weighed at Liverpool) without water.

"The usual load she took was eight wagons, engine, and tender, with persons, equal to 74 tons gross weight, in ten minutes, or nearly 6 miles per hour, the last quarter of a mile being at the rate of $9\frac{3}{4}$ miles per hour. Seven wagons, etc., equal to $67\frac{1}{4}$ tons gross weight, in about 9 minutes, or $6\frac{1}{2}$ miles per hour mean speed. Six wagons, etc., equal to 61 tons gross weight, in sometimes $5\frac{1}{4}$ and sometimes $6\frac{1}{2}$ minutes, say in 6 minutes average, or 9 miles per hour mean speed, the last quarter of a mile usually giving a speed of nearly 11 miles an hour. Five wagons, equal to about 53 tons gross, were usually taken at a speed of 13 miles per hour for the last half-mile up. The foregoing results occurred generally during fine weather, but sometimes the rails were partially wet, and this occasioned a difference of speed in the ascent of half a minute to a minute and a half. One day when showery the men walked over the rails with marl on their boots, rendering the way very greasy and slippery, also the lower part of the plane had been formed only a few hours, and was very soft and badly gauged.

Under these circumstances the 'Philadelphia' took five wagons, self, and tender, being a gross weight, including persons, of about 53 tons, up at a mean rate of rather more than 5 miles per hour, and the last quarter of a mile was passed at the rate of 8 miles per hour. Two wagons were then taken off, and the 'Philadelphia' took the remaining three wagons, self, and tender, being a gross weight, including persons, of 40 tons, up at a mean rate of 12 miles nearly per hour, her maximum speed being nearly 16 miles per hour."

Sir D. Gooch was not at all satisfied with the original broad-gauge locomotives, and in 1839 he obtained the sanction of the directors of the Great Western Railway to design two classes of locomotives for the railway. These engines were known as the "Firefly" class and the "Fury" class, the former having 7ft. driving wheels, cylinders 15in. diameter, 18in. stroke, and 700ft. of heating surface; the latter had 6ft. driving wheels, cylinders 14in. diameter and 18in. stroke, and 608ft. of heating surface.

One hundred and forty-two locomotives of the "Fury" and "Firefly" design were constructed. Sir D. Gooch states that the best

FIG. 32.—"HARPY," ONE OF GOOCH'S "FIREFLY" CLASS OF BROAD-GAUGE ENGINES

were built by Fenton, Murray and Jackson, of Leeds. The sixty-two of the "Firefly" class were built as follows:—Twenty, by Fenton, Murray and Jackson, Leeds; sixteen, by Nasmyth, Gaskell and Co., Manchester; ten, by Sharp, Roberts and Co., Manchester; six, by Jones, Turner and Evans, Newton; six, by Longridge and Co., Bedlington; two, by Slaughter and Co., Bristol; and two, by G. and J. Rennie, London.

It will be observed that most of these were built in the North of England, and it is a significant fact that these broad-gauge locomotives were conveyed on narrow-gauge trucks for some hundreds of miles to the Great Western Railway, thus showing that it would have been quite possible to widen the existing narrow-gauge railways, by simply decreasing the space between the two roads, comparatively at a small expense.

All these engines were built from the specifications and drawings supplied by the Great Western Railway to the makers, and thin iron templates were also supplied of those parts which were to be interchangeable. Fig. 32 illustrates the "Firefly" type.

The "Firefly," built by Jones and Co., Viaduct Foundry, Newton, was the first of these engines delivered. On March 28th, 1840, she made an experimental trip from Paddington to Reading, with a load of two carriages, containing 40 passengers, and a carriage truck; she performed the journey in 46 minutes 25 seconds from start to stop. A spring of one of the tender wheels broke on the journey, necessitating careful running. On the return trip, between the 26th and 24th mile posts, a speed of 56 miles an hour was reached, and the average speed from Twyford to Paddington was over 50 miles an hour. On the occasion of the Queen's accouchement in August, 1844, the news was brought to London by a special messenger travelling on one of these engines. The journey from Slough to Paddington, 18¼ miles, was accomplished in 15 minutes 10 seconds, or at the rate of 75 miles an hour. The illustration (Fig. 33) shows the interior of the old Paddington engine shed, and amongst the locomotives to be seen are the "Ganymede" and "Etna." All the engines had domed fire-boxes, and outside frames, the principal dimensions, in addition to those already given, being: Leading and trailing wheels, 4ft. diameter; boiler barrel, 8ft. 6in. long, 4ft. diameter; 131 tubes, 2in. diameter, 9ft. long; weight, in working order on leading 4¾ tons, driving 11 tons 13 cwt., trailing 7 tons 16 cwt.; total, 24 tons 4 cwt.

On November 20th, 1840, Daniel Gooch obtained a patent for steeled tyres, and the locomotives of the "Fury" and "Firefly" classes were fitted with these patent tyres. Although the tyres only contained one-fifth part of shear steel, yet the use of Gooch's tyres did not become general, as 56 years ago steel was an expensive commodity, and consequently railway rolling stock generally was not fitted with steel tyres; indeed, the Great Western Railway went no further than using the improvement for their locomotive and tender wheels. Many locomotives fitted with these patent tyres ran nearly 300,000 miles before new tyres were required.

FIG. 33.—INTERIOR OF PADDINGTON ENGINE HOUSE, SHOWING THE BROAD-GAUGE LOCOMOTIVES OF 1840

These first essays of Daniel Gooch as a locomotive designer at once placed him at the very head of locomotive engineers, and Gooch himself, usually so modest, says of these locomotives, "I may with confidence, after these engines have been working for 28 years, say that no better engines for their weight have since been constructed, either by myself or others. They have done, and continue to do, admirable duty." This candid eulogium of these engines by their designer certainly did not go beyond the truth in describing their good points. Gooch's first design of broad-gauge goods locomotives had six coupled wheels 5ft. in diameter, inside cylinders 16in. diameter, and a stroke of 24in. The fire-box was of the domed pattern. Fig. 34 ("Jason") represents one of these engines.

John Gray, who was in 1840 locomotive superintendent of the

Hull and Selby Railway, introduced a striking improvement into the construction of locomotives at that time. (Gray had, on July 26th, 1838, taken out a patent for his valve gear; and whilst on the subject of valve gears, it will be of interest to note that Dodds and Owen patented their wedge motion on September 16th, 1839.) In Gray's improvements in the Hull and Selby engines he adopted

FIG 34.—" JASON," ONE OF GOOCH'S FIRST TYPE OF GOODS ENGINES FOR THE GREAT WESTERN RAILWAY

inside bearings for the driving wheels, an extended base for the springs, and, of course, his patent valve motion and expansive working. Shepherd and Todd, of the Railway Foundry, Leeds, constructed the engines in question. The driving wheels were 6ft. diameter, cylinders 12in. diameter by 24in. stroke, fire-box 2ft. by 3ft. 6in. (inside), and 94 2in. tubes, 9ft. 6in. long. Two of these locomotives, "Star" and "Vesta," were tried in competition with other engines on Tuesday, November 10th, 1840. Sixteen trips were made by the "Star" and "Vesta," the average loads being 55.4 tons, or 1,718 tons over one mile; coke consumed, 465lb., or 0.271lb. per ton per mile; water evaporated, 2,874lb., or 1.62lb. per ton per mile.

Two other classes of locomotives were tried in competition with Gray's patent—viz., the usual kind of engines then in use, and the same with the addition of Gray's expansion gear

The result of the trials is shown in the following table:—

| Class of Engine | Load in Tons conveyed over 1 mile in lbs. | Elsecar Coke used per trip of 31 miles in lbs. | Coke used per mile in lbs. | Coke used per ton per mile in lbs. | Water used per trip of 31 miles in lbs. | Water per mile in lbs. | Water per ton per mile in lbs. |
|---|---|---|---|---|---|---|---|
| Patent    | 1649·4 | 446·98  | 14·41 | 0·271 | 2672   | 86·19  | 1·62 |
| Altered   | 1649·4 | 686·15  | 22·13 | 0·416 | 4601·6 | 148·43 | 2·90 |
| Unaltered | 1649·4 | 1007·78 | 32·59 | 0·611 | 6432·8 | 207·5  | 3·97 |

The financial annual result of the three classes of engines for coke and boilers, with such a traffic as that of the Hull and Selby line, was about:—

£4,500 for the unaltered engines.
£3,250 for the altered engines.
£2,000 for the patent engines.

We have now reached the era of another development of the locomotive—viz. the introduction of "long boiler" engines; but although the idea was well "boomed," it never was thought much of by competent locomotive engineers; indeed, many severely condemned the plan.

In 1841 Robert Stephenson patented a new form of valve gear, with a top and bottom gab fixed to the valve spindle, and the ends of the eccentric rods kept apart by a straight link. Here, again, Stephenson introduced nothing new, his gear being but a clumsy adaptation of Roberts's valve gear. An engine of this description (generally known as Stephenson's patent "long boiler" engine) was tried on the York and North Midland Railway in January, 1842, the dimensions being:—

```
Diameter of Cylinder     ...  ...  ...  ...  ...  14 inches
Length of Stroke         ...  ...  ...  ...  ...  20   ,,
Diameter of Driving Wheels ... ...  ...  ...  ...  5¼ feet
Diameter of small wheels ...  ...  ...  ...  ...  3    ,,
There are 150 tubes, giving a heating surface of  ... 765 ,,
Copper Fire Box, with a heating surface of  ...   ... 30  ,,
                                   Total heating surface 795 feet
Length of Boiler, including fire and smoke boxes ... ... 17   ,,
Weight of the Engine in working order   ...  ..  ...  15 tons
```

During a journey of 90 miles, a speed of 48 miles an hour was attained, but the train then consisted of only five carriages of light weight.

The consumption of fuel during the above experiment was 19.2lb.

per mile, with a load of eight coaches over half the distance (45 miles) and five coaches over the remaining half.

This consumption included the whole of the fuel used in lighting the fire and raising the steam.

R. Stephenson introduced tubes of wrought-iron instead of brass or copper, in order that the increased heating surface might be obtained without a corresponding augmentation in the price of the engine. This he did not adopt without making several experiments.
During the last twelve months he had several boilers working under his own eye with iron tubes, for the special purpose of determining how far he could recommend them for general adoption. The result was all that he could desire; and owing to this he introduced them with great confidence. The valve gear is thus eulogised: "In ordinary engines the mechanism for working the slide valves was very liable to derangement and considerable wear and tear.

"This part of the engine he so far simplified that it required only a simple connection between the eccentrics and slide valves, thus doing away with a considerable number of moving parts.

"This was attained by placing the slide valves vertically on the sides of the cylinders, instead of on the top as heretofore, so that the direction of the sliding motion of the valves and the central line of the valve-rods intersected the central line of the main axle at the point where the eccentrics were placed. In this case the eccentric-rods were connected immediately to the prolongation of the valve-rods, without the usual intermediate levers and weigh bars; the slide valves of both cylinders were placed in one steam chest, between the cylinders." Another improvement was in the working of the feed pumps; it consisted in connecting the pump-rods to the eccentrics used for reversing the engine. By this arrangement the velocity of the moving part of the pump was greatly diminished, by which was secured greater regularity of action.

Messrs. G. and J. Rennie, of Holland Street, Blackfriars, S.E., in 1841, constructed a locomotive named the "Lambro" for the Milan and Monzo Railway. The "Lambro" was built from the design of Mr. Albano, the engineer to the railway; the cylinders were 13in. diameter, 18in. stroke, driving wheels 5ft. 6in. diameter, steam pressure 50lb., weight 22 tons. Her average coke consumption with trains weighing 143 tons at 36 miles an hour, was only 22lb. per mile. The locomotive engineer of the railway reported that "no engine he had seen at all approached the locomotive engine 'Lambro' in any respect whatever, in the economy of fuel, in her immense dragging power, and in the excellency and solidity of her framing and working gear."

The particular evolution now about to be described occupies a foremost position in locomotive history. Like many other useful inventions, the link motion has been proclaimed as the production of different people.

Its popular title, the "Stephenson" link motion, is a well-known misnomer; indeed, Stephenson never appears to have put forward a claim in which he figured as the inventor of the curved link motion, perhaps, at first, he did not fully appreciate its value.

The germ of the idea belongs to Williams, of Newcastle, who, in 1842, designed a form of straight link coupling the two eccentrics together. Of course, such an arrangement was utterly impossible in practice, as the crank, in revolving, would soon place the two eccentrics in such a position that the link would be destroyed. The curved link, placed half-way between the valves and eccentrics, was soon evolved from Williams' crude idea, and up to 1846 it was most generally called Williams' motion. In an article describing expansion valves, in the *Practical Mechanics' Magazine* for April, 1846, it is so described; but in the May number of the magazine a letter appears from William Howe, a fitter employed by R. Stephenson and Co., Newcastle. In this communication Howe states that Williams proposed the straight link, previously mentioned, but that Howe saw its utter impracticability, and evolved the curved link. Williams made no reply to this communication; although he may not have seen Howe's letter claiming the invention. Be this as it may, Howe was thereafter given the credit for the curved link. It is, however, significant that he never patented it, and it is probable that at first neither he nor Stephenson saw its value as a means of effectually working the valves expansively, or one or the other would have protected the invention, seeing that Stephenson had then quite recently patented the top and bottom gab-gear. Then, again, Howe's supposed claim may have been a reason for not protecting it.

In the invention of the link motion, this country does not appear to have been forestalled by the Celestial Empire, as (it is asserted) is the case with so many useful discoveries. But the glory does not rest with us, for it has been shrewdly "guessed" that the idea originated with one of our American cousins, W. T. James, of New York, who, as early as 1832, constructed the "James" locomotive, which was provided with link motion. The invention at this period does not appear to have been considered of any value, for its use was not perpetuated in later locomotives in America until after it had been re-discovered by the Williams-Howe experiments of 1842-3.

In 1843, Mr. C. Beyer, then employed with Messrs. Sharp Brothers and Co., but afterwards of the well-known firm of locomotive builders, Messrs. Beyer, Peacock and Co., Manchester, introduced the single iron plate for locomotive frames.

Trevithick's son directed his attention to the evolution of the steam locomotive, and while chief engineer of the Grand Junction Railway, the now world-famous Crewe Works were erected, being opened in 1843. Mr. A. Allan became manager at Crewe, and under his superintendence a new class of engines was constructed, the novel points being the coupling of the driving and trailing wheels—Allan having, in 1863, publicly claimed this innovation as wholly and solely due to him.

The engines in question are usually described as "the old Crewe goods class," and had outside cylinders, 15in. by 14in. The coupled wheels were 5ft. diameter, and were placed one pair before and the other behind the fire-box; these wheels had inside bearings, and the small leading pair had outside bearings. The steam pressure was 120lb. These useful engines weighed 19½ tons, and were used for goods traffic for many years. Mr. Ramsbottom afterwards rebuilt several of them as tank engines, and some, as such, are still in use on the London and North Western Railway. Alexander Allan, who died as recently as 1891, was noted for his invention of a straight link motion in 1855.

The need of a powerful brake has always been one of the greatest necessities of locomotive engineers. For a long time they all agreed that it was not advisable to brake the driving wheels of locomotives; but Peter Robertson, the locomotive superintendent of the Glasgow and Ayr Railway, was of a different opinion, and in April, 1843, he fitted a locomotive on that railway with his patent steam brake. The apparatus consisted of a flexible metal band, of a semi-circular shape, surrounding the upper half of the driving wheel. One end of the band fastened to a hinge, and the other was fixed to a piston-rod. When "off," the piston-rod held the band away from the tyre of the driving wheel, but when steam was applied behind the piston the band was tightly pressed against the tyre. Such was the simple, but effective, application of Robertson's steam brake. A familiar example of its action can be seen in the hand brakes still fitted to cranes.

The Cowlairs incline at Glasgow is the *bête noir* of the North British Railway, and is situate just outside the Glasgow terminus of what was originally the Edinburgh and Glasgow Railway. When first opened this incline was (as is, indeed, at present the case) worked by stationary engines; but towards the end of 1843 Mr. Paton, the locomotive superintendent, and Mr. Millar, the engineer of the Edinburgh and Glasgow Railway, designed and built a powerful locomotive for working this two-mile incline of 1 in 42. The engine was

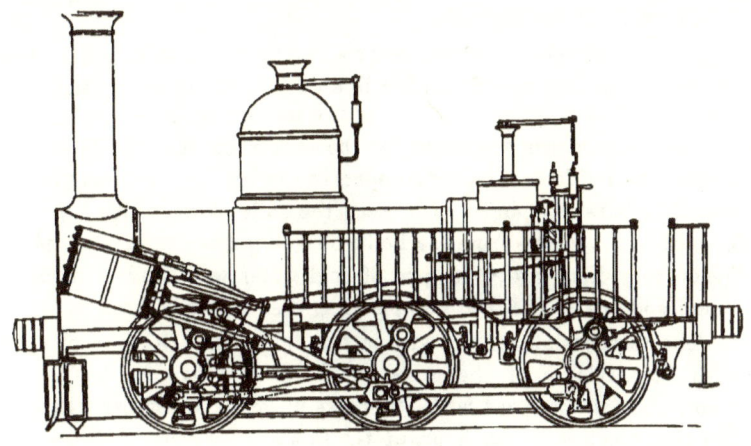

FIG. 34A.— PATON AND MILLAR'S TANK ENGINE FOR WORKING ON THE COWLAIRS INCLINE, GLASGOW

put to work in January, 1844, and during that year the cost of working the incline was, with the locomotive, one-third of the amount expended during the previous twelve months on the stationary engine.

Upon reference to the illustration (Fig. 34A) of this remarkable locomotive, the first detail that attracts notice is the immense steam dome. The engine was supported on six coupled wheels of 4ft. 3½in. diameter. The cylinders were "outside," fixed in an inclined position about half-way up the smoke-box, their diameter being 15½in. The stroke was 25in. These dimensions, it will be noticed, were considerably in advance of the general practice obtaining 55 years ago. The

## EVOLUTION OF THE STEAM LOCOMOTIVE

valve chests were above the cylinders, and the eccentrics were fixed on the driving axle, within the frames; the springs were underhung, and all the wheels were counterbalanced. Two lever safety valves were provided. The heating surface of the fire-box was 60 sq. ft., that of the tubes 748ft. The other principal dimensions of this engine were:—Fire-box, 4ft. long by 4ft. 6in. deep; smoke-box, 2ft. 6in. long by 4ft. 4in. deep; 136 tubes, 2in. diameter, and 10ft. 6in. long. This engine, it should be observed, was of the "tank" class, 200 gallons of water being stored in a tank below the smoke-box, that amount being sufficient for two trips. The water was supplied from a stand-pipe, and not from the usual columns.

The driving wheels were furnished with brakes, the levers of which were worked by a screw, the handle of the latter being placed within reach of the engineer.

The trailing pair of wheels had a steam brake, something like those applied to the engines of the Ayr line by Mr. Robertson. Sand-boxes were placed in front on each side of the water tank for dropping sand on the rails, which was done by the stoker on the footplate, by a handle and rod from valves or stoppers in the boxes. The most effectual remedy against slipping was to keep the rails clean, which was done by means of two jets from the boiler in going down the incline plane. When very dirty two other jets of cold water were used, a small air vessel and one of the feed pumps being used for that purpose.

The total weight of the engine was 26½ tons; the rate of speed with 12 carriages of the gross weight of 54 tons was 15 miles per hour; the rate of speed with 20 trucks of goods of a gross weight of 104 tons was 9 miles per hour, up the Cowlairs incline.

WORK OF ENGINE FOR THE MONTH OF NOVEMBER, 1844.
TOTAL WORK DONE ON INCLINE PLANE.

| Carriage. | | Trucks. | | Brakes. | | Gross Total. | Lifts. | Piloting. | Week end ing Nov. |
|---|---|---|---|---|---|---|---|---|---|
| No. | Tons. | No. | Tons. | No. | Tons. | Tons. | Total. | Hrs | |
| 335 | 1,675 | 694 | 3,817 | 298 | 1,937 | 7,429 | 121 | 120 | 7 |
| 344 | 1,720 | 673 | 3,701½ | 268 | 1,742 | 7,163½ | 107 | 118 | 14 |
| 375 | 1,875 | 658 | 3,619 | 248 | 1,612 | 7,106 | 104 | 118 | 21 |
| 376 | 1,900 | 640 | 5,525½ | 254 | 1,660 | 7,076½ | 103 | 118 | 28 |
| 1,430 | 7,170 | 2,665 | 14 663 | 1,068 | 6,951 | 28,775 | 435 | 474 | — |

The table on page 99 gives the results of one month's working of a second locomotive of similar design, the cylinders, however, being 16½in. diameter, and additional heating surface being provided by means of a water space dividing the fire-box. The second engine was put to work towards the end of 1844.

These engines were named "Hercules" and "Sampson," and were built at Cowlairs, whilst two others of the same general design, and named "Millar" and "Hawthorne," were constructed at Newcastle.

Mr. A. E. Lockyer states that these engines "had not run any length of time, however, before the foreman platelayer complained of the engines destroying the rails, which, it must be remembered, were only 58lb. per yard, with the sleepers 3ft. apart." In consequence of this report the incline was relaid, the distance between the sleepers being reduced to 2ft. between the centres. This did not much mend matters, and to crown all, the Forth and Clyde Canal began to leak, in consequence, no doubt, of the vibration induced by the constant passage of the heavy locomotives. A strategic movement to the rear then became necessary, and an eminent engineer (Mr. McNaught) was appointed by the directors to strengthen the land engine, and put it in proper working order, so as to reintroduce the haulage system for working the incline.

A Newcastle firm (R. S. Newall and Co., the original inventors and patentees of untwisted iron rope) supplied the railway company with one of their wire ropes. The land engine was finished by March 4th, 1847, and on trial under the new conditions the haulage system proved highly satisfactory, so much so that the four locomotives were removed altogether.

The Manchester and Sheffield line was, in 1845, supplied with four powerful goods locomotives, built on Bodmer's patent principle. The cylinders were 18in. diameter, stroke 24in.; the six coupled wheels were 4ft. 6in. diameter; but the weight of these engines was only 24 tons each. They are, however, stated to have been equal to hauling a gross load of over 1,000 tons. Bodmer's locomotives deserve recognition in the evolution of the steam locomotive, because of their curious construction, and also because other locomotive histories do not mention these peculiar engines.

The engines are described as "compensating," the whole strain being confined to the pistons, piston-rods, connecting-rods, and cranks. There were two pistons in each cylinder, one being connected with one crank and the other with the opposite crank of an axle with double cranks on each side, so that the driving axle was fitted with four cranks.

The steam was admitted alternately between the two pistons at the time the pistons met in the middle of the cylinders, also between the ends or tops of the cylinders and the pistons when the latter arrived at the other end of the stroke.

Bodmer claimed that by this arrangement the engine was perfectly balanced, and no oscillation or pitching of the engine resulted, no matter what speed was attained. Another engine of this description was supplied to the Sheffield and Manchester Railway, constructed by Sharp Bros. and Co. The cylinders were 14in. diameter, stroke 20in. (two strokes of 10in. each in both cylinders), driving wheels 5ft. diameter, steam pressure 90lb. per square inch. During November, 1844, the average coke consumption of this engine amounted to only 21.92lb. per mile.

A larger and more powerful engine on the same principle was supplied to the Joint Locomotive Committee of the South Eastern and London and Brighton Railways, and when the Committee was dissolved the engine was taken over by the South Eastern Railway in 1845, and was numbered 123. The cylinders were 16in. diameter and 30in. stroke, or rather, two pistons each working a stroke of 15in. Heating surface was: box 73 sq. ft.; tubes 769 sq. ft.; steam pressure 95lb.; weight 18 tons; coke consumption 15lb. per mile. The driving wheels were 5ft. 6in. diameter. Shortly after the South Eastern Railway took over this engine it broke down, and one of the men in charge was killed.

Bodmer also supplied the London and Brighton Railway with one of these patent reciprocating engines. This was in December, 1845; and she ran the first 5 p.m. express from London to Brighton. The locomotive in question was No. 7, and had single driving wheels, 6ft. diameter. The cylinders were 15in. diameter, and the 20in. of stroke was, of course, covered by two pistons in each cylinder working 10in. The fire-box was of the well-known "Bury" type. No. 7 was rebuilt in January, 1850, when Bodmer's reciprocating pistons were taken out, and ordinary ones put in. In later years No. 7 was named "Seaford."

Bodmer designed another engine on this plan, with outside cylinders 22in. diameter and 24in. stroke—*i.e.*, two pistons of 12in.

stroke each. The driving wheels were 7ft. diameter. The boiler pressure of this extraordinary engine was 100lb. and the coke consumption was estimated at 10lb. per mile, with trains of 12 coaches. This engine was fitted with cylindrical slides and expansion valves, under a patent obtained by Bodmer.

In 1845, J. E. McConnell, then locomotive superintendent of the Birmingham and Gloucester Railway, determined to construct a more powerful engine for working the Lickey Incline than the American engines previously described. The "Great Britain" was the result of his essay. She was a six-wheel coupled saddle-tank locomotive. The wheels were 3ft. 10in. diameter, and the cylinders 18in. by 26in. stroke. This powerful "iron-horse" easily hauled trains weighing 150 tons up the Lickey Bank. McConnell also rebuilt one of the American engines, as a saddle-tank locomotive, for working the Tewkesbury branch of the Birmingham and Gloucester Railway. This curious specimen of a saddle-tank engine had outside cylinders 10½in. diameter, 20in. stroke, single driving wheels 4ft. diameter, and a leading bogie.

Mr. Dewrance, of the Liverpool and Manchester Railway, about this time turned his attention to the experiments which were, ever and anon, being made towards the long-wished-for goal of a perfect coal-burning locomotive. In the "Condor" he tried the effect of two fire-boxes. The fuel was inserted in the usual manner into the exterior fire-box; the second, or combustion chamber being designed to consume the gaseous matter that escaped from the first furnace.

During the period of special attention to the working of the "Condor" this system of coal burning appears to have been of a fairly successful character. The idea of a combustion chamber as a solution of the vexed question of a successful smoke-consuming locomotive was afterwards tried by other locomotive designers. The division between the two fire-boxes of the "Condor" consisted of a transverse water-space, fitted with short tubes. Air was admitted to the combustion-chamber by means of a pipe, with a head perforated with small holes.

## CHAPTER VIII.

Stephenson's "long boiler" goods engines for the Eastern Counties Railway—Gray's prototype of the "Jenny Lind"—Hackworth builds twelve of the class for the Brighton Railway—Stephenson and Howe's three-cylinder locomotive not a success—The "Great A," another Stephenson absurdity—The competitive trials between broad and narrow-gauge locomotives—Gooch to the rescue!—The "Premier," the first engine constructed at Swindon—The "Great Western" the forerunner of the standard express engine of to-day—Trial trip of this "mammoth"—A notable run of the "Great Western"—The "Great Western" altered to an eight-wheel engine—Galloway's incline-climbing locomotive tried on the Great Western—Beyer's "Atlas" for the Manchester and Sheffield Railway—The Eastern Union "Essex" draws 149 loaded goods wagons—Stephenson's "White Horse of Kent"—Crampton, as a locomotive designer, the "Namur" constructed—Gooch's "Iron Duke" and "Lord of the Isles" make the broad-gauge still more popular—The "Jenny Lind," a "storm in a tea cup"—Trial of the "Jenny Lind" and "Jenny Sharp"—Trevithick's "Cornwall," a locomotive monstrosity—Exhibited at the 1851 Exhibition—Rebuilt in her present form, and still running—McConnell's "counter-balancing" experiments—The "most powerful narrow-gauge engine ever built"—"No. 185" of the Y.N. and B.R.—The oldest locomotive now running, "Old Coppernob," of the Furness Railway—"Lablache," another locomotive freak—"Cambrian" locomotives, and the peculiarities of their construction—The "Albion," of 1848—Half a century later, the writer unearths the working drawings of this engine and her sisters.

During 1845 R. Stephenson and Co. built seven of their "long-boiler" engines, with outside cylinders, for working the goods traffic of the Eastern Counties Railway. Fig. 35 is an illustration of one of these ungainly specimens of locomotive construction. The boiler barrel was no less than 13ft. 6in. in length, all the axles were beneath the barrel, the leading wheels were 3ft. diameter, and the driving and trailing (coupled) wheels 5ft. 9½in. diameter. The cylinders were 16in. diameter, the stroke being 21in. In working order, these locomotives weighed 23 tons 12 cwt. After looking at the illustration, it is scarcely necessary to add that these engines were very unsteady when travelling, the oscillation being excessive.

In the arrangement of inside and outside bearings to the various wheels of the patent engines, designed by John Gray for the Hull and Selby Railway (previously described), we make acquaintance with the embryo design, afterwards perfected, and known the whole world over as the "Jenny Lind" class.

In 1846 Gray had become locomotive superintendent of the Brighton Railway, and he prepared another design of express engines for that line, in which the type now known as "Jenny Lind" was

FIG. 35.—STEPHENSON'S "LONG-BOILER" GOODS ENGINE, EASTERN COUNTIES RAILWAY

further developed. J. Hackworth and Co. obtained the contract for the supply of twelve of these locomotives, and in November, 1846, they delivered the first pair, numbered 53 and 54. Fig. 36 represents No. 49, one of these engines. The leading and trailing wheels were 3ft. 6in. diameter, the drivers being 6ft. diameter. Cylinders 15in. by 24in. stroke. Heating surface: tubes, 700 sq. ft.; fire-box, 79 sq. ft. Inside bearings were provided to the driving, and outside to the leading and trailing wheels; the engines

FIG. 36.—GRAY'S PROTOTYPE OF THE "JENNY LIND," No. 49, LONDON AND BRIGHTON RAILWAY

were fitted with Gray's "horse-leg motion," and several of the dozen had two square-seated steam domes, one located on the centre of the boiler barrel, the other over the fire-box. Each dome was provided with a steam safety-valve. The steam pressure was 100lb. per square inch.

These engines were found to be good at hauling heavy loads (as computed 50 years ago) at speeds up to and slightly exceeding 40 miles an hour.

In 1846 Stephenson and Howe obtained a patent for a three-cylindered engine. Z. Colburn, in his "Locomotive Engineering," exposes the fallacy of the idea that the action of the steam admitted alternately to cylinders whose centres are far apart, sets up a dangerous sinuous motion. The object of Stephenson and Howe's three-cylinder engine was to overcome this winding motion. Colburn states that a "few pounds of counterweight would have served a better purpose than the extra cylinder and working parts." Two engines appear to have been built on this plan before the true cause of the rocking motion and the real way of overcoming it, were fully grasped by the patentees. The outside cylinders were only $10\frac{1}{2}$in. diameter and 22in. stroke; whilst the centre or inside cylinder was $16\frac{3}{8}$in. diameter, but the stroke in this case was restricted to 18in. It is needless to add that these three-cylinder locomotives were not successful.

Passing reference must be made to the celebrated gauge experiments which took place during the last days of December, 1845, and resulted so greatly in favour of the broad-gauge, despite the fact that the Great Western Railway had no new engines prepared for the competition, but used those regularly in work on the broad-gauge railways.

The narrow-gauge experiments were made on the Great North of England Railway, a special engine being built for the purpose by R. Stephenson and Co., and called "A." The "A" was a six-wheel long-boiler engine, with outside cylinders and 6ft. 6in. driving wheels. Hot water for supplying the boiler was used on the narrow-gauge in place of cold on the broad-gauge. The latter started from a state of rest, but the narrow-gauge approached the starting-point at as great a velocity as possible; yet, notwithstanding these sharp practices of the narrow-gauge officials, they were completely beaten in the experiments.

The Swindon Works commenced to build locomotives early in 1846; and, as its name implies, the "Premier" was the first engine constructed at these now world-famous locomotive shops.

She was a six-coupled goods engine, with wheels 5ft. diameter.

Numerous engines of this type, with slight modifications, were built at Swindon; "Hero" (Fig. 37) is a good example of the G.W. standard goods engine at the time.

The narrow-gauge engineers having made frantic efforts to produce locomotives as powerful as those in use on the Great Western

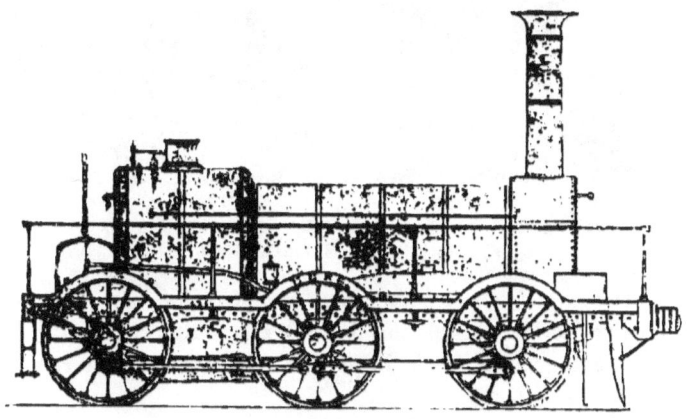

FIG. 37.—"HERO," A GREAT WESTERN RAILWAY SIX-COUPLED BROAD-GAUGE GOODS ENGINE

Railway, the directors of the latter company decided to have a larger and more powerful engine constructed, and Mr. Gooch received orders to construct a colossal locomotive, and to have it in work before the commencement of the Parliamentary Session of 1846. From the time the decision was arrived at, until the "Great Western" was at work, only 13 weeks elapsed, during which short period the design of the engine had to be decided upon, the drawings made, the patterns prepared, and the whole of the complex machinery made and put together; yet those three months were sufficient to produce this most famous locomotive.

As originally constructed, the "Great Western" (Fig. 38) was a six-wheel engine, the dimensions being:—Cylinders, 18in. diameter and 24in. stroke; driving wheels, 8ft. diameter; leading and trailing wheels, 4ft. 6in. diameter; 278 tubes, 9ft. long, 2in. diameter; fire-box (outside), 5ft. 6in. by 6ft., inside 4ft. 10in. by 5ft. 4in., with partition through the centre; heating surface, tubes 1,591 sq. ft.; fire-box, 160 sq. ft.; grate area, 20ft.; height, from level of rail to top of boiler, 9ft. 6in.; the chimney was 5ft. 2in. high; length of engine, 24ft.; weight (empty), 36 tons. In this engine Gooch retained the Gothic fire-box, as supplied to the engines he had previously designed. By the way, a picture, purporting to be an illustration of this engine, was given in a book on locomotive history, with the flush

FIG. 38.—THE "GREAT WESTERN" BROAD-GAUGE ENGINE AS ORIGINALLY CONSTRUCTED

top fire-box and four leading wheels! The "Great Western" continued to work trains on the Great Western Railway until the end of 1870, having run a total distance of 370,687 miles during the 23¾ years she was in work.

On Saturday, June 13th, 1846, the "mammoth" locomotive (as the "Great Western" was usually called) made a sensational trip from London to Bristol and back, and, but for the failure of one of the six-feed pumps, necessitating slower running, even better results would have been attained. But, despite the accident, the result of the trip came like a "bolt from the blue" upon the narrow-gauge engineers.

The train weighed 100 tons, and consisted of ten first-class carriages, seven of which were ballasted with iron, the other three being occupied by the directors and those interested in the experiment. The train started from Paddington at 11 hours 47 minutes 52 seconds; at Didcot a stop of 5¼ minutes was made; Swindon was reached in 78 minutes. After staying there 4 minutes 27 seconds, the journey was continued to Bristol, the whole distance of 118½ miles being covered in 2 hours 12 minutes, or at the rate of 54 miles an hour, or, excluding the 9¾ minutes spent in the two stoppages, at about 59 miles an hour for the complete journey, including the slowing down and getting up speed again on three occasions. The maximum speed was obtained between the 82nd and 92nd mile-posts (from the 80th to the 85th mile there is a falling gradient of 8ft. per mile, and from the 85½th to about the 86½th mile there is a falling gradient of about 1 in 100, and a fall of 8ft. per mile then reaches to about the 90½th mile-post; a rising gradient of 8ft. per mile then succeeds and extends beyond the 92nd mile-post), performing the ten miles in 9 minutes and 8 seconds, or at an average speed of nearly 66 miles an hour. The 87th and 88th miles, on a falling gradient of 8ft. per mile, were run over at a rate of 69 miles per hour.

One Monday early in June, 1846, the "Great Western" was attached to the 9.45 a.m. express Paddington to Exeter, the crack train of that time, which, indeed, continued to be the fastest ordinary passenger train until the establishment of the "Flying Dutchman" many years later. When it was advertised that this train would perform the journey between London and Exeter in 4½ hours, people said it was impossible; what, then, must have been thought of the run performed by the "Great Western" and chronicled below? The 193¾ miles from Paddington to Exeter were covered in 214 minutes (3 hours 34 minutes) running time, being an average rate of 55¼ miles per hour. The actual running time on the journey was as follows:—

| | | | | |
|---|---|---|---|---|
| From | Paddington to Didcot ... ... ... | 53 miles | ... | 55 minutes |
| ,, | Didcot to Swindon ... ... ... | 24 ,, | ... | 30 ,, |
| ,, | Swindon to Bath ... ... ... | 29¾ ,, | ... | 33 ,, |
| ,, | Bath to Bristol ... ... ... | 11½ ,, | ... | 14 ,, |
| ,, | Bristol to Taunton ... ... ... | 44¾ ,, | ... | 45 ,, |
| ,, | Taunton to Exeter ... ... ... | 30¾ ,, | ... | 37 ,, |
| | | 193¾ | | 214 |

The return journey was performed in less time, and could have been accomplished with ease at a rate exceeding 60 miles an hour. The actual running time, exclusive of stoppages, was as follows:—

| | | | | |
|---|---|---|---|---|
| From | Exeter to Tauntou ... ... ... | 30¾ miles | ... | 34 minutes |
| ,, | Taunton to Bristol ... ... ... | 44¾ ,, | ... | 43 ,, |
| ,, | Bristol to Bath ... ... ... | 11½ ,, | ... | 14 ,, |
| ,, | Bath to Swindon ... ... ... | 29¾ ,, | ... | 34 ,, |
| ,, | Swindon to Didcot ... ... ... | 24 ,, | ... | 26 ,, |
| ,, | Didcot to Paddington ... ... ... | 53 ,, | ... | 56 ,, |
| | | Miles, 193¾ | | Minutes, 208 |

After the engine had been running a short time, Gooch found the weight on the leading axle too much to be safely carried by one axle, and he fitted another pair of leading wheels to the "Great Western" (Fig. 39), making her an eight-wheeled engine, having a group of four wheels in front of the driving wheels. It must be remembered that these four wheels were not affixed to a bogie frame. So well satisfied were the directors of the Great Western Railway with the "Great Western" that 29 more engines of almost similar design (except the domed fire-box) were constructed during the next eight years, and these engines, with a few of the same design, built at a more recent period, worked the famous broad-gauge expresses between London and Newton Abbot until the abolition of the broad-gauge in May, 1892.

In March, 1847, the Great Western Railway laid down a length of line at Maidenhead for the purpose of testing Elijah Galloway's

FIG. 35.—THE ORIGINAL "GREAT WESTERN" AS REBUILT WITH TWO PAIRS OF LEADING WHEELS

system of locomotive propulsion with horizontal driving wheels. The horizontal wheels gripped a centre rail, and the engine not being dependent upon the weight placed upon the driving wheels for adhesion, was enabled to ascend inclines that were impossible for ordinary locomotives; whilst the fact that the two horizontal driving wheels were pressing one on either side of the centre rail enabled the engine to safely pass round curves of extremely short radii, such as would be impossible with ordinary locomotives. The line put down at Maidenhead was on an incline of 1 in 19, but a model engine and train successfully ascended an incline of 1 in 6. Mr. D. Gooch gave the following account of the experiments:—

"Engineer's Office, Paddington,
"March 25th, 1847.

"The following is the result of the experiment I made with Mr. Galloway's locomotive engine, in which the driving wheels are placed horizontally, and act against the sides of a centre rail:—

| Weight of engine | ... | ... | ... | ... | ... | 20 tons. |
| Weight of load | ... | ... | ... | ... | ... | 13¼ " |
| | | | | | | 33¼ tons. |

"This weight was taken at a slow speed up an incline of 1 in 19, with a pressure on the boiler of 60lb. on the inch, and calculating the power of the engine and actual duty performed, we have as follows:— With steam at 60lb. in the boiler, the average effective pressure on the pistons, after deducting back pressure, will be about 50lb. on the inch, then the area of the two cylinders 308 × 50 = 15,400in., and double stroke of piston equals 32in., and circumference of driving wheel 116in.

"Therefore, as 116in. : 15,400 : : 32 : 4,248 tractive power on the rim of the wheel,

And gravity per ton, 1 in 19 = 118 lb,
Friction ditto ... 7 lb.

125 × 33·5 tons = 4,187·5 lb.
resistance overcome.

therefore, 4,248 − 4,187 = 61lb., the total loss from the friction of the working parts of the engine, which I think, is as small a loss as can be hoped for in any class of engines, and from the facility of applying screws to increase the weight on the driving wheels to any required amount, there is no difficulty from slipping.

"(Signed) DANIEL GOOCH."

The "Atlas," constructed for the Manchester and Sheffield Railway, deserves notice. She was built by Sharp Bros. and Co., from the designs of Mr. Beyer, their then chief engineer, but afterwards head of the well-known firm of locomotive builders, Beyer, Peacock and Co., of Manchester.

The "Atlas" commenced work in May, 1846, and during the succeeding 17 months she travelled 40,222 miles, with a coke consumption of 36.53lb. per mile, although engaged in hauling heavy goods trains. The engine had inside cylinders, 18in. diameter, 24in. stroke; the whole of the framing and bearings were inside the wheels; the boiler was 13ft. 6in. long and 3ft. 6in. diameter, and contained 175 brass tubes of 1⅝in. external diameter; the wheels were cast-iron, 4ft. 6in. diameter; a copper fire-box was provided, its inside measurements being 3ft. 8in. long, 3ft. 3½in. wide, and 3ft. 4½in. from the fire-bars to the top. The water space around the fire-box was 3in., and a mid-feather, 4in. wide, divided the fire-box.

The cylinders were secured to each other by internal flanges, which formed the bottom of the smoke-box, and also the chief cross-stay between the frames. The valves were in one chest, located below the cylinders, and inclined towards each other. The weight of the valves was carried by spindles working through stuffing-boxes. The regulator was provided with two perforated discs, so that the steam was admitted very gradually, the volume increasing as the two sets of perforations came opposite each other.

The weight of the "Atlas" was 24 tons, and five other engines of exactly similar designs were supplied to the Manchester and Sheffield Railway.

Another engine of the same description was supplied to the Manchester and Birmingham Railway, and on October 3rd, 1836, "No. 30" hauled a train of 101 wagons, weighing 597 tons, from Longsight to Crewe, a distance of 29 miles, at the average speed of 13.7 miles an hour.

The mention of a powerful engine and a record train on one railway naturally suggests a better one on another line, so we have the "Essex" going "one better" than "No. 30."

This time we have a load of 149 loaded wagons (probably equal to 890 tons), and forming a train nearly half a mile long. The "Essex" is also stated to have hauled a train of 192 empty trucks. The engine in question was built for the Eastern Union Railway by Stothart, Slaughter, and Co., Bristol, in 1847, and had wheels 4ft. 9in. diameter, cylinders 15in. by 24in. stroke, weight 22 tons.

In 1846, Stephenson and Co. supplied the South Eastern Railway with an engine called the "White Horse of Kent" (the "White Elephant of Newcastle" would have been a far more descriptive name). This engine probably exhibited the "long boiler" folly in a more marked manner than any other engine of that notorious class. She was 21ft. 10in. long, with a wheel base of only 10ft. $3\frac{1}{4}$in.! She had cylinders 15in. by 22in. stroke, 5ft. 6in. driving wheels, and weighed $18\frac{3}{4}$ tons. Gooch says this engine was so unsteady that it was necessary to be tied on to make experiments on the smoke-box temperature, and that the tubes were so long that one end of the engine was actually condensing the steam generated at the other end!

At this time Mr. T. R. Crampton turned his attention to locomotive construction, and patented a design of locomotive. He claimed for his design the following advantages—viz., a reduction of the rocking and vibrating motion, obtained by lowering the centre of gravity, and by locating the greater portion of the weight between the supports; an increased heating surface; and a superiority of arrangement of the working parts, the whole of which were placed immediately under the eye of the driver.

The first engine constructed on this principle was the "Namur" (Fig. 40), built under Crampton's patent by Tulk and Ley, of the Lowera Works, Whitehaven, for the Namur and Liège Railway.

The illustration shows that the chief peculiarity of the "Namur" was the position of the driving wheels, the axle of which was behind the fire-box, so that the axle extended across the foot-plate. One spring, formed of plates, also extended across the back of the fire-box, parallel with and above the driving axle, and acting upon it at the bearings.

The chimney was 6ft. 6in. high; the smoke-box was very narrow, being no wider than the diameter of the chimney; all the wheels had inside bearings; the cylinders were outside, and horizontal; the valve-chests were on the outer side of the cylinders, so that the eccentrics were at the extreme ends of the axles, beyond the wheels, and quite exposed.

FIG. 40.—THE "NAMUR," THE FIRST ENGINE BUILT ON CRAMPTON'S PRINCIPLE

The boiler barrel was surmounted by an immense fluted dome, which was fitted with two lever safety valves, whilst a third one, of the spring pattern, was provided on the fire-box casing.

The following are the principal dimensions of the "Namur":—

Diameter of driving-wheels, 7ft.; diameter of leading and middle wheels, 3ft. 9in.; total wheel base, 13ft.; cylinders, 16in. diameter, 20in. stroke; number of tubes, 182—length 11ft., external diameter 2in.; fire-box, 4ft. 3in. long, 3ft. 5in. wide; area of fire-tube, 14ft. 6in.; heating surface: fire-box 62ft., tubes 927ft., total 989ft.

# EVOLUTION OF THE STEAM LOCOMOTIVE

The engine was completed early in February, 1847, and previous to its exportation, it was tried for several weeks on the London and North Western Railway, running over 2,300 miles. All classes of traffic were hauled by the engine, and she gave general satisfaction. A speed of 75 miles an hour was attained between Willesden and Harrow, when running "light." On another occasion, 50 miles an hour was attained on a trip from Camden Town to Wolverton with a coke train, weighing 50 tons, between Tring and Wolverton.

The "Namur" weighed 22 tons, of which 7½ tons were on the leading wheels, 4 tons on the centre wheels, and 10½ tons on the driving wheels.

The L. and N.W.R. were so satisfied with the "Namur" that Tulk and Ley were instructed to build a Crampton engine for that railway; and the "London" (Fig. 41) was produced in 1848 in response to this

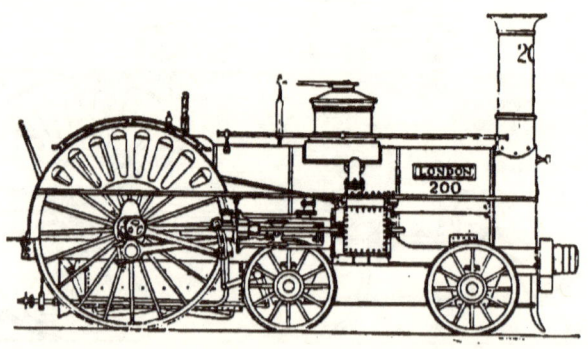

FIG. 41.—CRAMPTON'S "LONDON," THE FIRST ENGINE WITH A NAME ON THE SOUTHERN DIVISION OF THE L. & N.W.R.

order. She was the first engine on the southern division of the L. and N.W.R. to have a name. The driving wheels were 8ft. diameter, the cylinders 18in. diameter and 20in. stroke. The boiler was oval in shape, its vertical diameter being 4ft. 8in., and its horizontal diameter 3ft. 10in. The heating surface was 1,350 sq. ft. The fire-box extended below the driving axle.

In April, 1847, Mr. D. Gooch's famous broad-guage express engine, "Iron Duke," commenced to run. Fig. 42 represents an engine of this class. She was the first of a set of twenty-nine locomotives of almost similar construction, designed to work the

114    EVOLUTION OF THE STEAM LOCOMOTIVE

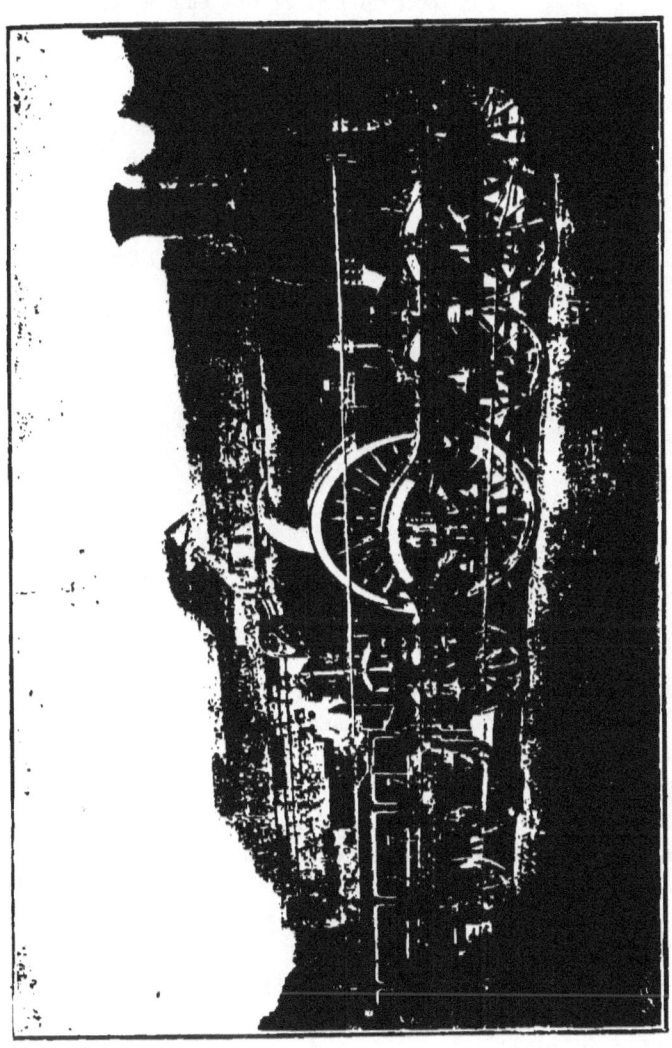

FIG. 42.—"GREAT BRITAIN," ONE OF GOOCH'S FAMOUS 8FT. "SINGLE" BROAD-GAUGE ENGINES FOR THE G.W.R.

# EVOLUTION OF THE STEAM LOCOMOTIVE 115

Great Western express trains. The "Iron Duke" was an improvement on the celebrated "Great Western," previously described; the most noticeable difference was the absence of the domed fire-box in the "Iron Duke." The total mileage of this engine, up to October, 1871, when it was withdrawn from service, amounted to 607,412 miles. The best-known engine of the class is "Lord of the Isles," built at Swindon in 1850, and exhibited at the International Exhibition, London, 1851; she commenced to run July, 1852, and continued in active service on the Great Western Railway for 29 years, during which time 789,300 miles were covered by the "Lord of the Isles." This famous

FIG. 43.—"No. 61," LONDON AND BRIGHTON RAILWAY

broad-gauge locomotive is still preserved by the Great Western Railway.

The next point in the evolution of the locomotive that deserves attention is the famous class of engines known as the "Jenny Lind" design.

Much has been written concerning these engines during recent years, and many uncorroborated and absurd statements have been made; but it was most clearly demonstrated that to Mr. David Joy was due the chief honour of designing the successful class of locomotive known far and near as "Jenny Linds." Such a design was elaborated from the adoption of the best features of the several descriptions of locomotives then in use.

The first of the type of engine afterwards known as the "Jenny Lind" class was constructed for the London and Brighton Railway by E. B. Wilson and Co., Railway Foundry, Leeds, and was commenced building in November, 1846, and completed in

May, 1847. The principal features of the engines may be summarised as follows:—Steam pressure 120lb. per square inch, inside bearings to driving and outside bearings to the leading and trailing wheels, outside frames, outside pumps located between the driving and trailing wheels, and worked by cranks fixed on the outside of the driving axles. The engine had a raised fire-box; the dome was fluted and had a square seating; the safety valve was enclosed within a fluted column, and fixed on the fire-box.

Polished mahogany lagging was used for both the boiler and fire-box, the same being secured by bright brass hoops. The tops of the safety valves and dome were bright copper. The first trip of the "Jenny Lind" was from Leeds to Wakefield and back. Ten engines of this class were supplied to the London and Brighton Railway, and were numbered 61 (Fig. 43) to 70. The principal dimensions were:— Driving wheels 6ft. diameter; leading and trailing wheels, 4ft. diameter; cylinders (inside), 15in. diameter, 20in. stroke; boiler, 11ft. long, 3ft. 8in. diameter; 124 tubes, 2in. diameter. A water space of 3in. was left between the inner and outer shells of the fire-box. Heating surface, tubes 700 sq. ft., fire-box 80 sq. ft.

It is significant to note that in the original description of the "Jenny Lind," published in 1848, we are informed that "in establishing this class of engine Messrs. Wilson have studied less the introduction of dangerous novelties than the judicious combination of isolated examples of well-tried conveniences." This statement exactly agrees with those recently made by Mr. Joy.

The great success of the "Jenny Lind" type caused Sharp Bros. and Co. to introduce a rival class of engines nicknamed "Jenny Sharps."

The engines were provided with a mid-feather in the fire-box for the purpose of augmenting the heating surface. The principal dimensions of the "Jenny Sharps" were as follow:—Steam pressure, 80lb.; cylinders, 16in. diameter, 20in. stroke; driving wheels, 5ft. 6in. diameter; heating surface, tubes (of which there were 161, each 10ft. long and 2in. diameter) 847 sq. ft., fire-box, 72 sq. ft.; total, 919 sq. ft. Mr. Kirtley, the locomotive superintendent of the Midland Railway, arranged a trial between the rival "Jennies," and the event came off on May 4th, 5th, and 6th, 1848.

Sharp's engines were Nos. 60 and 61, and Wilson's Nos. 26 and 27. The first trip was with a load of 64 tons, made up of nine carriages and two brake-vans, weighted with iron chairs to 64 tons.

EVOLUTION OF THE STEAM LOCOMOTIVE 117

Sharp's No. 60 took the first train, the weight being, engine 21 tons 9 cwt., tender 12 tons 11 cwt., load 64 tons; total, 98 tons, or, including officials, etc., about 100 tons.

The journey was from Derby to Masborough, 40¼ miles, the line rising for the first 20 miles at about 1 in 330, and falling for the remainder of the distance at about the same rate. The weather was fine, the metals dry, and there was no wind.

William Huskinson drove the train, which left Derby at 3h. 39min. 5½sec. p.m., and arrived at Masborough at 4.28 p.m. Among the passengers were Messrs. Kirtley, locomotive superintendent; Marlow, assistant locomotive superintendent; Harland, carriage superintendent; E. B. Wilson and Fenton, of the firm of E. B. Wilson and Co.; and T. R. Crampton.

The first 18 miles up the bank of 1 in 330 were covered in 25 minutes 12½ seconds, being at an average speed of nearly 43 miles an hour. Before starting, the water in the tender had been heated to nearly boiling point; 16 cwt. of coke were consumed, or 44.8lb. per mile; 10,290lb. of water were evaporated, equal to 5.7lb. of water to 1lb. of coke.

Wilson's engine, No. 27, was next tried. She weighed 24 tons 1 cwt., and her tender, loaded, 15 tons 13 cwt., the total load with train thus being 103 tons 14 cwt. William Carter drove the train, which left Derby at 7h. 10min. 20sec., and arrived at Masborough at 7h. 56min. 42sec., the speed averaging 52 miles an hour. The first 18 miles were negotiated in 22 minutes 44¾ seconds, or at nearly 47 miles an hour. Only 13 cwt. of coke was used, equalling 36.4lb. per mile.

The following table shows the working of the two engines up the bank to the seventeenth mile-post:—

| Mile Post. | "Jenny Sharp." Miles per hour. | "Jenny Lind." Miles per hour. | Mile Post. | "Jenny Sharp." Miles per hour. | "Jenny Lind." Miles per hour |
|---|---|---|---|---|---|
| 1 | 21·6 | 21.9 | 10 | 46·8 | 52 0 |
| 2 | 39·6 | 44·5 | 11 | 45·9 | 51·4 |
| 3 | 42·0 | 51·0 | 12 | 45·9 | 52·3 |
| 4 | 42·5 | 51·4 | 13 | 45·6 | 52·7 |
| 5 | 45·4 | 51·4 | 14 | 46·6 | 51·8 |
| 6 | 46·8 | 51·2 | 15 | — | 51·8 |
| 7 | 44·5 | 48·9 | 16 | 48·0 | 51·4 |
| 8 | 46·2 | 50·0 | 17 | 47·0 | 51.8 |
| 9 | 47·0 | 52·5 | | | |

Trials were then made with trains of 17 coaches, weighted to 99 tons 16 cwt. Twenty passengers were carried, including Captain Symmons, the Government Inspector. The gross load was 101 tons.

William Mould drove the Sharp engine, and William Barrow the Wilson engine (No. 26).

The coke consumption was—Sharp's, 16 cwt., or 44.8lb. per mile; Wilson's, 12 cwt., or 33.6lb. per mile.

Water evaporated—Sharp's, 10,840lb., equal to 27.1lb. per mile, or 6lb. of water by 1lb. of coke; Wilson's, 10,116lb., equal to 25.29lb. per mile, or 7.5lb. of water by 1lb. of coke.

The first 18 miles up the bank were covered in 26 minutes 19 seconds by the "Jenny Lind," and in 27 minutes 55 seconds by the "Jenny Sharp."

The tables show the speeds at which the posts were passed:—

| Mile Post. | Sharp's Engine. Miles per hour. | Wilson's Engine. Miles per hour. | Mile Post. | Sharp's Engine. Miles per hour. | Wilson's Engine. Miles per hour. |
|---|---|---|---|---|---|
| 1 | 15·0 | 18·3 | 10 | 43·9 | 44·5 |
| 2 | 36·5 | 40·9 | 11 | 44·5 | 44·5 |
| 3 | 48·0 | 45·6 | 12 | 43·9 | 45·0 |
| 4 | 42·4 | 46·8 | 13 | 43·4 | 45·0 |
| 5 | 43·9 | 46·8 | 14 | 43·4 | 44·5 |
| 6 | 43·9 | 46·2 | 15 | — | 45·0 |
| 7 | 41·9 | 43·4 | 16 | 42·9 | 43·9 |
| 8 | 42·4 | 43·4 | 17 | 42·9 | 42·4 |
| 9 | 43·9 | 44·5 | 18 | 41·9 | 41·4 |

Beyond the thirtieth mile-post Wilson's engine, which had been considerably in advance, according to the time taken, began to lose ground, in consequence of the driver allowing the fire to get low, and upon arrival at Masborough he had scarcely sufficient steam to shunt the train.

Mr. Kirtley considered the trial unsatisfactory for this reason, and a second one was arranged for the next day, but with no more satisfactory result, as upon this occasion, after travelling a mile, a joint cover of one of the cylinders worked loose, consequently a great deal of steam escaped during the remaining 39 miles of the trip. We have given the real facts in connection with the original "Jenny Linds" at some length, for the purpose of placing on permanent record the details of these capital locomotives, and so prevent our readers and students of locomotive history generally from being misled by the absurdly inaccurate romances that have, for some obscure purpose, been recently circulated concerning the "Jenny Lind." (Fig. 44.)

The original design of the locomotive now to be described is so singular that we are reminded of the extravagant examples of locomotive construction appertaining to 1830, or thereabouts, rather than to the year now under review. Yet, strange as it may appear, the "Cornwall" (Fig. 45) is still running express trains, although it must be confessed it has undergone a complete metamorphosis since it was built at Crewe in 1847. The engine in question was designed by Mr. F. Trevithick, son of the famous "father of the locomotive," and was intended to be a narrow-gauge improvement on Gooch's famous "Great Western," as Trevithick wished to build a locomotive that would be able to attain a higher rate of speed than the renowned broad-gauge engine. To do this, he considered an increase of the diameter of the driving wheels, a *sine qua non*. He therefore constructed the "Cornwall" with driving wheels 8ft. 6in. in diameter. His next

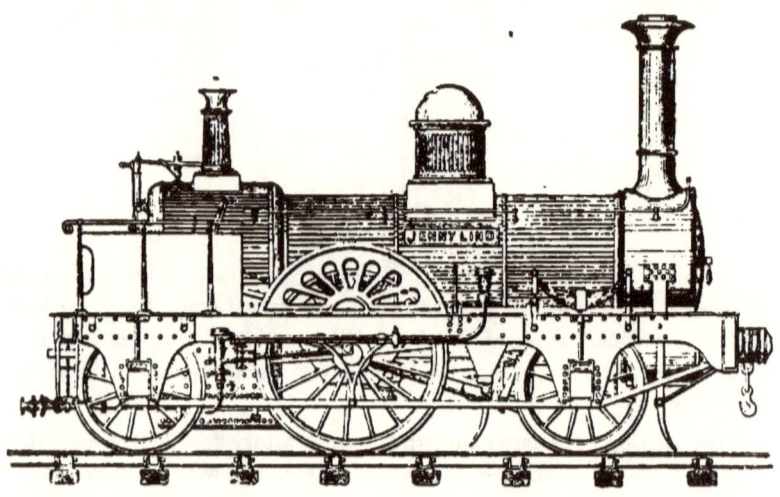

FIG. 44.—THE "JENNY LIND," A FAMOUS LOCOMOTIVE BUILT BY WILSON AND CO., LEEDS, IN 18'6

proposition was that as 8ft. was then considered the limit of size for driving wheels on the broad-gauge, with the boiler above the driving axle, it was necessary to place the boiler below the driving axle with wheels 8ft. 6in. diameter on the narrow gauge. And, therefore, Trevithick constructed the "Cornwall," with underhung boilers, i.e., beneath the driving axle. The cylinders were outside, 17½in. diameter, with a stroke of 24in. The heating surface was 1,046 sq. ft. The locomotive was carried on eight wheels—a group of four leading wheels, the driving, and a single pair of trailing wheels. Weight of engine in working order, 27 tons. The "Cornwall" was very successful in attaining high rates of speed, and, indeed, far exceeded Trevithick's expectations in this respect.

It has been stated that she attained a speed equal to 117 miles an hour down the Madeley Bank. Such a statement must be accepted with reserve—not that the *bonâ fides* of the engineer who made it are doubted, but rather because of the difficulty of obtaining correctly the exact speed of engines when travelling at a great rate, even when proper instruments are employed. We know that

FIG 45.—TREVITHICK'S "CORNWALL," WITH 8FT. 6IN. DRIVING WHEELS, AND BOILER BELOW THE DRIVING AXLE

with an ordinary watch correct results are almost impossible, and an error of a second or two when calculating a quarter of a mile will make a very great difference when arriving at the approximate rates in miles per hour. However, be this as it may, it is generally acknowledged that the "Cornwall" attained speeds that may fairly be called phenomenally high.

On November 9th, 1847, the "Cornwall" was hauling a goods train from Liverpool, and upon rounding the curve near Winsford Station, ran into a coal train, the result being the death of the driver of the "Cornwall," the engine being thrown across both lines, whilst the tender and trucks were projected over the engine, and did not come to a standstill for several yards.

The "Cornwall" was one of the features of the first International Exhibition (held in Hyde Park, London, in 1851). In 1862 Mr.

Fig. 46. TREVITHICK'S 8FT. 6IN. "SINGLE" LOCOMOTIVE, "CORNWALL," AS NOW RUNNING ON THE L. & N.W.R. BETWEEN LIVERPOOL AND MANCHESTER

J. Ramsbottom rebuilt the "Cornwall," and placed her new boiler over the driving wheels. She was numbered "173," and still works the three-quarter-of-an-hour express trains between Liverpool and Manchester. She completed her jubilee of active service last year, and is still running. The present number of the "Cornwall" is "3020," and she is now only a six-wheeled engine.

McConnell made an experiment in counterbalancing a locomotive on the London and North Western Railway in 1848. The engine in question was the "Snake," No. 175, built by Jones and Potts on Stephenson's long-boiler principle. McConnell's plan was to provide a connecting-rod attached to a block working between slide bars, on the opposite side of the driving axle to that on which the piston, etc., were located. By this method he considered that, providing his extra rod-block, etc., weighed the same as the pistons and other reciprocating parts, he had attained a perfect method of counterbalancing. The result was a rude disillusion of the idea, and a complete wreckage of both the theory and the "Snake," the engine breaking down on its first trip, after being fitted with this reciprocating counterbalance. The only result of such an addition to the "Snake" was an increase in the weight of the engine and an augmentation of the friction and axle strains.

In the spring of 1848 McConnell built an engine which he expected "to prove the most powerful narrow-gauge engine ever yet built."

It had outside cylinders 18in. diameter, and 7ft. 6in. between centres. The driving wheels were 6ft. diameter, leading and trailing 3ft. 10in. The boiler was 4ft. 3in. external diameter, 12ft. 7in. long, and contained 190 tubes of 2in. diameter. Height of top of boiler from rail level, 7ft. 9in.

The fire-box was 5ft. 9¼in. wide, by 5ft. 5in. long, and of the same height. The wheel base was as follows:—Leading to driving, 6ft. 8in.; driving to trailing, 10ft. 6in.

Another combination design in locomotive practice is to be found in engine "No. 185," delivered to the York, Newcastle, and Berwick Railway on October 3rd, 1848, by R. Stephenson and Co.

This engine had inside cylinders, but outside valve gearing and eccentrics. The cylinders were 16in. diameter, with 20in. stroke. The boiler was 3ft. 10in. diameter and 11ft. long; there were 174 tubes, 1⅞in. outside diameter, 11ft. 5in. long, the heating surface being: tubes, 964 sq. ft.; fire-box, 82 sq. ft. The driving wheels were 6ft. 6in.

diameter, the leading and trailing being 3ft. 9in. diameter. Inside bearings were provided for the driving wheels and outside bearings for the leading and trailing wheels. Inside and outside iron-plate frames, 1in. thick and 8in. deep, were provided. This engine weighed 22 tons in working order, and consumed 18lb. of coke per mile with express trains of four carriages. The peculiar feature of "No. 185" was the vertical valves, worked by eccentrics outside the driving wheels; the pumps were also worked off the same eccentrics, and were consequently outside, as in the "Jenny Lind" design. The exhaust ports were below the cylinders, the pipes from which united at the blast orifice.

FIG. 47.—"OLD COPPER-NOB," No. 3, FURNESS RAILWAY, THE OLDEST LOCOMOTIVE NOW AT WORK

Locomotives that attain their "jubilee" of active service are indeed very few and far between, and it redounds much to the honour of the late firm of Bury, Curtis and Kennedy, of the Clarence Foundry, Liverpool, that locomotives constructed by them in the year 1846 are still engaged in hauling trains on an English railway.

This firm of builders ceased to exist 46 years ago, but engines Nos. 3 (Fig. 47) and 4 of the Furness Railway are continuing monuments of the good material and sound workmanship of Bury, Curtis and Kennedy. The locomotives in question are mounted on four wheels (coupled) of 4ft. 9in. diameter, the cylinders are 14in. diameter, and stroke 24in., the valves being between the cylinders. The wheel base is 7ft. 6in. The boiler is 11ft. 2in. long, with a mean diameter of 3ft. 8in., and contains 136 tubes of 2in. diameter, the total heating surface being 940 sq. ft. Steam pressure, 110lb. The tenders are carried on four wheels of 3ft. diameter, the wheel base being 6ft. 9in. The tank holds 1,000 gallons of water, and the coal space is 100 cubic feet. The engines weigh 20 tons each, and the tenders 13 tons each.

The prominent "Bury" features—bar framing and round back fire-boxes with dome tops—are, of course, *en évidence*.

The chimneys appear abnormally high when viewed side by side with modern engines; whilst the pair of Salter safety valves with long horizontal arms, the one reaching from the centre to the back of the fire-box, and its fellow continuing to the front, are also noticeable objects. These engines are usually employed in shunting goods trains in the Barrow Docks and goods yards, and are locally called the "old copper nobs."

Two further peculiarities of these Bury engines are worth recording —viz., the splashers, which are extended in a curious way over the rear of the wheels, and reach within a few inches of the rails, and the round "old copper nobs."

The period under review was a time of considerable competition between the rival gauges, and this competition naturally led to the projection of various extraordinary designs in locomotive construction, such designs being the results of the efforts made by the narrow-gauge engineers to equal the splendid broad-gauge locomotives then recently introduced.

During the first weeks of 1848 E. Wilson and Co., of the Railway Foundry, Leeds, turned out a remarkable specimen of locomotive construction; the engine in question was named "Lablache" (after a celebrated singer). This locomotive had two inside cylinders 16in. diameter, 20in. stroke, and was supported on four wheels each 7ft. diameter; the wheel base was 16ft.

It is necessary to describe the mode of working introduced into the "Lablache." Between the two pairs of wheels was a straight bar, or shaft, extending under the boiler, parallel with the axles, and projecting on each side beyond the frames. Between the frames two levers were attached to this shaft, and the other extremities of these levers were attached to the pistons by the usual piston-rod and connecting-rods. Now comes the difference in working; the driving axle, it will be observed, was not cranked, but provided with arms. The axle did not revolve, but simply vibrated backwards and forwards. Outside the frames were double-ended levers, one end being coupled to a crank on the leading wheel, and the opposite end connected in a similar manner to the trailing wheels. The wheels on both sides of the locomotive were connected in the same way that a rotary motion is communicated to a lathe by a treadle. When first constructed india-rubber springs were provided for this engine's bearings.

Another engine of a similar design was built, but much lighter. It ran upon the York, Newcastle, and Berwick line for some years. We may say that no other engine on this system was ever built. With a train of three carriages, an average speed of 75 miles an hour is said to have been maintained between Rugby and Leicester. This was, however, due to the high pressure of the steam. Upon another occasion 80 miles an hour was attained; and the engine hauled a train of 53 loaded wagons, weighing 430 tons, at an average speed of 30 miles an hour. After some little time, the fire-box of the "Lablache" was destroyed, and she was then returned to the Railway Foundry, and altered into a four-coupled engine of the usual type, and sold to a railway contractor.

Another locomotive of peculiar design now deserves notice. At a first glance it might be supposed that the "Albion" was propelled on the same principle as the "Lablache" previously described. Such is not, however, the case, the machinery being of an entirely different character. We have been fortunate enough to secure the original working drawings of the "Albion" and the three other engines constructed on the same method, designated the "Cambrian" system. A patent for this method of working steam engines was obtained in 1841 by Mr. John Jones, of Bristol, and applied to stationary engines.

Broadly speaking, the *modus operandi* is as follows:—A central shaft is provided, extending under the boiler of the locomotive and projecting beyond the frames on both sides. Between the frames the shaft passes through a segmental cylinder, within which and fitted to the shaft was a species of disc piston, made to vibrate throughout the length of the hollow segment of the cylinder. It will, therefore, be observed that the motion was obtained from a vibrating disc engine, the blades of which were fixed on the driving shaft; the difference between Wilson's locomotive and the ones we are now describing being that the former was actuated by two horizontal engines working a rocking shaft by connecting-rods, whilst the latter were driven by a disc engine, fixed directly upon the rocking shaft. The arrangement for connecting the driving wheels with the shaft was very similar in both classes of engines.

The premier "Cambrian" locomotive was named "Albion" (Fig. 48), and was built in 1848 by Messrs. Thwaites Bros., of the Vulcan Foundry, Bradford. She was a six-wheel engine, the leading and middle pairs of wheels both receiving motion by means of the connecting-rods from the outside levers attached to the driving shaft. The top of the fire-box was considerably above the level of the top of the boiler barrel. Upon this raised fire-box was fitted a steam dome with a square seating, above the dome was an enclosed Salter pattern safety valve.

The principal dimensions of the "Albion" were:—Leading and driving wheels, 5ft. 6in. diameter, and trailing, 3ft. 9in. diameter; wheel base—leading to driving, 9ft. 6in.; driving to trailing, 5ft. 8in.; boiler, 12ft. long, containing 149 tubes; throw of cranks, 20in.

It should be observed that the "Albion" was fitted with the "link" motion.

The patentee claimed the following advantages for locomotives built on the "Cambrian" system—viz., perfect balance of working parts, thus entirely doing away with the centre pressure and strain; the complete avoidance of all dangerous oscillation; the ends of the oscillating levers, in passing through the greater part of a circle, gained increased power at the extremities of the stroke, and so compensated for the loss of power in the cranks as they approached the dead centres.

This is explained by observing that as the lever approaches the extremities of the stroke the actual length diminishes, and becomes from 18in. to 17¾in., 16in., 15½in., 14in., and 13¼in. at the centres, so that the power of the lever increases in proportion to its diminution in length.

The wear and tear of the machinery was less than in an ordinary locomotive, there being fewer working parts, whilst the centre of gravity was considerably lowered.

The above advantages summarised amounted to the advantages of the long-stroke crank without a long-stroke cylinder, and consequently the absence of a high-piston velocity.

# EVOLUTION OF THE STEAM LOCOMOTIVE 127

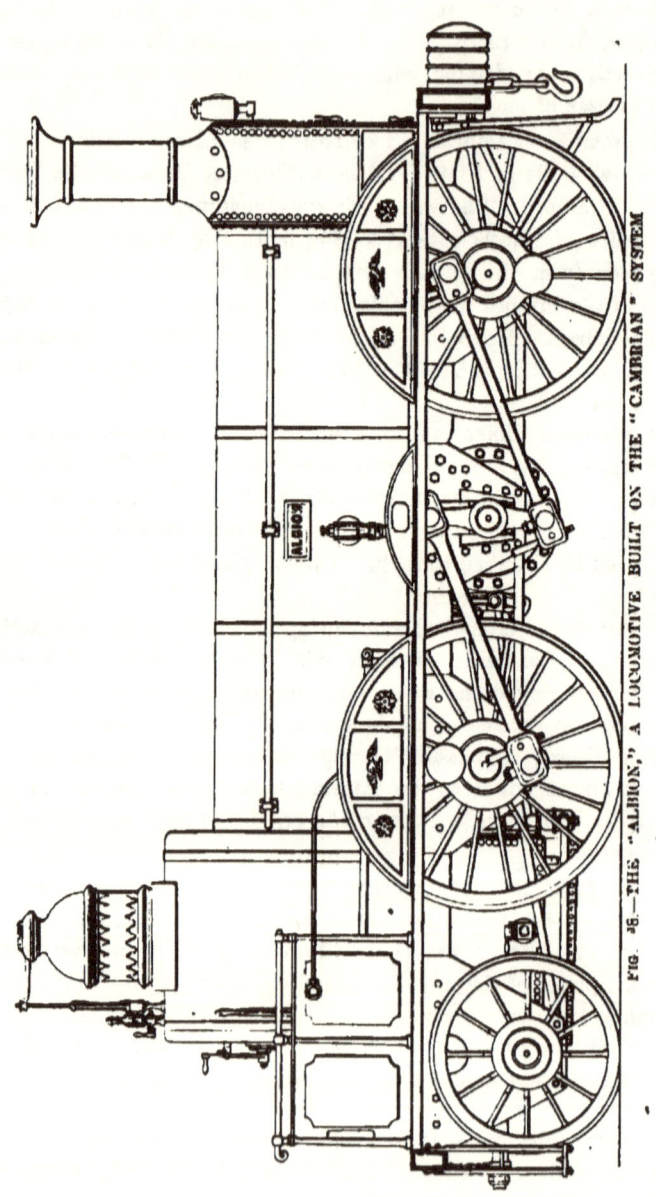

FIG. 8.—THE "ALBION," A LOCOMOTIVE BUILT ON THE "CAMBRIAN" SYSTEM

The "Albion" made its initial trip in June, 1848, the length of line selected being from Bradford to Skipton, on the Leeds and Bradford Railway; the distance was about 18 miles. The speed attained and the low fuel consumption are stated to have more than satisfied the builders and others concerned. The "Albion" was afterwards tried on the Midland Railway between Derby and Birmingham, and the result of these trials showed that the coke consumption was 5lb. per mile less than with the ordinary locomotives, although the trains hauled were of greater weight than usual. We have been unable to obtain further details of the working of this interesting locomotive. The patentee appears to have sent details of the duties performed by the "Albion" to the Institution of Mechanical Engineers in 1849; but these were not printed in the "Proceedings," nor is the Secretary of the Institution now able to find any trace of the papers in question among the archives of the Institution. Messrs. Thwaites Bros., the builders, inform us that about 30 years ago the engine in question was working at Penistone, near Sheffield, and that she was afterwards taken over by the Manchester, Sheffield and Lincolnshire Railway.

Unfortunately, the locomotive department of that railway does not appear to have preserved any particulars relating to the "Cambrian" locomotive after it came into the possession of the Manchester, Sheffield and Lincolnshire Railway.

The other three engines with "Cambrian" machinery were tank locomotives. Two of these were propelled in a similar manner to the "Albion," the segmental cylinder being below the frames, and located between the driving and leading wheels, both pairs of which were 5ft. 3in. diameter, the trailing wheels being 3ft. 9in. diameter. One of these two tank engines had a raised fire-box, similar to that of the "Albion"; but the other had a "Gothic" fire-box, with the wood lagging exposed to view. The other features of the former were a boiler 12ft. long, and a steam dome on the fire-box, fitted with two Salter safety valves, placed side by side. This engine had the "link" motion. Three water-tanks were provided, one beneath the foot-plate, the second below the frames between the leading and driving wheels, and the third extended from the front of the leading axle under the smoke-box, and terminated at the buffer beam. The wheel base was, L. to D. 9ft. 6in., D. to T. 5ft. 8in. The engine had inside frames and bearings.

The locomotive with the "Gothic" fire-box was fitted with a gab reversing gear, worked off the leading axle; the throw of the cranks was 19in. The boiler was 12ft. long and 3ft. 5in. diameter, and contained 121 tubes. Two water-tanks were provided—one beneath the foot-plate, the other below the frames between the leading and driving wheels. The wheel base of this engine was, L. to D. 11ft., D. to T. 5ft. 8in.

The third Cambrian tank engine of which we possess the drawings was a six-wheel locomotive, with single driving wheels 5ft. 6in. diameter, the leading and trailing wheels being 3ft. 9in. diameter. The wheel base was 15ft. 5in., equally divided.

This engine also had a "Gothic" fire-box, and was provided with a sledge brake, which acted on the rails between the driving and trailing wheels. The reversing gear was of the fork pattern. The water tanks were fixed—one below the foot-plate, the other beneath the frames, between the driving and trailing wheels. The boiler was 11ft. 2in. long, and contained 126 tubes. The machinery in this engine was arranged in an entirely different manner, the segmental cylinder being below the smoke-box. The driving shaft passed through the cylinder, and projected beyond the frames on either side of the engine, and vibrated in an arc, as did that of the "Albion"; but instead of a lever being attached to each end of the cranks, the latter only extended in one direction, so that at one end the crank was fixed on the driving shaft, while to its other extremity was pivoted a connecting-rod, 4ft. long, the other end of which was pivoted on a vertical arm, the upper end of this arm being attached to the frame by a horizontal bolt, on which it hung. It is very difficult to explain the method of propulsion without a drawing, but it will be understood that the connecting-rod from the driving shaft to the hanging-rod only vibrated. Another crank, 6ft. long, was also attached to the bottom end of the vertical swinging-rod; the other end of this crank was connected with the driving wheel by means of the usual outside pin. It will, therefore, be seen that by means of the hanging-rod the vibrating motion was transformed into a rotary one. The feed-pumps were worked off the vertical rod, the motion of which was similar to that of a pendulum, with the connecting-rods fastened to its bottom end. The drawings of these four remarkable locomotives are on a large scale, and are well executed; parts of them being coloured, they are also mostly in a good state of preservation.

K

## CHAPTER IX.

The era of "light" and combination locomotives—Samuel's "Lilliputian" and "Little Wonder"—The broad-gauge "Fairfield," constructed by Bridges Adams—Samuel's "Enfield"—Original broad-gauge "singles" converted into tank engines—The rise of "tank" engines, "saddle," and "well"—Adams' "light" engines on Irish railways—The Norfolk Railway adopts them—England's "Little England" exhibited at the 1851 Exhibition—Supplied to the Edinburgh and Glasgow, the Liverpool and Stockton, Dundee and Perth, and Blackwall Railways—Hawthorne's "Plews" for the Y.N. and B.R.—Crampton's monster "Liverpool"—Taylor's design for a locomotive—Pearson's prototype of the "Fairlie" engine—Ritchie's non-oscillating engine—Timothy Hackworth again to the front—His celebrated "Sanspareil, No. 2"—His challenge to Robert Stephenson unaccepted—Bury's "Wrekin"—Caledonian Railway locomotive, No. 15—"Mac's Mangle" on the L. and N.W.R.

MANY curious contrivances were introduced into the construction of the locomotive about the period now under review. Among these early proposals for the improvement of locomotion, few are more interesting than the combined locomotive and carriage introduced some fifty years ago by Mr. W. Bridges Adams.

Mr. Adams had a wide experience of every section of railway construction. Indeed, in the preface to one of his books, in writing of his experience, he says that he had "years of practical utility in planning the construction of nearly all machines that run on roads and rails also —from navvy's barrow up to a locomotive engine."

Nor are Mr. Adams's contributions to railway literature inconsiderable, for, besides writing several books between 1838 and 1862, he was at one time editor of a periodical, and also wrote voluminously under the pseudonym of "Junius Redivivus."

Having thus briefly mentioned Mr. W. B. Adams as being entitled to a far more important position in the evolution of our locomotives than is usually accorded him, we will now proceed to discuss the subject of combined locomotives and railway carriages, of which Mr. Adams was the chief advocate. The first machine of the kind, however, appears to have been constructed by Mr. Samuels, of the Eastern Counties Railway, for the purpose of quickly and economically conveying the officials of the railway over the system.

This engine was apparently called both the "Lilliputian" and the "Little Wonder." It was constructed in 1847, and made its first trip to Cambridge on Saturday, October 23rd, leaving London at 10.30 a.m., and reaching the University town at 2.45 p.m. Stops were made at three intermediate stations for water, etc., which occupied about half an hour, so that the $57\frac{1}{2}$ miles were covered in about 105 minutes' running-time.

The total length of the "Little Wonder" was 12ft. 6in., in which space was included the boiler, machinery, water-tank, and seats for seven passengers. The frame was hung below the axles, and carried on four wheels 3ft. 4in. diameter.

The floor was 9in. above rail level. The machinery consisted of two cylinders, $3\frac{1}{2}$in. diameter, and placed one on each side of the vertical boiler; the driving axle was cranked. The stroke was 6in. The boiler was cylindrical in shape, 19in. diameter and 4ft. 3in. high; it contained 35 tubes, 3ft. 3in. long and $1\frac{1}{2}$in. diameter; the tube heating surface being 38 sq. ft. The fire-box was circular in shape, 16in. diameter and 14in high, its heating surface being $5\frac{1}{2}$ sq. ft.

The link motion, feed pumps, etc., were provided. The water-tank held 40 gallons, and was placed under the seats. The usual speed of the "Little Wonder" with a full load was 30 miles an hour; and as high a rate as 44 miles an hour was often attained. The coke consumption was only $2\frac{1}{4}$lb. per mile. The weight of the whole vehicle, including fuel and water, was only $25\frac{1}{2}$ cwt.

Samuels' initial effort with light locomotives having been so successful, it occurred to him that branch traffic could be much more cheaply worked by means of a combined engine and carriage, instead of the usual locomotive and train of carriages.

Mr. Adams also had for some time been in favour of a combination of the kind, and Mr. Gregory, the engineer of the Bristol and Exeter Railway, was also in favour of the system being tried on the short branches of that railway, the passenger carriages on one at least of which were at that time drawn by horses. Acting upon the advice of Mr. Gregory, the directors of the Bristol and Exeter Railway ordered Mr. Adams to construct a vehicle and engine for working the traffic on the Tiverton branch. The machine was completed in December, 1848, and a satisfactory trial of it was made upon the broad-gauge metals of the West London Railway. This combination, which was constructed by Mr. Adams at Fairfield Works, Bow, E., was called the "Fairfield" (Fig. 49), and was brought into use on the Tiverton branch on December 23rd, 1848.

Its length was 39ft., and the boiler was placed in a vertical position. The driving wheels were 4ft. 6in. diameter, and were originally made of solid wrought iron. The middle and trailing wheels, 3ft. 6in. diameter, were of wood, and loose on their axles as well as their journals, the middle wheels having a lateral transverse of 6in.

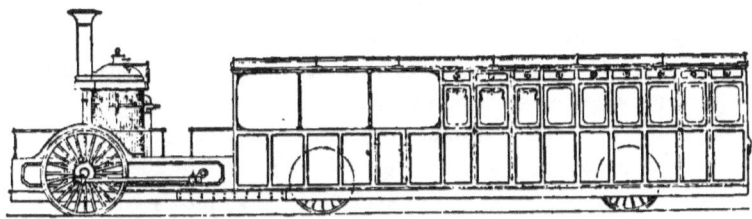

FIG. 49.—THE "FAIRFIELD," ADAMS'S COMBINED BROAD-GAUGE ENGINE AND TRAIN, FOR THE BRISTOL AND EXETER RAILWAY

The boiler was vertical, 3ft. in diameter and 6ft. high, and contained 150 tubes; the fire-box was 2ft. high and 2ft. 6in. in diameter. The cylinder was 8in. diameter, with 12in. stroke. The connecting-rods worked on a separate crank shaft, which communicated with the driving wheels by side rods, the axle of the driving wheels being straight, with crank pins on the outside.

The boiler was placed behind the driving axle, the tank, capable of holding 200 gallons of water, being in front of it; and the coke-box was attached to the front part of the carriage behind the driver. The working pressure was 100lb.

The bottom of the framing was within 9in. of the rails, so that by keeping the centre of gravity low greater safety might be ensured at high speed, and freedom from oscillation obtained.

The first-class carriage was in the form of a saloon, and accommodated sixteen passengers; whilst the second-class compartment seated thirty-two. The entire weight of the machine was about 10 tons, and when occupied with forty-eight passengers it amounted to about 12½ tons.

On the experimental trip, on December 8th, 1848, the "Fairfield" left Paddington Station at 10.30 a.m. for Swindon, 77 miles down the line, with a party of gentlemen connected with various railways. Mr. Gooch officiated as driver on both the up and down journeys.

Though the rails were greasy from the prevailing rain, in addition to a head wind—and, what was worse, a leak in the boiler—the machine soon attained considerable speed, and for a portion of the way reached the rate of 49 miles an hour. On arriving at Swindon the fire was extinguished, the leak partially repaired, and, after a reasonable sojourn, the party returned to town. The run back was exceedingly satisfactory, the speed of 49 miles being maintained for a considerable part of the way, the passage from Slough to Paddington being performed in 30 minutes.

As previously stated, the crank-shaft was unprovided with wheels, the motion being conveyed to the driving wheels by means of cranks fixed on the outsides of the driving axle, and connected to similar cranks on the driving wheels by means of connecting-rods.

This method has erroneously been called "Crampton's system," but it should be noticed that Adams used it for several years previous to Crampton adopting the plan in question. These combined engines and carriages were, in fact, built under a patent obtained by Mr. Adams in 1846, and, therefore, some time before Crampton adopted the inside cylinder and intermediate driving shaft.

It was found in practice that the vertical boiler of the "Fairfield" was not a success, so after some nine months' trial it was replaced by a horizontal tubular boiler. Then, after further experience, several drawbacks to the efficient working of branch line traffic by means of the combined engine and carriage were evident. So the engine was disconnected from the carriage and given an extra pair of wheels, and became, in fact, a miniature four-wheeled tank locomotive, a style of engine Adams afterwards became noted for building.

Mr. Samuel having obtained the sanction of the directors of the Eastern Counties Railway, Mr. Adams constructed a locomotive carriage for the Enfield branch traffic. The "Enfield" (Fig. 50), in appearance resembled a four-wheel tank engine and a four-wheel carriage, built together on a continuous frame, instead of being connected by couplings and buffers.

The whole framing, with the exception of the two buffer bars, was of wrought-iron, and was 8ft. 6in. in width, bound together by deep cross-bars.

The engine was of the outside cylinder class. The cylinders were 7in. in diameter, with a 12in. stroke. They were simply bolted down to the surface of a stout wrought-iron plate, in the middle of which the boiler was placed.

The driving wheels were 5ft. in diameter, and, as well as the front pair of wheels of the carriage, were without flanges, those of the leading engine wheels and the hind pair of the carriage being sufficient to retain the engine on the rails, whilst greater freedom was thus obtained for passing around curves. The boiler was constructed in the usual manner, and was 5ft. in length by 2ft. 6in. in diameter, and had 115 1½in. tubes 5ft. 3in. long, giving 230ft. of tube-heating surface. The dimensions of the fire-box were 2ft. 10½in. by 2ft. 6in., being an area of 25 sq. ft., making the total heating surface 255 sq. ft. The water was carried below the floor of the carriage in wrought-iron tubes 12in. in diameter and 12ft. long.

The coke was carried in a chest placed behind the foot-plate of the engine and immediately in front of the carriage head. The side frames were ingeniously trussed by diagonal bars of iron, and were thus rendered of great strength without adding much weight to the machine.

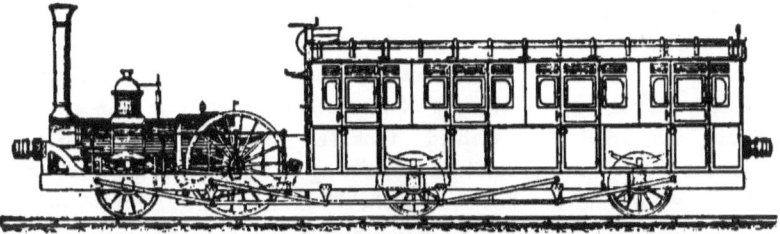

FIG. 50.—THE "ENFIELD," COMBINED ENGINE AND TRAIN FOR THE EASTERN COUNTIES RAILWAY

The leading engine wheels, together with the running wheels of the carriage, were 3ft. in diameter. The carriage was divided into four compartments, the two middle ones being for first-class and the two external ones for second-class passengers. The guard's seat was on the top of the carriage head. A vertical shaft with a hand-wheel on its upper end passed down the side of the head, and was connected beneath the framing with two transverse rocking shafts, carrying the brake blocks, placed one on each side of the driving wheels, thus giving the guard a ready means of control over the speed of the engine.

To bring up the buffers to the line of those of ordinary carriages, separate timber beams were passed across each end of the carriage, the front one being supported by neat wrought-iron brackets, rising from the framing. The total weight of the whole was not more than 10 tons, including its supply of coke and water, and accommodation was afforded for 42 passengers, to convey which, at 40 miles per hour, the calculated consumption of coke was 7lb. per mile.

Mr. Samuel stated that the accommodation provided by the combined engine and carriage was not sufficient for the traffic, so two additional carriages (one with a guard's compartment) were added, the train thus having accommodation for 150 passengers. The "Enfield" worked this train regularly at 37 miles an hour speed.

From January 29th to September 9th, 1849, the train travelled 14,021 miles, and was in steam 15 hours daily, but only five of which were spent in running. The total time in steam during the above period was 2,162 hours, the total coke consumed being 1,437 cwt., of which 743 cwt. was consumed in running, 408 in standing, and 286 in raising the steam. The average coke consumption per mile was 11.48lb., but a considerable portion of this was spent in standing, the actual consumption for running being only about 6lb. per mile.

In addition to the passenger traffic, the "Enfield" hauled all the goods and coal traffic on the branch, which, during the period under review, amounted to 169 tons of goods and 1,241 tons of coal. On June 14th, 1849, the "Enfield" took the 10 a.m. train from Shoreditch to Ely, 72 miles, the train consisting of three passenger carriages and two horse-boxes; but the "Enfield" arrived eight minutes before time, and the coke consumed only amounted to $8\frac{3}{4}$lb. per mile for the trip, including that used in raising steam.

When tried between Norwich and London, the "Enfield" performed the journey of 126 miles in 3 hours 35 minutes, including stoppages. An ordinary train had, at that time, never made the journey so quickly.

Although the "Enfield" appeared to use so little fuel, the broad-gauge "Fairfield" does not seem to have been an economical machine. A special trial was made between Gooch's famous 8ft. single "Great Britain" and "Fairfield," between Exeter and Bristol. A loaded wagon weighing 10 tons was drawn by the "Fairfield," making a total weight of $26\frac{1}{2}$ tons, of which the engine portion can be reckoned at $9\frac{1}{2}$ tons and 17 tons for the weight of the train. The distance is 76 miles, and the time allowed for the 8 a.m. train, including ten stops, was 2 hours 35 minutes; but the "Fairfield" took 3 hours 17 minutes to cover the distance, and consumed 13lb. of coke per mile, only 6.3lb. of water being evaporated for each pound of coke.

The duty performed by the two locomotives is thus tabulated:—

|  | Load in tons. | Coke per mile. | Consumption of Coke per ton per mile. |
|---|---|---|---|
| "Great Britain" | 100 | 26 lb. | 0·26 lb. |
| "Fairfield" | 17 | 13 lb. | 0·76 lb. |

But, in comparison with the old "Venus," the "Fairfield" comes out no better.

The "Venus," it will be remembered, was one of the original broad-gauge engines built for the Great Western Railway by the Vulcan Foundry Company, with 8ft. driving wheels. This engine had her driving wheels reduced to 6ft. diameter, and a small water-tank fitted on the foot-plate in place of a tender, thus being converted into a six-wheel "single" tank engine. The "Venus" only used 14lb. of coke per mile in working the Tiverton branch; while the "Fairfield" consumed 19lb. of coke per mile on the same work. The evaporating powers of the "Venus" had been greatly improved since N. Wood's experiments in 1838, as at that time she consumed 52.7lb. of coke per mile run.

FIG. 51.—" RED STAR," A 7FT. SINGLE BROAD-GAUGE SADDLE TANK ENGINE. CYLINDERS, 15in. BY 18in.

In addition to "Venus," several other of the early broad-gauge locomotives were reconstructed as tank engines. Fig. 51 ("Red Star") is a good example of the peculiar tank locomotives on the G.W.R. 60 years ago.

In addition to the "Fairfield" and "Enfield," combined engines and carriages were constructed by Mr. Adams for several other railways. One for the Cork and Bandon Railway had cylinders 9in. diameter, and accommodation for 131 passengers. This engine was constructed in such a manner as to enable it to run independently of the carriage. Another engine and carriage was built for a Scotch railway, and was guaranteed to work at 40 miles an hour. But the advantage of having the engine separate from the carriage was so great that Mr. Adams soon ceased to build the combination vehicles, and instead constructed his celebrated "light" locomotives; these, and the somewhat similar "Little England" engines, built by England and Co., were at one time very popular.

Fig. 52, representing "No. 148," one of the first batch of outside cylinder engines on the Southern Division of the L. and N.W.R., shows also a good example of Stephenson's "long boiler" locomotive. "148" was built by Jones and Potts, of Newton-le-Willows in 1847. The cylinders were 15in. diameter, the stroke being 24in. The driving

FIG. 52.—"No. 148," LONDON AND NORTH WESTERN RAILWAY; AN EXAMPLE OF STEPHENSON'S "LONG BOILER" ENGINES

wheels were without flanges, and were 6ft. 6in. in diameter. The leading wheels were 4ft. diameter. This engine was destroyed in a collision at Oxford on January 3rd, 1855, in which accident seven people lost their lives.

At this period a fashion for "tank" engines had become prevalent, and most of the locomotive builders produced designs, each having characteristic features. Thus Sharp Brothers and Company's "tank" engines had outside cylinders, with the tank between the frames and below the boiler, whilst the coal was carried in a bunker affixed to an extension of the foot-plate. Somewhat similar "single tank" engines were made by the same firm for the Manchester and Birmingham Railway (London and North Western Railway). The two engines in question were Nos. 33 and 34, and were used in working the traffic between Manchester and Macclesfield, the daily duty of each averaging 114½ miles. These engines commenced work in May, 1847. They weighed 21 tons in working order; the driving wheels were 5ft. 6in. diameter, and the leading and trailing 3ft. 6in. Two water-tanks were provided, one between the leading and driving wheels, the other under the coal bunker, at the rear of the trailing wheels. The two tanks contained 480 gallons of water. A woolen float attached to a vertical rod was fitted to show the amount of water in the tanks! The bunker contained half a ton of coals. These engines were fitted with sand-boxes; but these were placed in front of the leading wheels only, although the locomotives were specially constructed for running either bunker or chimney in front. However, the introduction of the sand-box was a step in the right direction; yet Tredgold only mentions the innovation in an apologetic manner. He says (after describing the working of the apparatus) that "it is very seldom required on the Macclesfield line, owing to the ballast between the rails being mostly sand; but when the rails are moist it is necessary in starting a heavy train to open the sand-cock." Tredgold then proceeds to give a detailed explanation of "how it is done."

In September, 1849, Walter Neilson, of Glasgow, obtained a patent for his design of tank engine.

The tank was of the now well-known "saddle" kind, and covered the whole boiler, barrel, and smoke-box; the bottom of the saddle tank rested on the frames on either side of the boiler, so that the tank was semi-circular in shape, instead of being but an arc, as is the practice with modern "saddle tanks." Neilson was, however, sufficiently ingenious not to limit the design of his saddle tank, for we find that "the tank may be supported from the boiler, instead of the framing, if necessary, and its length may be made shorter than that of the boiler, if required." The boiler was fed with water drawn from the smoke-box end of the tank, to obtain the advantage of the escaping heat. The coal bunkers were placed at the sides of the fire-box, and extended some distance towards the back buffer beam, but a bunker was not provided at the end, so as to allow "of ready access to the couplings of the wagons behind." The engine in question had inside frames, underhung springs, outside cylinders, single driving wheels, unprovided with flanges, and small leading and trailing wheels. A short cylindrical dome was placed over the fire-box, and on this were fixed two "Salter" pattern safety valves, covered by a brass casing.

# EVOLUTION OF THE STEAM LOCOMOTIVE 139

"Light locomotives" was the popular name of tank engines when the general use of such engines was being urged as a method of reducing the working expenses of unremunerative railways. We have previously alluded to Mr. W. Bridges Adams and his combined engines and carriages. This gentleman and Mr. England were the principal advocates of the "light" locomotive, and both attained some success in connection therewith.

The engines in question would now be considered absurdly light, but nearly fifty years ago far different ideas of "light" and "heavy,", as applied to locomotive engines, obtained.

The practice of Adams and England regarding "light" locomotives differed considerably. The former was a firm advocate of four wheels

FIG. 53.—ADAMS'S "LIGHT" LOCOMOTIVE FOR THE LONDONDERRY AND ENNISKILLEN RAILWAY

and a long wheel base. England, on the other hand, preferred his light locomotives to be supported by six wheels. In 1847, Adams built a light locomotive (Fig. 53) for the Londonderry and Enniskillen Railway (Ireland), with outside cylinders 9in. in diameter, the stroke being 15in. The driving wheels were 5ft. in diameter, and located in front of the fire-box; the other pair of wheels were 3ft. diameter, and were placed beneath the smoke-box. The fire-box was 2ft. 9in. long, the boiler 2ft. 3in. diameter and 10ft. 3in. long; height of top of boiler from rails, 5ft. 8in. The connecting-rods were 5ft. 3in. long; the steam pressure was 120lb. The water-tank was placed beneath the boiler, and reached to within a few inches of the surface of the rails. Mr.

Adams built a similar engine for the St. Helen's Railway. In November, 1849, a broad-gauge light locomotive was built at Mr. Adams's Fairfield Works, for service on the Holyhead breakwater. The engine in question was from designs prepared by Mr. Thos. Gray, resident engineer of C. and J. Rigby, the contractors for the breakwater. This engine had cylinders 8in. diameter, the stroke being 18in.

In July, 1849, Adams supplied two of his light engines to the Cork and Bandon Railway. These differed from those already described, as the driving wheels were the leading ones, the smaller pair of wheels being at the rear. The Irish names of the engines signified "Running Fire" and "Whirlwind."

In August, 1853, the engineer of the Cork and Bandon Railway reported that "the cost of repairs to the engines was very small, more particularly on the light engines, which have worked all the fast passenger trains in a satisfactory manner, and with the same consumption of coke as heretofore—viz., about 10lb. per mile. These engines were put upon the line in July, 1849, since which period they have been daily working the passenger traffic. The principal item of cost in their repairs during the four years has been a new crank axle to each of the two light engines, as also a new set of tyres on the driving wheels. The light special trains conveyed by these engines generally occupy about 26 minutes between the two termini of Cork and Bandon." These two light locomotives continued to work traffic over the Cork and Bandon Railway for several years.

On May 1st, 1851, Mr. Peto, the chairman of the Norfolk Railway, provided four light engines with 12in. cylinders, and weighing 10 tons each, to work the branch traffic of that railway under the following circumstances.

The Norfolk Railway was worked by the Eastern Counties, and the branch or local trains of the former were supposed to meet the main line trains of the latter line at the junctions.

But the Eastern Counties trains had a habit of being behind-hand, putting in an appearance at the junctions any time between thirty minutes and an hour after the times given in the time-tables. As a result, the traffic on the Norfolk branch lines was thoroughly disorganised; indeed, so little could it be depended upon that local passengers almost completely neglected the line. Then the Eastern Counties Railway worked the Norfolk branches with the main line engines, and charged the Norfolk Railway the average expense per mile incurred in working with these engines.

Such a method did not meet with the approval of the chairman of the Norfolk Railway, so Mr. Peto obtained the sanction of the Eastern Counties Railway to allow the Norfolk Company to work the local branch traffic itself, and independent of the arrival and departure of the main line trains. Mr. Peto's new system met with instantaneous and complete success, a great saving being effected. Thus the coke consumption of Adams's light engines, introduced by Mr. Peto, only averaged 10lb. per train mile; but the Norfolk Railway had been paying the Eastern Counties Railway at the rate of 27lb. per mile, that being the average coke consumption of the Eastern Counties Railway main line engines. A large and remunerative local passenger traffic was built up by reason of the improved local services.

The advantages claimed by Mr. Adams for his light engines were as follows:—Less dead weight, less friction, and less crushing and deflecting of the rails.

We will now proceed to give some account of England's light locomotives, popularly called "Little Englanders"; but this cognomen then had a very different meaning, as applied to locomotives, than the words have at the present time in their application to certain individuals. England constructed his premier light engine in 1849, and the "Little England" (Fig. 54) was exhibited at the Exhibition of 1851. The chief dimensions were:—Driving wheels, 4ft. 6in. diameter, located in front of the fire-box; leading and trailing wheels, 3ft. diameter; inside cylinders, placed between the leading and driving wheels, and not under the smoke-box; the frames were outside. The fire-box was of the Bury type, with safety valves, similar to those previously described as on the Bury engine still at work on the Furness Railway. A dome was placed on the boiler barrel over the cylinders, so that the steam-pipes proceeded in a curved vertical line from the dome to the cylinders. The dome was on a square seating. An auxiliary pipe for the escape of the steam was provided at the back of the chimney, but was only about one-half as high as the chimney. At the rear of the foot-plate was a well-tank, holding water sufficient for a 50-mile trip. A prize medal was awarded to this engine at the Exhibition.

England and Co. in August, 1850, sent one of their light engines to the Edinburgh and Glasgow Railway on the following conditions: A guarantee that the engine should work the express trains between Edinburgh and Glasgow, consisting of seven carriages, and keep good time as per time-bill, while the fuel consumption was not to exceed 10lb. of coke per mile. If the light engine performed these conditions to the satisfaction of the railway company's engineer, the Edinburgh and Glasgow Railway was to purchase the locomotive for £1,200. But if the work done and the quantity of fuel consumed were not as guaranteed, England and Co. were to remove the engine and pay all expenses of the trial.

This "Little England" was tried in competition with the "Sirius," the coke consumption of the former being 8lb. 3oz. per mile against 29lb. 1oz. of the "Sirius," both performing exactly the same work. The "Little England" so frequently ran in before her time that the driver had to be ordered to take longer time on the trips for fear of an accident happening in consequence of the train arriving before it was expected. The speed of this light engine frequently exceeded 60 miles an hour, and during the heavy winds and gales of January, 1851, the "Little England" was the only locomotive on the line that

FIG. 54.—ENGLAND & CO.'S "LITTLE ENGLAND" LOCOMOTIVE, EXHIBITED AT THE PREMIER INTERNATIONAL EXHIBITION, LONDON, 1851

kept time. With a train of five carriages the coke consumption only amounted to 6½lb. per mile. On the Campsie Junction line, the "Little England" hauled a train of seven carriages and a brake-van, all of which were overloaded with passengers, over the several gradients of Nebrand, at 30 miles an hour. Although the train stopped at a station on the incline, the light engine successfully started from the station and continued the ascent. An ordinary engine was sent to assist the train at the rear, in case the "Little England" proved unequal to the task, but it is said that the bank engine was unable to keep up with the train!

The following table shows the result of the trial of the "Little England" on the Edinburgh and Glasgow Railway:—

| Number of Carriages per Train. | Daily Mileage (47⅜ each way). | No. of Stoppages (8 each way). | Time on each trip. | Coke Consumed. | |
|---|---|---|---|---|---|
| | | | | While running. Per Mile. | Including lighting up and standing 4 hours between each trip. |
| | | | | lbs. oz. | lbs. oz. |
| 7 7 5 4 | 95 miles. | 6 | 90 min. | 8  3<br>8  3<br>7  4<br>6  5 | 9  7<br>9  7<br>9  7<br>8  5 |

On September 7th, 1850, another "Little Englander" commenced service on the Liverpool and Stockport Railway, under guarantee to haul a train of seven carriages up an incline of 1 in 100, stopping and starting upon it, at a speed of 25 miles an hour, and consuming not more than 10lb. of coke per mile; on the level the speed was to be 45 miles an hour. This engine frequently drew ten carriages under the conditions laid down for only seven. In June, 1849, a "Little Englander" had been supplied to the Dundee and Perth Railway for working the mail train of four carriages. This the engine did successfully for a considerable time.

After the abolition of rope traction on the Blackwall Railway "Little Englanders" were used for the passenger trains.

England and Co. guaranteed these light engines to haul trains of six carriages at a speed of 40 miles an hour on gradients of 1 in 100, at a coke consumption of only 10lb. per mile. These engines cost £1,200 each, and the builders were willing to back them for 1,000 guineas a-side, with a load in proportion to the weight of any other engine, or the amount of fuel consumed. We do not think anyone ever cared to accept this challenge.

In March, 1848, a patent was granted to McConochie and Claude, of Liverpool, for various improvements in the locomotive. The cylinders were inside, behind the leading wheels, the valve gearing being outside the frame and worked by eccentrics on the naves of the driving wheels. It will be remembered that the valve gearing of Stephenson's "No. 185" was on this plan. The pumps were worked off the driving wheels, as in the "Jenny Linds." A double-beat safety valve was provided.

To enable a low-pitched boiler to be employed, the axle was cranked at the extreme ends, so that at each extremity of the axle only one return crank-arm was provided, the wheel itself forming the second one, and a pin connecting the wheel and axle-crank formed the shaft upon which the connecting-rod worked.

To increase the weight upon the driving axle, a toggle joint was placed between the bearing of the trailing axle and the springs; a rod connected the knuckle of the toggle joint with the piston of a small steam cylinder.

When the driver wished to obtain additional adhesion for the driving wheels, he admitted steam to this auxiliary cylinder, which drove the toggle joint into an upright position, thereby removing the weight from the trailing wheels and placing it upon the driving wheels. Several other novel proposals were included in the specification in question.

In 1848, Hawthorne, of Newcastle, built an engine named "Plews," No. 180, of the York, Newcastle, and Berwick Railway (makers' number of engine, 711). The locomotive had a copper fire-box. The boiler was 10ft. 8in. long, of oval shape, and consequently had to be stayed with four plates; 229 brass tubes of 1¾in. external diameter were provided; two lever safety valves were fixed on a raised fire-box and enclosed in a brass casing; the steam pressure was 120lb. A very large cast-iron dome placed on the centre ring of the boiler was a characteristic of the "Plews."

The cylinders were placed between the outside and inside frames, diameter 16in. and stroke 20in.; whilst the slide valves were outside the cylinders, being worked by four eccentrics, on the outside of the wheels, but within the outside frames. The driving wheels were 7ft. diameter, the leading and trailing being 4ft. diameter; the whole of the bearings were outside.

When at rest, the steam was turned into the tender for the purpose of heating the feed-water. The tender was carried on six wheels of 3ft. 6in. diameter, and was capable of holding 1,400 gallons of water.

Brake blocks were provided for both sides of the six wheels, and an ingenious arrangement of tooth wheels and rack applied the whole of the blocks by means of a few turns of the brake handle.

Crampton's engine, "Liverpool" (Fig. 55), has been described as the "ultimatum for the narrow-gauge." Why, we are at a loss to understand; many other narrow-gauge engines have been constructed of greater power, and certainly of more compact and pleasing design. The "ultimatum of locomotive ugliness" would have been a correct title for the "Liverpool."

The engine in question was built by Bury, Curtis, and Kennedy, for the London and North Western Railway, in 1848. The one good point about the engine was the immense heating surface, which amounted to 2,290 sq. ft. When our locomotive superintendents make up their minds to construct express locomotives with such an amount of heating surface, we shall hear no more of "double engine running," and our

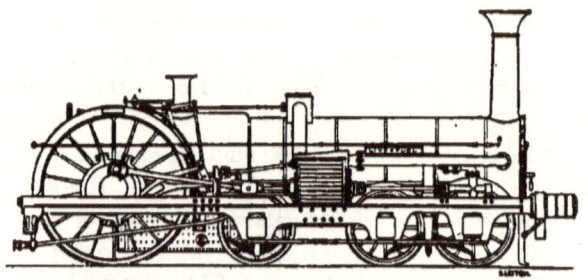

FIG. 55.—CRAMPTON'S "LIVERPOOL," L. AND N.W.R.

express trains may be expected to average a speed of over 50 miles an hour from start to finish (including stops) on all trips.

The general arrangement of the "Liverpool" was similar to the engines on Crampton's system already described—viz., the driving wheels at the back of the fire-box and outside cylinders fixed about the centre of the frames. This engine had three pairs of carrying wheels under the boiler, in addition to the driving wheels. The cylinders were outside, fixed upon transverse bearers, formed of iron plates 1¼in. thick, curved to the shape of the boiler and passing below it. The cylinders were 18in. diameter, the stroke being 24in. Metallic packing, consisting of two concentric rings of cast-iron, each with a wedge and circular steel spring, was used for the purpose of making the pistons steam-tight. The valves were above the cylinders, and were

L

inclined, the eccentrics being of large size and outside the driving wheels. The regulator was located in a steam-box on the top of the boiler barrel; the steam reached the valves by means of curved vertical copper pipes outside the boiler, whilst the exhaust was conveyed to the smoke-box by similar horizontal "outside" pipes. The two exhaust pipes united within the smoke-box beneath the bottom of the chimney, the blast orifice being $5\frac{1}{2}$in. diameter.

The leading wheels were 4ft. 3in. diameter, the two intermediate pairs 4ft., and the driving wheels 8ft. in diameter. The area of the fire-grate was 21.58ft. The tubes were of brass, 12ft. 6in. long; 292 were 2 3-16th in diameter, the remaining eight being $1\frac{3}{4}$in. diameter. The heating surface was:—Tubes, 2,136.117 sq. ft.; fire-box, 154.434 sq. ft. The pumps were horizontal, fixed on the frames over the leading wheels; they were worked by extension piston-rods, worked through the covers of the cylinders.

The engine weighed (loaded) 35 tons, of which weight 12 tons were on the driving axle. The tender weighed 21 tons. With a light load the "Liverpool" attained a speed of nearly 80 miles an hour, whilst on one occasion she hauled the train conveying Franconi's troupe and horses, consisting of 40 vehicles, from Rugby to Euston under the schedule time. Three engines had been engaged to haul the same train from Liverpool to Rugby, when time was lost. The power of the "Liverpool" would, therefore, appear to have exceeded that of three of the usual London and North Western Railway locomotives of that date.

Adams's idea of a straight driving shaft connected by means of outside rods with the driving wheels soon attracted attention, and in 1849 Crampton incorporated the principle in his patent locomotive specification of that year. But it was some two years later before any engines were built under this particular patent of Crampton's. These locomotives will be described in due sequence.

We will now give a few details of some engines that would have been most interesting had we knowledge that they were ever built. We possess drawings of the engines in question, but lack authentic details of their performances, so we will mention the principal features of the designs, as given in the patent specifications. George Taylor, of Holbeck, Leeds, obtained his patent on June 3rd, 1847. The drawing shows the boiler to be hung below the wheels, of which there are only four; these were to be 15ft. diameter, and in addition the wheels were geared up 2 to 1, so that one revolution of the cogged driving wheel would have propelled the engine six times the distance of a driving wheel of 5ft. diameter. The cylinders were inside the frames, over the boiler, and, of course, at the rear of the smoke-box; the connecting-rods were attached to cranks on either side of a central cog-wheel, which engaged with a cog-wheel of half its diameter, fixed on the centre of the rear axle. The motion being conveyed to the centre of the axle, instead of alternately on each side, as is usual, practically abolished the oscillating motion so apparent in two-cylinder engines. An examination of the drawing of this locomotive design of George Taylor shows with what ease and slight alteration it was possible for the two geared engines supplied to the Great Western Railway by the Haigh Foundry to have been altered to ordinary direct action engines.

Large wheels were also to be used for the tender, the axles passing through the water-tank, so that the centre of gravity was lowered.

James Pearson, the locomotive superintendent of the Bristol and Exeter Railway, obtained a patent on October 7th, 1847, for a double locomotive. Fairlie's "Little Wonder" narrow-gauge engines were probably suggested by Pearson's design of 1847; whilst the latter's famous broad-gauge double-bogie tanks were decidedly evolved from his earlier form of locomotive.

The boiler was to have the fire-box in the centre, the latter being divided into two parts, connected below the furnace doors; the driving axle was across this central foot-plate, to allow of very large wheels and a low centre of gravity. Each boiler (there being practically two, one each side of the central double fire-box) was carried on a four-wheel bogie, so that the locomotive was carried on ten wheels, as in the later design. The bogie frames were connected by tension-rods, passing outside the fire-box. India-rubber springs were employed, their use being to allow each bogie to adjust itself to any inequality of the road, and to bring the bogies back to the straight position on an even road. The coke was to be stowed in bunkers over the boilers, and the water could be either in tanks between the tops of the boilers and the coke bunkers, or a separate tender could be provided. The steam domes were on the fire-box, and were to be of abnormal height, and connected over the head of the foot-plate, thus forming the roof of the cab. An exhaust fan was fixed in the smoke-box to draw the heated air through the tubes and discharge it up the chimney, or it could be used again as a hot blast for the furnace, and a chimney and a smoke-box were provided for each boiler. The fans were to be driven by pulleys off one of the axles, and it was claimed that, as the exhaust steam was not required for the purpose of creating a blast, extra large exhaust pipes could be used, and the cylinders thereby relieved of "back pressure." The cylinders were outside, and the valves were beyond the cylinders. These were fixed between the wheels of one of the bogies. The general design of this engine, as shown in the drawings, was very ingenious, and is certainly the most symmetrical "double-ended" type of engine we have seen illustrated. Pearson for some reason did not construct an engine after this style, but produced the well-known 9ft. "single" (double-bogie) tanks instead.

The third patent now to be described had also for its leading feature extra large driving wheels. The specification is that of Charles Ritchie, of Aberdeen, the patent being granted to him on March 2nd, 1848. The principal feature was the providing of two piston-rods to each piston, one on each side. Four driving wheels were proposed, one pair placed in front of the smoke-box and one pair behind the fire-box. The cylinders were outside, and were, of course, fixed at an equal distance between the two pairs of driving wheels. One pair of carrying wheels was to be used, placed below the cylinders. It was claimed that this arrangement of pistons and connecting-rods exactly balanced the reciprocating parts of the machinery, and therefore abolished oscillation. Another improvement related to the slide-valves, the starting, stopping, and reversing of the engine, together with the expansive working of the steam, the whole to be controlled by a wheel on the foot-plate, connected by cogs with the link of the valve-gear.

Other improvements were compensating safety-valves, an "anti-primer," and an improved feed-water apparatus. The last is described as follows:—"Upon steam being admitted from the boiler into the cylinder, through the steam-port, the piston will be acted upon, and the ram be withdrawn; the water will then raise the valve and enter the barrel, to occupy the space previously occupied by the ram. By this time the piston will have acted upon a lever, so as to cause the slide-valve to uncover one port and cover the other, thereby allowing the steam on the other side of the piston to escape through the exhaust pipe.

"The piston will now be impelled in a contrary direction, and the ram entering the barrel will cause the one valve to be closed and the other to be opened by pressure of the water therein, which, as the ram advances, will be forced into the boiler."

Another part of the specification related to an "anti-fluctuator." A partition-plate was to be fixed between the tube-plate and the fire-box, and the water was to be let into the boiler at the fire-box end, and would only reach that portion of the boiler beyond the fire-box by flowing over the top of the partition-plate. By this means the fire-box would always be covered with water. It will be seen that the specification contained several useful propositions, which, however, do not appear to have been put into practice.

FIG. 56.—TIMOTHY HACKWORTH'S "SANSPAREIL No. 2"

We have previously, upon more occasions than one, shown the important position occupied in the evolution of the steam locomotive by the engines built or designed by Timothy Hackworth. We now have to give an account of his last locomotive, the "Sanspareil No. 2."

A comparison of the drawings of this engine (copies of which are in our possession) with Hackworth's earlier efforts of 20 years before, clearly discloses the remarkable strides made in the improvement of the locomotive during that period, and also most clearly shows that in 1849 Hackworth was still in the very van of locomotive construction, even as he had been in the days of his "Royal George."

The "Sanspareil No. 2" (Fig. 56) was constructed by Timothy Hackworth at his Soho Engine Works at Shildon. The patent was obtained in the name of his son, the late John Wesley Hackworth. We are indebted to the executor of the will of Timothy Hackworth for many of the following details concerning the engine now under review.

The locomotive was of the six-wheel "single" type, with outside bearings to the L. and T. wheels, and inside bearings of the driving wheels. The cylinders were inside. A cylindrical steam dome was placed on the boiler barrel close to' the smoke-box. The fire-box was of the raised pattern, and on it was an encased Salter safety-valve. Cylindrical sand-boxes were fixed on the frame-plates in front of the driving wheels. The principal dimensions of the engine were:— Driving wheels, 6ft. 6in. diameter; leading and trailing, 4ft. diameter; cylinders, 15in. diameter, 22in. stroke. Weight in working order:—L., 8 tons 6 cwt.; D., 11 tons 4 cwt.; T., 4 tons 5 cwt. Total, 23 tons 15 cwt.

It would be well if we mentioned the principal novelties in construction—viz.: Welded longitudinal seams in boiler-barrel; the boiler was connected to the smoke-box and fire-box by means of welded angle-irons, instead of the usual riveted angle-irons; the lagging of the boiler was also covered with sheet-iron, as is now general, instead of the wood being left to view, as was at that time the usual practice.

A baffle-plate was fitted at the smoke-box end of tubes, as well as at the fire-box end.

The pistons and rods were made of wrought-iron in one forging.

The valves were constructed under Hackworth's patent, and were designed to allow a portion of the steam required to perform the return stroke to be in the cylinder before the forward stroke was completed, and thus to form a steam cushion between the piston and cylinder covers. Such working was said to economise 25 to 30 per cent. of fuel.

The engine conveyed 200 tons 45 miles in 95 minutes, consuming 21 cwt. of coke, and evaporating 1,806 gallons of water. She also drew a train of six carriages over the same distance without a stop, in 63 minutes, with an expenditure of 13 cwt. of coke and 1,155 gallons of water.

Upon the* completion of this engine, J. W. Hackworth sent the following challenge to Robert Stephenson:—

"Sir,—It is now about 20 years since the competition for the premium of locomotive superiority was played off at Rainhill, on the Liverpool and Manchester Railway. Your father and mine were the principal competitors. Since that period you have generally been looked to by the public as standing first in the construction of locomotive engines. Understanding that you are now running on the York, Newcastle and Berwick Railway a locomotive engine which is said to be the best production that ever issued from Forth Street Works, I come forward to tell you publicly that I am prepared to contest with you, and prove to whom the superiority in the construction of locomotive engines now belongs.

"At the present crisis, when any reduction in the expense of working the locomotive engine would justly be hailed as a boon to railway companies, this experiment will no doubt be regarded with deep interest as tending to their mutual advantage. I fully believe that the York, Newcastle and Berwick Railway Company will willingly afford every facility towards the carrying out of this experiment.

"Relying upon your honour as a gentleman, I hold this open for a fortnight after the date of publication.

"I am, Sir, yours, etc.,

JOHN W. HACKWORTH."

We do not think Robert Stephenson accepted the challenge; at all events, no records of such a competition have ever been made public, and had it taken place the victor would have doubtless well published the result.

The "Sanspareil" frequently attained a speed of 75 miles an hour on favourable portions of the line. She was sold to the North Eastern Railway by the executors shortly after the death of Timothy Hackworth, something like £3,000 being obtained for the engine, which continued to work upon the North Eastern Railway until recent years, having, of course, been rebuilt during the long time it was in active service.

We have now to describe another specimen of the locomotives constructed by the celebrated firm of Bury, Curtis and Kennedy. This locomotive was one of the last engines built by the firm before its final dissolution. The "Wrekin" was a six-wheel engine with inside bar frames and inside cylinders, and was constructed for the Birmingham and Shrewsbury Railway in 1849.

The special points noticeable in the construction of the engine in question are the width of the framing, which was arranged horizontally instead of vertically, and only two bearings to each axle. The axleboxes of the leading wheels were bolted to the frames, those of the other wheels being welded to the frames, and the cylinders were also directly affixed to the framing. An advantage claimed by the builders, as resulting from the method of construction employed, was that the weight being placed entirely within the wheels, such weight had a tendency to press down the axle between the bearings, and so counteract the constant tendency arising from the flanges of the wheels, when pressing against the edge of the rails, especially in passing round curves.

The cylinders were 15in. diameter, the stroke being 20in The driving wheels were 5ft. 7in. diameter, the leading 4ft. 1in., and the trailing 3ft. 7in.

The boiler contained 172 brass tubes, 11ft. 6in. long and 2⅛in. external diameter. The heating surface was: Tubes, 1,059 sq. ft.; fire-box, 80 sq. ft.; total, 1,139 sq. ft. Grate area, 15 sq. ft.

No steam dome was provided, the main steam-pipe being of iron, with a longitudinal opening 3-16th inches wide along the top; this -pipe extended to the smoke-box, at which end of it the regulator valve was placed; the actuating-rod passing through the main steam-pipe from end to end. Two encased Salter safety valves were fixed on the fire-box. The wheel base of the "Wrekin" was: leading to driving, 8ft. 1in.; driving to trailing, 6ft. 11in.

In 1849 the Vulcan Foundry Company supplied the Caledonian Railway with an engine known as "No. 15." In general appearance the locomotive was very similar to Allan's "Velocipede" engine on the London and North Western Railway.

"No. 15" (Fig. 57) was a six-wheel engine, with inclined outside cylinders, 15in. diameter and 20in. stroke. The driving wheels were 6ft. diameter, leading and trailing wheels 3ft. 6in. diameter. The boiler barrel was 9ft. 9in. long and 3ft. 6¾in. diameter, containing 158 brass tubes of 1¾in. external diameter. Wheel base, L. to D., 6ft.; D. to T., 6ft. 6½in. The chimney was 6ft. 6in. high; on the centre of the boiler was a man-hole, surmounted by a column safety-valve of Salter's pattern, the blowing-off steam pressure being 90lb. The steam dome was of brass, placed on the raised fire-box, and surmounted with a second Salter's safety-valve. The driving and leading wheels were provided with underhung springs, but the trailing wheels

EVOLUTION OF THE STEAM LOCOMOTIVE 153

had the springs over the axle-boxes. These latter springs were of elliptic shape, and were provided with a screw device fixed on the foot-plate, by means of which the weight was taken off the trailing wheels and thrown upon the driving wheels.

In addition to the semi-circular brass name-plates (*i.e.*, Caledonian Railway) affixed to the splashers of the driving wheels, brass number-plates of diamond shape (12in. long by 6in. diameter) were fixed on the buffer beams of " No. 15." The tender was supported on four wheels, 3ft. 6in. diameter, and held 800 gallons of water.

During June, 1849, " No. 15 " made a number of trial trips between Glasgow and Carlisle, with seven, eight, and nine coaches of an average weight of five tons each, the weight of the engine and tender

FIG. 57.—CALEDONIAN RAILWAY ENGINE, "No. 15"

being 28 tons. On the trips to Glasgow the Beattock Summit had, of course to be climbed. This consists of 10 miles of stiff gradients, varying between 1 in 75, 80, and 88. The run of 13½ miles from Beattock to Elvanfoot, consisting of the 10 miles just described and of 3½ down at 1 in 100, was negotiated by " No. 15 " in 33 minutes, with a train of six coaches; with seven coaches the time was 41 minutes, and with a pilot and eleven coaches, 30 minutes, or at the rate of 27 miles an hour. These were considered exceptionally good specimens of hill-climbing performances 48 years back, but are, of course, entirely out of comparison with modern Caledonian records over the same line with much heavier trains.

McConnell, the locomotive superintendent of Wolverton, turned out several remarkable locomotives for the London and North Western Railway, and No. 227, or, as she was generally called, " Mac's Mangle," (Fig. 58), was one of these peculiar specimens of McConnell's design. The cylinders were of large size, being 18in. diameter, with a 24in. stroke;

they were outside, as were also the axle bearings—a very uncommon combination. No. 227 was a six-wheel "single" engine, the driving wheels being 6ft. 6in. diameter, and the leading and trailing wheels 4ft. diameter. The fire-box was of the raised pattern, and a Salter

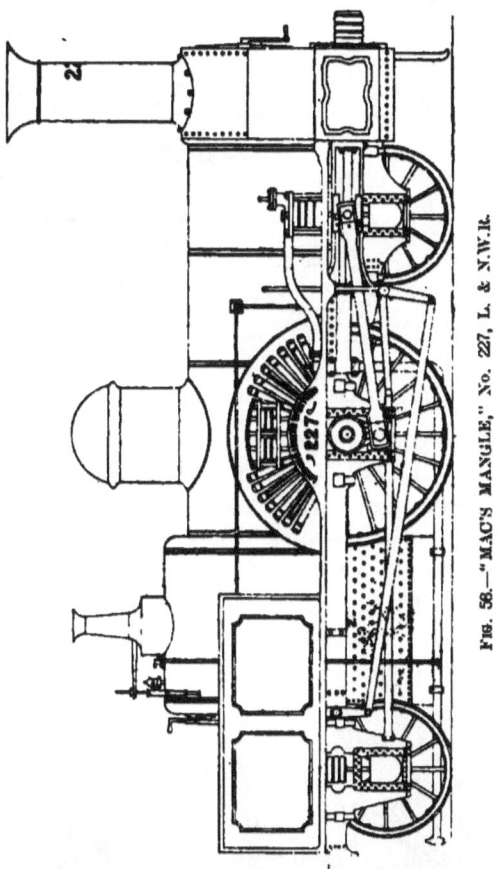

Fig. 58.—"MAC'S MANGLE," No. 227, L. & N.W.R.

safety valve (encased) was fixed on it. A huge steam dome was provided, located, originally, close to the smoke-box end of the boiler barrel, but afterwards (in 1850) placed near the fire-box end, over the driving wheels. The boiler-heating surface of "Mac's Mangle" was 1,383 sq. ft. No. 227 enjoyed but a short locomotive career, being built in April, 1849, and "scrapped" in May, 1863. It is stated that in consequence of the extreme width of this engine, caused by outside cylinders being employed in conjunction with outside axle-boxes, it became necessary to set back the platforms at some of the stations, so that the engine could clear these erections without coming to grief.

FIG. 59.—" PRESIDENT," ONE OF McCONNELL'S "BLOOMERS," L. & N.W.R. AS ORIGINALLY BUILT

In 1850 McConnell designed a very powerful class of passenger engines for the L. and N. W. R. These are generally called the "Bloomers." "President" (Fig. 59) illustrates this favourite class of L. and N. W. R. locomotive, when built. The cylinders were inside, 16in. diameter, with a stroke of 22in. The driving wheels were 7ft. in diameter. The heating surface was 1,152 sq. ft. These engines weighed 28¾ tons. (Fig. 60) is from a photo of a "Bloomer" as rebuilt by Ramsbottom.

FIG. 60.—ONE OF McCONNELL'S "BLOOMERS" AS REBUILT BY RAMSBOTTOM

## CHAPTER X.

The locomotive exhibits of 1851—The "Hawthorn"—Wilson's two-boiler engine, the "Duplex"—Fairbairn's tank engine—The S.E.R. "Folkestone" on Crampton's system—Sharp's "single" engines for the S.E.R.—J. V. Gooch's designs for the Eastern Counties Railway—The "Ely," Taff Vale Railway—Beatties "Hercules"—A much-vaunted locomotive, McConnell's "300." L. & N.W.R.—London and Birmingham in two hours—The chief features of "300"—Competitive trials with other engines—Coal v. coke—An earlier "recessed" boiler—Dodd's "Ysabel"—The first compound locomotive—Another Beattie design—Pasey's compressed air railway engine—Its trial trips on the Eastern Counties Railway—The original Great Northern engines Sturrock's masterpiece, "No. 215," G.N.R.—Pearson's famous 9ft. "single" double bogies, Bristol and Exeter Railway—Rebuilt with 8ft. drivers, and a tender added by the G.W.R.—More old Furness Railway engines—Neilson's outside cylinder locomotives—A powerful goods engine on the Maryport and Carlisle Railway—Gooch's 7ft.-coupled broad-gauge locomotives—His first narrow-gauge engines.

THE premier International Exhibition, which, as all the world well knows, was held in Hyde Park, London, 1851, brought together quite a respectable collection of railway appliances. The British exhibitors showed the following locomotives:—

London and North Western Railway's "Cornwall" and "Liverpool."
Great Western Railway's "Lord of the Isles."
Hawthorne's express, "Hawthorn."
Adams's combined engine and carriage, "Ariel's Girdle," built by Wilson and Co., Leeds.
England's light locomotive, built by Fairbairn.
Fairbairn's tank engine.
South Eastern Railway's "Folkestone."
E. B. Wilson and Co.'s double boiler tank engine.

Several of these have been described in an earlier chapter, whilst details of other types (such as the "Lord of the Isles" type) have also been given, so that it is not necessary to describe such designs again. We have, however, to give particulars of Hawthorne's express, Fairbairn's tank, the "Folkestone," and Wilson's "double boiler" tank engine. The dimensions of the first are: cylinders, 16in. diameter,

22in. stroke; driving wheels, 6ft. 6in.; leading and trailing wheels, 3ft. 9in. diameter; heating surface of fire-box, including water bridge, 110 sq. ft.; tubes, 865.4 sq. ft. The tubes were of brass, of 2in. external diameter, and 158 in number.

The " Hawthorn " had inside cylinders and double sandwich frames, a raised fire-box, with an enclosed safety-valve, no dome, but a perforated steam-pipe for the collection of the steam was provided. The engine was designed for running at 80 miles an hour; the special features of the engine being double compensating beams for distributing the weight uniformly on all the wheels, equilibrium slide valves, and an improved expansion link suspended from the slide-valve rods. Instead of fitting a spring to each wheel, two only were placed on each side of the engine between the wheels. These springs were inverted, and sustained by central straps attached to the framing. Their ends were connected by short links to the wrought-iron double-compensating beams placed longitudinally on each side of the engine, inside and beneath the framing.

The two inner contiguous ends of these beams were linked by a transverse pin to an eye at the bottom of the axle-box of the driving axle, whilst the opposite ends of the beams were respectively linked in a similar manner to eyes on the top of the leading and trailing axle-boxes. The action of these beams was obvious. By them a direct and simultaneous connection was given to all the axle-bearings, and consequently a uniform pressure was always maintained on all the wheels, irrespective of irregularities on the permanent-way. The slide valves were placed on vertical faces in a single steam-chest, located between the two cylinders. One slide valve had a plate cast on its back, and the other had an open box cast on its back to receive a piston, which had its upper end parallel with the valve face. This piston was fitted steam-tight in the box, and its planed top bore against the face of the plate in working. By this arrangement the slides were relieved from half of the steam pressure; and to assist a free exhaust, a port was made in the back plate of one of the slides, so providing an additional exit for the spent steam by means of the piston and the exhaust ports of the opposite valve.

The expansion link was placed in such a position as to allow the bottom of the boiler to be quite near the axle. The link, instead of being fixed to the ends of the eccentric-rods, so as to rise and fall with them when the reversing lever was moved, was suspended from its centre, by an eye, from the end of the slide-valve spindle. This removed the weight of the link, etc., from off the reversing gear. The eccentric-rods were jointed to the opposite ends of the link slide-block, to secure steadiness and durability of the parts. It was claimed that this method of a fixed link-centre as fitted to the "Hawthorn" ensured a more correct action of the valves.

Wilson and Co., of the Railway Foundry, Leeds, exhibited a curious tank engine at the Exhibition of 1851, called the "Duplex," in consequence of it being provided with two boilers. The idea of the designer was to obtain sufficient steam from an engine of light weight to haul a heavy train. The original drawings of this engine are still in the possession of Mr. David Joy, who designed it; and at first it was proposed to build the "Duplex" with three cylinders and six-coupled wheels, but afterwards fresh drawings were prepared, and it was from these latter ones that the engine was built. The two boilers were placed side by side, and these each measured 10ft. 6in. long by 1ft. 9in. diameter, and together contained 136 tubes of 1¾in. diameter, the heating surface of which was 694 sq. ft., that of the fire-box being 61 sq. ft., making a total of 755 sq. ft. The cylinders were outside, their diameter being 12½in., and the stroke 18in. The leading wheels were 3ft. 6in. diameter; the driving and trailing (coupled) 5ft. diameter. Some other dimensions were:—Total length, 24ft. 3in.; breadth, 5ft. 3in.; height from rail to top of chimney, 13ft. 6in.; weight, empty, only 16 tons, with fuel and water 19 tons 17 cwt. The capacity of tank was 520 gallons, sufficient for a journey of 25 miles; coke bunker, 42 cubic feet, equal to 26 bushels, or 15 cwt. The "Duplex" was sold to a Dutch railway after the Exhibition, and its further career is, therefore, unknown to those interested in it.

Fairbairn's tank locomotive was of the "well" type, supported on six wheels, the driving pair being 5ft. diameter, and the L. and T. each 3ft. 6in. diameter. The cylinders were inside, measuring 10in. by 15in. stroke. The boiler was 8ft. long by 3ft. diameter, and contained 88 brass tubes of 2in. diameter. The heating surface amounted to 480 sq. ft. The internal fire-box was of copper, and measured 2ft. 5in. long, 3ft. wide, and 3ft. 5in. deep. The tank behind and under the foot-plate held 400 gallons of water. The coke consumption of this little engine was only 10lb. per mile with trains of six carriages, the weight in working order only 13 tons; and it may interest our readers to know that this diminutive locomotive was described as "a fair specimen of the *heavier* class of tank engine".

FIG. 61.—THE "FOLKESTONE," A LOCOMOTIVE ON CRAMPTON'S SYSTEM. BUILT FOR THE S.E.R. 1851

The engine calling for the greatest attention at the Exhibition of 1851 was the "Folkestone" (Fig. 61), exhibited by the South Eastern Railway. This was an engine built by R. Stephenson and Co., under one of Crampton's patents, but the principal feature in its design was an intermediate driving axle, connected by means of outside cranks, and coupling-rods to the driving wheels, which were (under Crampton's patent) behind the fire-box, the axle extending across the foot-plate. It will be well, perhaps, if we at this point reiterate the fact that the method of working locomotives by means of an intermediate crank-shaft was not introduced by Crampton, it having been used some years previously by W. B. Adams, not to mention some of the early Stockton and Darlington Railway engines, where the same arrangement was employed, but with vertical cylinders. Readers will, therefore, see it is incorrect to describe locomotives with this system of machinery as "Crampton's patent," although it is quite possible for a "Crampton patent" locomotive to be provided with an intermediate driving shaft, as was the case with the "Folkestone."

Eight engines of this type were built by Stephenson and Co. for the South Eastern Railway, and were numbered 136 to 143, the first of which was named "Folkestone." These engines were supported by six wheels, a group of four being arranged close together at the smoke-box end. Their diameter was 3ft. 6in. The driving wheels were 6ft. in diameter, the wheel base 16ft. These engines weighed $26\frac{1}{4}$ tons each, of which only 10 tons were on the driving wheels, the remainder of the weight being supported by the four leading wheels. The cylinders were inside, 15in. diameter, and the stroke 22in. The fire-box top was flush with the boiler barrel, the straight lines of which were unrelieved by a dome, but an encased safety valve was fixed near the back of the fire-box top. The boiler contained 184 tubes, of 2in. diameter and 11ft. in length.

The "Folkestone" ran its trial trip on Monday, March 31st, 1851, when Mr. McGregor, the chairman of the South Eastern Railway, Mr. R. Stephenson, the builder of the engine, Mr. Barlow, the South Eastern engineer, and Mr. Cudworth, the South Eastern locomotive superintendent, were present. From London Bridge to Redhill no great speed could be attained, as a Brighton train was in front; but beyond the latter station, and with a train of nine carriages, the $19\frac{1}{2}$ miles to Tonbridge were covered in $19\frac{1}{2}$ minutes, a maximum speed of 75 miles an hour being attained. After a short stop, the journey to Ashford was resumed, and that town was reached in $20\frac{1}{2}$ minutes after leaving Tonbridge. The times and distances were as follow:—Redhill to Tonbridge, 19 miles 47 chains, start to stop in $19\frac{1}{2}$ minutes; Tonbridge to Ashford, 26 miles 45 chains, start to stop in $20\frac{1}{2}$ minutes, or at the rate of 78 miles an hour; the whole 46 miles 12 chains being covered in 40 minutes, running time, or, including the stop at Tonbridge, in 43 minutes. It must be remembered that the line between Redhill and Ashford is, perhaps, the most level and straight in England for so long a distance.

These eight engines did not prove very successful in general working, and they were afterwards rebuilt as four-coupled engines, an ordinary cranked axle with wheels being provided in place of the intermediate driving shaft.

It will not be out of place if we here mention eight "single" engines built by Sharp Bros. in 1851 for the South Eastern Railway, and numbered 144 to 151. The general dimensions were similar to the Cramptons, except that the wheel base was only 15ft., and that the heating surface was 1,150 sq. ft. The admission of the steam to the cylinders was controlled by a hand lever, with catch and notches, similar to and placed by the side of the ordinary reversing lever. Six eccentrics were on the driving axle, two of them working the pumps. The framing and springs of these engines were afterwards perpetuated by Cudworth in his later and better known types of South Eastern locomotives.

FIG. 62.—ONE OF J. V. GOOCH'S "SINGLE" TANK ENGINES, EASTERN COUNTIES RAILWAY

The locomotives of the despised "Eastern Counties," that were designed about 1850 by Mr. J. V. Gooch, will now be concisely described. They were of three kinds—viz., "single" tanks, "single" express, and four-wheels-coupled tender engines. Of the tanks, three sizes were constructed, chiefly at the "Hudson Town" (or Stratford Works). The largest of these were provided with outside cylinders, 14in. diameter and 22in. stroke, the boiler being 10ft. 6in. long, and containing 164 tubes of 1 3-16th in. diameter. The leading and trailing wheels had outside bearings, the driving wheels being provided with inside bearings only. A steam dome was placed over the raised fire-box, and a screw-lever safety valve on the boiler barrel. The water was stored in two tanks, fixed between the frames, one below the boiler and the other beneath the foot-plate. These tank-engines were known as the "250" class, and some of our readers may recollect

M

that when Peto, Brassey and Betts leased the London, Tilbury and Southend Railway, engines of this design were used to work the traffic on that railway. We understand it is now 20 years since the last of them (No. 08) reached the final bourne of worn-out locomotives— the "scrap heap."

The dimensions of the smallest class of these tanks (Fig. 62) were: Cylinders, 12in. diameter, 22in. stroke; boiler, 10ft. long and 3ft. 2in. diameter, 127 tubes of 1 7-8th in. diameter; the total heating surface was 709 sq. ft.; grate area, 9.7 sq. ft. The driving wheels were 6ft. 6in. diameter, and the L. and T. 3ft. 8in. The total weight of these engines was 23 tons 19 cwt., of which 9 tons 14 cwt. was on the driving axle. The wheel base was: L. to D., 6ft. 3in.; D. to T., 5ft. 9in.

J. V. Gooch's four-coupled, or "Butterflies," had leading wheels 3ft. 8in. diameter, and driving and trailing (coupled) 5ft. 6in. Wheel base, L. to D., 6ft. 3in.; D. to T., 7ft. 9in. The cylinders were 15in. diameter, the stroke being 24in. The boilers of this class, and also of the singles, next to be described, were of the same dimensions as those of the "250" class of tanks.

The "single" expresses were provided with 6ft. 6in. driving wheels, and cylinders 15in. diameter and a 22in. stroke; in this class also the leading and trailing wheels were 3ft. 8in. diameter. The wheel base was 14ft., the driving wheels being 6ft. 9in. from the leading and 7ft. 3in. from the trailing wheels. Ten engines of this design were constructed, some at Stratford, and others at the then recently opened Canada Works of Brassey and Co. at Birkenhead. Their official numbers were from 274 to 283.

The "Ely" (Fig. 63) represents the type of 6-wheel passenger engine in use on the Taff Vale Railway at this period. She was built in 1851 by Messrs. Kitson and Company, from Taff Vale designs. She had 13in. cylinders, with 20in. stroke, and four wheels coupled, of 5ft. 3in. diameter. She carried a pressure of 100lbs., she had a four-wheel tender, carrying 900 gallons of water, and as the gross weight of the tender was about 11 tons in working order, the gross weight of the engine and tender would be 33 tons. The "Ely" could not take a train of three carriages, weighing only 21 tons, up the Abercynon bank of 1 in 40 without the assistance of a "bank" engine.

In 1851 Mr. Beattie, the locomotive superintendent at Nine Elms, built for the London and South Western Railway the four-wheels-coupled engine, "Hercules," No. 48. The frames of this engine were of the "lattice" type, examples of which can be still seen on some of the older Great Northern Railway tanks.

The diameter of wheels was: L., 3ft. 6in.; D. and T., 5ft. 6in.; tender, 3ft. 6in.; wheel base, L. to D., 7ft. 1in.; D. to T., 6ft. 6in.; T. to leading tender, 7ft. 3½in.; the tender wheel base being 10ft. 3in. equally divided.

FIG. 63.—"FLY," A TAFF VALE RAILWAY ENGINE, BUILT IN 1851

The weight was distributed as follows:—Engine, L. axle, 8 tons 17 cwt.; D., 9 tons 17 cwt.; T., 9 tons 16 cwt.; tender, L., 4 tons 19 cwt.; M., 5 tons 19 cwt.; T., 7 tons 10 cwt. The cylinders were 15in. by 22in.; tractive force on rail, 7,500lb.; 1,800 gallons of water could be carried in the tender tank. The "Hercules" had a flush top boiler, and a raised fire-box surmounted by a large inverted, urn-shaped dome. This design of locomtive was a favourite one on the London and South Western Railway for many years, but the last engine of the kind has now been scrapped.

Having favoured the London and South Western Railway, to equalise matters, we cannot do better than give a description of a locomotive belonging to its cousin-german, the London and North Western Railway. The latter was indeed the more famous, being no other than McConnell's notorious "No. 300," (Fig. 64) which, being introduced with a vast amount of publicity, became a nine days' wonder, then sank into quiescent mediocrity, and after a brief locomotive career, was seen no more—a rather different fate, be it observed, to that of the London and South Western Railway's "Hercules."

It has been stated that only one drawing of this engine exists.

M 2

This is incorrect; the writer possesses a complete set of drawings relating to "No. 300," together with the whole of the specifications from which the engine was constructed. To reproduce this specification in detail would give too technical a character to this narrative, and would try the patience of even the most ardent locomotive enthusiast.

The directors of the London and North Western Railway in 1851 expressed their determination to run their express trains from London to Birmingham in two hours, and gave instructions to McConnell, the locomotive superintendent at Wolverton, to design the necessary locomotives. The salient features of the design were: Inside cylinders, 18in. by 14in.; six wheels, with inside and outside frames; driving wheels, 7ft. 6in. diameter; leading, 4ft. 6in.; and trailing, 4ft. diameter.

The boiler was 11ft. 9in. long and 4ft. $3\frac{1}{4}$in. external diameter. The tubes were of brass, 303 in number, only 7ft. in length, and $1\frac{3}{4}$in. outside diameter. The crank axle bearings were—outside, 7in. deep and 10in. in length, the inside ones being 7in. and $4\frac{1}{4}$in. respectively. The leading and training axles were hollow, the metal being $1\frac{1}{2}$in. thick, and the hollow centre $4\frac{1}{2}$in. diameter, thus making the total diameter of the straight axles $7\frac{1}{2}$in. The slide valves had an outside lap of $1\frac{1}{4}$in. The principal innovations were: Coleman's patent india-rubber springs, fitted below the driving axle and above the leading and trailing axles, and also to the buffers. McConnell's patent dished wrought-iron pistons, forged in one piece with the piston-rod, and encased with continuous undulating flat metal packing. The steam-pipe was of flat section, and passed through a superheating chest in the smoke-box; the steam was thus dried during its journey from the dome to the cylinders. The great feature of the design was the arrangement of the fire-box, with a mid-feather, a combustion chamber, hollow stays for a free supply of air to the fire-box, and the cutting away of the bottom of the fire-box to obtain clearance for the cranks and yet retain a low centre of gravity with large driving wheels. Assertion to the contrary notwithstanding, it should be observed that so much did McConnell insist upon a low centre of gravity that he specially mentioned it in his patent specification of February 28th, 1852.

A more particular description of the fire-box, etc., is requisite. It extended into the cylindrical portion of the boiler a distance of 4ft. 9in., so that the boiler tubes were only 7ft. long. The whole length of the fire-box was 10ft. 6in.; depth at front-plate 6ft. 5in., at door-plate 6ft. 10in.; length on fire-bars 5ft. 10¼in., thus leaving 4ft. 7¾in. for the portion over the axle and the combustion chamber. At its narrow part (directly at the top of the recess above the driving axle) the fire-box was only 2ft. 3in. in height; height at tube-plate 3ft. (beyond the cut away portion); width at tube-plate 3ft. 9in. It will be noticed that Webb's "Greater Britain" class of locomotives is designed with the long fire-box and combustion chamber; but as Mr. Webb, unlike McConnell, does not object to the high-pitched boiler, the former does not recess the boiler barrel for the purpose of obtaining a low centre. Webb also divides his tubes into two sets by having the combustion chamber between them. McConnell's combustion chamber was a continuation of the fire-box. We must now describe the general appearance of this engine.

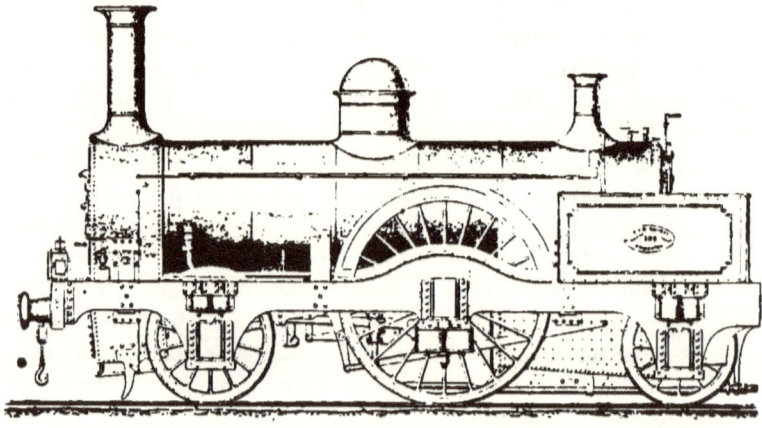

FIG. 64.—McCONNELL'S "300," LONDON AND NORTH WESTERN RAILWAY.

The cylinders were inclined upwards from the front, and the valve-chests were above them, below the smoke-box. Two Salter safety valves were provided, encased within a sheet-brass covering of Stirling's Great Northern pattern. The steam pressure was 150lbs. The dome was also of brass, with a hemispherical top surmounting the cylindrical lower part. The steam regulator was at the mouth of the steam-pipe, which was placed at the top of the dome (inside, of course).

The heating surface was: Tubes, 980 sq. ft.; fire-box, 260 sq. ft. Wheel base, 16ft. 10in. Sufficient steam could be raised in 45 minutes after lighting the fire to move the engine. Two of these engines were built about the same time—one (No. 300) by Fairbairn and Co., Manchester, the other by E. B. Wilson and Co., Leeds. The orders were given early in July, 1852, and the engines delivered the second week in November, Wilson and Co. having occupied but eight weeks in the construction of the one given to them.

Both engines were delivered at Wolverton on the same day, and on Thursday, November 11th, 1852, Wilson's engine was tried for the first time, when on her first journey to Euston she attained a speed of 60 miles an hour.

It was soon found that "No. 300" and her sister engine were unable to cover the 111 miles—Euston to Birmingham—in two hours, as was confidently predicted, and the failure to do so was—perhaps justly—attributed to the inferior condition of the permanent-way. On March 8th, 1853, "No. 300" hauled a train of 54 carriages, weighing 170 tons, from Birmingham to London in three hours eight minutes, including five stoppages. A similar train drawn by the "Heron" and "Prince of Wales" took ten minutes longer to perform the same journey. These two engines had cylinders 15in. by 20in., and 6ft. driving wheels. The results of this trial are thus tabulated:—

|  | Coke. | Coke per mile. | Average speed per hour. | Maximum speed per hour. |
|---|---|---|---|---|
| No. 300 | 4,529 lb. | 40·8 lb. | 36·4 miles | 54 |
| "Heron" & "Prince of Wales" | 4,851 lb. | 43·7 lb. | 34·5 miles | 48 |

Upon the result of this run it was claimed that McConnell's patent engines were considerably superior to two of the ordinary London and North Western Railway locomotives, and one of Stephenson's "long boiler" abortions was altered by McConnell, being fitted up with his patent combustion chamber, short tubes, and the other innovations, as mentioned in our description of "No. 300."

The "long boiler" originally had 1,013 sq. ft. of tube-heating surface; when altered, the length of the tubes was reduced to 4¾ft., and some additional ones were fixed diagonally across the combustion chamber. By this alteration the tube-heating surface was reduced to 547 sq. ft., and the engine is stated to have drawn 170 tons at 60 miles an hour, and to have attained a speed of 70 miles an hour with light trains. From the working of this locomotive the following table (by which a reduction of 23 per cent. in the amount of fuel consumed was claimed for the altered engine) was prepared:—

|  | Miles run. | Average load. | Coke consumed. | Coke per mile. | Coke per ton per mile. |
|---|---|---|---|---|---|
| Original | 29,442 | 115 tons | 1,715,952 lb. | 58·28 lb. | ·5ˆ4 lb. |
| Altered | 12,060 | 144 tons | 519,120 lb. | 43·04 lb. | ·298 lb. |

But D. K. Clark's paper on "Locomotive Boilers," read before the Institution of Civil Engineers, soon placed a very different complexion upon the result of the trials between the ordinary and patent engines, resulting in the "air tubes" to the combustion chamber being speedily abandoned. The attention of the directors of the London and North Western Railway was called to the failure of these engines, with the result that they ordered Messrs. Marshall and Wood to report on the two classes of engines—viz., the ordinary London and North Western type and McConnell's patent locomotives. This report was ready in August, 1853, but for some reason its publication was suppressed at the time, but the directors countermanded the construction of other engines already ordered on McConnell's patent principle.

In the summer of 1854 Marshall and Wood conducted another set of experiments for the directors of the London and North Western Railway, with the object of determining the relative value of coke and coal as fuel for the locomotives.

The engines chosen were McConnell's patent "No. 303" and the "Bloomer," No. 293. Double trips were run between Rugby and London daily for six consecutive days, coal being burnt on three days and coke on the three alternate days. The trains chosen were the 12.55 p.m. up and 5.45 p.m. down.

It was found that 1lb. of coal evaporated 5.83lb. of water, and 1lb. of coke 8.65lb. of water; but the monetary saving was 6s. 9d. per ton in favour of coal.

McConnell's patent engines were again condemned. Marshall and Wood's report concluded as follows: "Although we consider the experiments we made with No. 303 engine satisfactory in point of smoke burning, we cannot resist the belief that the consumption of coal is in excess of what it ought to be, and that there is room for considerable improvement in this respect, by means which shall tend to utilise the heat which is at present wasted."

The whole report is of great interest to the technical reader; it is, however, too long to reproduce *in extenso*.

It is abundantly evident that there is no great pecuniary gain from locomotive designing, or we should be treated to great law-suits regarding the validity of the patents, such as have recently been the case with pneumatic tyres and incandescent gas-burners. We have already, upon more occasions than one, pointed out that certain patented locomotive designs had previously been anticipated, although the later patentees were probably unaware of the fact. We find this to have been the case with McConnell's "recessed" boiler locomotives just described, for on December 2nd, 1846, W. Stubbs and J. J. Grylls, of Llanelly, enrolled a design of locomotive. The specification in question not only mentioned the recessing of the boiler for the purpose of allowing the use of a large driving wheel and yet retaining a low centre of gravity, but it even anticipated McConnell's combustion chamber between the fire-box and tubes. An adaptation of Bodmer's double piston motion was also specified by Stubbs and Grylls. The two cylinders were placed below the boiler, four wheels being connected by means of side-rods with the cross-heads of the two cylinders in such a manner that from each cylinder two wheels were driven, by means of a crosshead, and each cross-head, by means of two connecting-rods, rotating the wheels. Another claim under this patent related to driving a locomotive by eccentrics fitted with antifriction rollers as a substitute for the ordinary cranks.

Although in the "Evolution of the Steam Locomotive" it is only intended to describe locomotives for British railways, it may not be out of place to mention an engine for a foreign railway, for two reasons— first, because it was built by an English firm in England, and, secondly, because it was tried on an English railway before exportation. The "Ysabel" was constructed in 1853 by Dodds and Sons, of Rotherham, for the "Railway of Isabella II. from Santander to Abar del Rey," and was tried on the Lickey incline of 1 in 37 for two miles, under the direction of Mr. Stalvies, the locomotive superintendent at Broomsgrove. The "Ysabel" had four-coupled wheels 4ft. 6in. diameter; cylinders, 14¼in. by 20in. stroke; 137 tubes, 1⅞in. diameter, and 11ft. 3in. in length, and was fitted with Dodds' patent wedge expansive motion, which required only two eccentrics. For the purpose of easy transportation, the "Ysabel" was so constructed that when disconnected no single portion weighed more than six tons; in addition to the fittings necessary to secure the boiler, the only connections between it and the frames, machinery, etc., were the steam-pipe and the two feed-pump connections. When tried upon the Lickey bank this locomotive hauled six trucks weighing 45 tons $12\frac{3}{4}$ cwt. up the two miles one furlong in 12 minutes 12 seconds, and with a train weighing 29 tons $4\frac{1}{4}$ cwt. the incline was negotiated in seven minutes five seconds.

The compound locomotive is not quite so modern an invention as is popularly supposed, for, putting aside the suggestion emanating in 1850 from John Nicholson, an Eastern Counties Railway engine-driver, whose plan of continuous expansion is generally accepted as the foundation of the compound system, we find that in 1853 a Mr. Edwards, of Birmingham, patented a "duplex" or in other words a compound engine, the steam, after working in a high-pressure cylinder, being used over again in a low-pressure one. The cylinders were so placed that the dead centre in one occurred when the other piston was at its maximum power.

In 1853 Beattie constructed for the London and South Western Railway at Nine Elms Works, the "Duke," No. 123, a six-wheel "single" express engine; driving wheels, 6ft. 6in. diameter; L. and T. 3ft. 6in. diameter; cylinders, 16in. by 21in. stroke. The weight was arranged in an extraordinary manner, 10 tons 9 cwt. being on the leading axle, only 9 tons 9 cwt. on the driving axle, and 5 tons 11 cwt. on the trailing axle. The wheel base was, L. to D., 6ft. $8\frac{1}{2}$in.; D. to T., 7ft. 6in. The "Duke" had a raised fire-box, surmounted by a large dome similar to that of the "Hercules," whilst another dome was located on the centre of the boiler barrel. The shape of this centre dome resembled a soup-tureen turned upside down.

At this point we take the opportunity to briefly describe a railway locomotive which, although not propelled by steam, deserves to be mentioned as an initial attempt at railway haulage by means of compressed air.

The engine in question was constructed by Arthur Pasey, and was tried on the Eastern Counties Railway in July, 1852. This machine was, in point of size and power, nothing more than a model, the dimensions being: Cylinders, $2\frac{1}{2}$in. diameter, 9in. stroke; driving wheels, 4ft. diameter; weight, $1\frac{1}{2}$ tons; air capacity of reservoirs, 39 cubic ft.

170  EVOLUTION OF THE STEAM LOCOMOTIVE

By reference to the illustration (Fig. 65) it will be seen that this curious little locomotive had the six wheels of 4ft. diameter within the frames, and the horizontal cylinders outside the frames, and actuating the centre pair of wheels. Above the frames was placed a cylindrical air reservoir, with egg shaped ends. This extended from the buffer beam at one end of the vehicle to the leading axle, a distance of about 12ft. The remainder of the space, about 4ft., was occupied by the pressure-reducing and other apparatus, and afforded a place of vantage for those in charge of the machine. The reservoir was constructed to withstand a pressure cf 200lb., but the engine was only pressed to 165lb., and this at the time

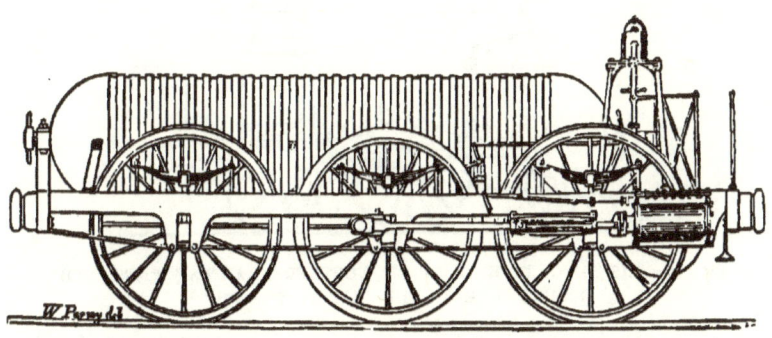

FIG. 65.—PASEY'S COMPRESSED AIR LOCOMOTIVE, TRIED ON THE EASTERN COUNTIES RAILWAY IN 1852

of the trial at Stratford was reduced to 20lb. working pressure. With a load of eight people, the engine ran the four miles, Stratford to Lea Bridge and back, in 30 minutes. The incident of the trial so aroused the curiosity of the men engaged at the Stratford Works, that they all left their employment for the purpose of witnessing the trial of so great an innovation as Pasey's compressed air locomotive. For this reason no further trials could be held at Stratford, but on July 2nd a second trip was made at Cambridge, and on this occasion, with six passengers, the following results were recorded : Starting from the 60th mile-post near the Waterbeach Junction, with a working pressure of 15lb. per sq. in., the first mile was covered in five minutes. By increasing the pressure on the pistons, the second mile was covered in four minutes; the pressure was then reduced to 18.85lb., and $3\frac{1}{2}$ additional miles were covered in ten minutes. The designer of this little machine gives eight reasons by which he apparently succeeds—at all events to his own satisfaction—in proving the great superiority of compressed air traction over that of steam. Unfortunately for Mr. Pasey's theory, steam is still triumphant, and compressed air dead—or nearly so—for tractive purposes.

The opening of the Great Northern Railway next claims our attention. The first locomotives were supplied by contract, an order for 50 passenger engines having been given to Sharp Bros. and Co. These

FIG. 66.—THE FIRST TYPE OF GREAT NORTHERN RAILWAY PASSENGER ENGINE, ONE OF THE "LITTLE SHARPS"

were six-wheel single engines (Fig. 66), the driving wheels being 5ft. 6in. diameter. The cylinders were 15in. by 20in. stroke. Weight of engine, loaded, 18 tons 8½ cwt. These engines were called "Little Sharps," and (Fig. 66) is an illustration of one of them

We will now describe the famous "No. 215" (Fig. 67) of the Great Northern Railway, designed towards the end of 1852 by Mr. Archibald Sturrock, constructed by Hawthorn and Co., Newcastle, and delivered to the Great Northern Railway on August 6th, 1853.

Fortunately, Mr. Sturrock has supplied the writer with complete and authentic details, together with a drawing, of this engine, so that readers may rely upon the information being strictly accurate, although it should be noted that it does not correspond in several particulars with other statements concerning "No. 215" that have been published.

It is a matter of railway history that in 1852 the "Gladstone" award settled the great rivalry existing at that period between the London and North Western and Great Northern Railways. The competition had been carried on in a manner still in favour in American railroad warfare—viz., the cutting of rates and fares; but Mr. Gladstone having decided this point, the Great Northern Railway introduced the method of rivalry now universally recognised as English railway competition—that is, trial of speed. Mr. Sturrock, with the experience gained under the daring broad-gauge leaders, was, of course, conversant with what a locomotive could do, and his published reasons for the construction of "No. 215" are as follow:—

"This engine was constructed to prove to the directors of the

FIG. 57. – STURROCK'S MASTERPIECE, THE FAMOUS G.N.R. "215"

Great Northern Railway that it was quite practicable to reach Edinburgh from King's Cross in eight hours, by only stopping at Grantham, York, Newcastle, and Berwick. This service was not carried out, because there was no demand by travellers for, nor competition amongst, the railways to give the public such accommodation."

Although delayed for 35 years, the demand for such a service arose in 1888, and Mr. Sturrock then had the satisfaction of seeing runs such as he had built "No. 215" to perform become daily accomplished facts. It should be noted that when "No. 215" was originally built, she was fitted with a leading bogie, such an arrangement being a principal feature of Mr. Sturrock's original design for the engine. The bogie and trailing wheels were 4ft. 3in. diameter, the driving wheels being 7ft. 6in. diameter; the cylinders were inside, and had a diameter of 17½in., with a stroke of 24in. The heating surface was large, this being another of the strong points in Mr. Sturrock's design. Tubes, 1,564 sq. ft.; fire-box, 155.2 sq. ft.; total heating surface, 1,718.2 sq. ft. The weight was, empty, 32 tons 11 cwt. 2 qr.; in working order, 37 tons 9 cwt. 2 qr. Wheel base, 21ft. 8½in. Water capacity of tender, 2,505 gallons. The frames and axle bearings were outside; the latter were curved above the driving axle, as in the broad-gauge "Lord of the Isles" type.

The boiler and raised fire-box were also after the same pattern. The engine had no dome, but an encased safety valve on the fire-box —a further evidence of attention to the Swindon practice. Compensation beams connected the two pairs of bogie wheels, and the underhung springs of the driving wheels were also connected with the trailing axle springs by means of compensation levers. "No. 215" frequently ran at 75 miles an hour. She appears to have been broken up about 29 years back, for in 1870 Mr. Stirling built an engine, "No. 92," in which he used the 7ft. 6in. driving wheels of Mr. Sturrock's famous "215." Engine No. 92, is still at work, so that the driving wheels must be 45 years old. A comparison of Mr. Sturrock's "215" with McConnell's "300" will show the immense superiority of the former, especially with regard to the amount of heating surface, the pitch of the boiler, and the bogie in place of the rigid wheel base.

In the last chapter, Mr. Pearson's initial patent for a locomotive was described, and a description of his famous double-bogie tank engines, with 9ft. "single" driving wheels, is given below. The design (Fig. 68), which was brought out in 1853, was a modification of the patent specification already alluded to. The engines were constructed by Rothwell and Co., Union Foundry, Bolton-le-Moors, and were famous for the low average cost for repairs and fuel consumption per mile run; indeed, a feature of most of the broad-gauge locomotives was the low average cost of maintenance and working. The ends of the frames were supported on a four-wheel bogie, the wheels of which were 4ft. in diameter, and the driving wheels 9ft. diameter; these latter had no flanges. The cylinders (the ends of which projected beyond the front of the

174  EVOLUTION OF THE STEAM LOCOMOTIVE

FIG. 68.—PEARSON'S 9FT. "SINGLE" TANK ENGINE, BRISTOL AND EXETER RAILWAY

smoke-box) were 16½in. diameter and 24in. stroke; the driving axle was above the frame. The boiler was 10ft. 9in. long and 4ft. ½in. diameter; it contained 180 brass tubes of one and thirteen-sixteenths inch external diameter. The steam pressure was 130lb. No dome was provided, and the Salter safety valves were located on the top of the fire-box and enclosed by a brass casing. The weight of the engine, in working order, was 42 tons. The water was stored in three tanks, one beneath the boiler, another below the fire-box, and the usual well tank, behind the foot-plate. The two suspended tanks were connected by means of a stuffing-box jointed pipe, which was continued to the bottom of the wheel-tank, so that the water in the three tanks was thus able to pass from one tank to any other one. The feed-pumps were worked from the piston-rod cross-head, and the feed-pipes passed along behind the splashers to the boiler. To steady the suspended tanks, link-rods were passed between the two. There were also "bogie safety links" connecting the bogie frames with the main frame at each end, and similar links connected the suspended tanks with the other ends of the bogie frames.

These links were each fitted with india-rubber disc buffers, to allow of the necessary elastic working. The parts were thus so strongly linked together, that should a bogie centre-pin break, or should the bogie movement fail in any way, the wheels would still remain in their right position. The whole of the springs were of the india-rubber disc kind. Those of the driving axle presented some remarkable peculiarities.

They were double, an elastic connection being formed between the boiler and the axle-boxes by large plate brackets projecting from the boiler barrel, and carrying centre studs for a short double-armed lever; each end of this lever had a separate spring-box attached to it by a long link.

The inner spring-box worked down behind the disc plate of the driving-wheel splashers, whilst the outer one worked parallel to it, outside the driving wheel.

The springs for the other wheels were all beneath their axles, and were very compact and neat in appearance. The brake action was confined to the after bogie, all four wheels being used for the frictional effect, the sliding bars carrying the brake blocks being actuated in reverse directions by a screw spindle, which carried a winch to be worked by the driver.

176  EVOLUTION OF THE STEAM LOCOMOTIVE

FIG. 69.—ONE OF PEARSON'S 9FT. "SINGLE" TANKS, TAKEN OVER BY THE GREAT WESTERN RAILWAY

The regulator valve was a slide, worked in a simple and certainly a convenient manner by a short lever, set on a pillar stud on the front of the fire-box, and passing through a slot in the end of the slide spindle. This was a far more effective plan of working the valve than the ordinary rotatory handle.

These engines were remarkable for their steady running at high speeds, 80 miles an hour and over being a daily performance of the engines on certain portions of the main line between Exeter and Bristol.

One reason for the freedom from excessive oscillation for which these engines were famous was attributable to the 9ft. driving wheels, and the slow piston velocity arising therefrom; thus with 6ft. wheels at a speed of 60 miles an hour, the pistons have to make no less than 280 double strokes per minute without making allowance for "slip." With the 9ft. driving wheels the double piston strokes per minute at 60 miles an hour fall to 186, and consequently with so considerable a reduction in the movements of the reciprocating and rotating machinery of the locomotive, it is only reasonable to expect and obtain a much more steady movement of the machine.

In the matter of coal consumption the engines were no less successful. Writing in August, 1856, Mr. Pearson reported: "Engine No. 40 has run 81,790 miles since her delivery in October, 1853, and has consumed 794 tons 17 cwt. 2 qr. of coke, or 21.76lb. per mile; the repairs as yet have been very trifling, consisting chiefly of returning the tyres. This engine has been working passenger trains on the main line almost the whole of the time since she was delivered. Our mileage is rather heavy, each engine averaging 750 miles per week."

After 1876, when the Bristol and Exeter Railway was amalgamated with the Great Western Railway and the former company's locomotive stock became the property of the latter, 4 of the 8 original 9ft. tank engines then in existence were rebuilt, and their character and design entirely remodelled. The diameter of the driving wheels was reduced to 8ft., and tyres fitted to them, a pair of trailing wheels were provided in place of the rear bogie, and a separate tender was added, the tanks being done away with. The B. and E.R. numbers of these engines were 39 to 46. The G.W.R. numbered the four taken over 2001 to 2004. The latter was hauling the "Flying Dutchman" when the Long Ashton accident happened on July 27th, 1876. It was

178  EVOLUTION OF THE STEAM LOCOMOTIVE.

FIG. 70.—A BRISTOL & EXETER RAILWAY TANK ENGINE, AS REBUILT (WITH TENDER) BY THE G.W.R.

in consequence of this disaster that the engines were rebuilt with 8ft. wheels. In concluding this sketch of Pearson's famous broad-gauge double tanks, we may state that until recent years, when phenomenall; high locomotive speeds have been recorded, these engines held the "blue ribbon" in that respect with an authenticated speed of 81 miles an hour. Figures 69 and 70 represent them as rebuilt.

The Furness Railway Company is certainly notorious for the manner in which it preserves its locomotives; not only has it the two old Bury engines (already described) yet in active service, but there are still at work on the same Company's iron roads other engines manufactured as long ago as 1854. These locomotives are first cousins to Bury's four wheel (coupled) goods engines; they were built by Fairbairn, of Manchester, and have cylinders 15in. diameter, with a stroke of 24in. Of course, they are technically inside cylinder—*i.e.*, of the Bury "inside" type, with the cylinders within the frames, but below the smoke-box, instead of within it. The cylinders are, in fact, but a few inches above rail level; they incline upwards, and the connecting-rods pass beneath the leading axle and actuate the trailing axle; the four wheels are 4ft. 9in. diameter, and are coupled by means of round section side-rods; the wheel base is 7ft. 9in.; the frames are of the inside bar pattern; the fire-box is round, with circular top, and surmounted by a double Salter safety valve. The boiler is 11ft. 2in. long and of 3ft. 11in. mean diameter; it contains 148 tubes, 2in. diameter. The total heating surface is 940 sq. ft.; steam pressure, 120lb.; weight of engine in working order, 22½ tons. There is no dome on the boiler; but some modern attachments have been fixed on the upper portion of the round fire-box, the steam pressure gauge being very noticeable. The tender is supported on four wheels of 3ft. diameter, the wheel base being 8ft., capacity of tank 1,000 gallons, and coal space 100ft.; weight in working order, 14½ tons. The tender has outside frames, and the brake actuates blocks to both sides of the four wheels. These engines are used for working goods and mineral traffic over the Furness Railway. The particular engine we have been describing is "No. 9."

"Ovid" (Fig. 71) represents a type of bogie saddle-tank engines, with four-coupled wheels, designed by D. Gooch for working the passenger trains on the steep inclines of the South Devon Railway. The cylinders were 17in. diameter, with a stroke of 24in. The coupled wheels were 5 ft. in diameter. Weight, in working order, 38½ tons,

Steam pressure, 120 lbs. per square inch. "Ovid" was built by Hawthorn in 1854.

"Plato" (Fig. 72) was one of the six-coupled banking engines, designed by Gooch for the South Devon Railway. She was built at Swindon in 1854. The steam pressure, cylinders, stroke, and weight were the same as in the "Ovid" class. The wheels were 5ft in diameter. The tanks contained 740 gallons of water. The rectangular projection in front of the smoke-box is the sand-box!

Neilson and Co., of the Hyde Park Works, Glasgow, produced in 1855 a type of outside cylinder goods engine. Readers will remember that at that period goods locomotives were not necessarily

FIG. 71.— "OVID," A SOUTH DEVON RAILWAY SADDLE TANK ENGINE, WITH LEADING BOGIE

of the six or eight wheels coupled description; they more generally had but the leading and driving wheels coupled. This type of engine, it will be remembered, is now usually described as "four-coupled in front" or a "mixed traffic" engine. The locomotive in question was built for the Edinburgh and Glasgow Railway, and was numbered "353" in Neilson and Co.'s books.

The boiler was of considerable length, and appeared longer from the fact that the fire-box top was not raised, so that a long, unbroken line of boiler top met the eye, relieved at the extremity of the fire-box end by being surmounted by an immense steam dome, on the top of which was fixed an enclosed Salter safety valve. The horizontal outside cylinders were below the foot-plate side frames, located as usual at the smoke-box end. Their diameter was 16in.

and stroke 22in. The coupled wheels were 5ft. and the trailing wheels 3ft. 6in. diameter.

The frames were "inside," and the driving and leading wheels were provided with inside bearings only, but by a curious practice of bolting on to the main frames—at about the middle of the fire-box—an elongated portion, which curved outwards, the trailing wheels were provided with outside bearings. The rams actuating the boiler feed-pumps were simply extensions of the piston-rods, the pumps being

FIG. 72.—"PLATO," A SIX-COUPLED SADDLE TANK BANKING ENGINE, SOUTH DEVON RAILWAY

fixed between the leading and driving wheels. The engine was provided with a steam-pressure gauge, fixed on a vertical pillar over the top of the fire-box—indeed, in much the same position the steam gauge still occupies, save that "No. 353" had no cab or weatherboard, and it therefore appeared singular to see the gauge in the place indicated.

Rotatory valves for locomotives are almost annual "inventions," and as old friends as the "biggest gooseberry" and "sea serpent," which appear regularly year by year. Under such circumstances, we may be excused for giving an account of Locking and Cook's patent rotatory valve, fixed to the York and North Midland Railway engine, "No. 48," on January 26th, 1854, and taken out in May of the same year, the locomotive in the interim having run 10,000 miles. "No. 48" was used on the Hull and Bridlington branch; and although she was an old engine, having been built for the Hull and Selby Railway in 1840, yet with the rotatory valve, good old "48" is stated to have consumed 20 per cent. less coke than a modern engine doing the same work on the same branch; we also read that when the valve was removed no perceptible wearing was to be noticed. We are not, however, aware that "No. 48" or any other of the York and North Midland Railway locomotives were afterwards fitted with Locking and Cook's patent rotatory valves.

Mr. G. Tosh, locomotive superintendent of the Maryport and Carlisle Railway, designed in 1854 a powerful goods engine to work the heavy mineral traffic over the railway. This engine had six coupled wheels, 4ft. 7in. diameter; cylinders, 16¾in. by 22in. stroke; heating surface—tubes, 1,181ft.; fire-box, 84ft.; total, 1,265 sq. ft.; steam pressure, 120lb.; weight, 26 tons 12 cwt.; cost, £2,175. She hauled a train of 100 loaded wagons, weighing 445 tons, for a distance of 28 miles in 1¾ hours. The line is of a very undulating character, including an ascent nine miles long, one mile of which is 1 in 192. The wagons were borrowed from the Newcastle and Carlisle Railway, and the 100 only weighed 172 tons, or an average of less than 1¾ tons each.

The dead weight of mineral wagons has largely increased since 1854, although it is to be feared their carrying capacity has not increased in the same proportion.

FIG. 73.—THE FIRST TYPE OF NARROW-GAUGE PASSENGER ENGINES ON THE GREAT WESTERN RAILWAY

About this time, the growth of narrow gauge lines in the districts served by the G.W.R., together with the amalgamations and alliances of narrow gauge railways with the G.W.R., made it necessary for the latter railway to provide narrow-gauge engines. Fig. 73 represent one of the first narrow-gauge Great Western locomotives. It will be

EVOLUTION OF THE STEAM LOCOMOTIVE     183

seen that Daniel Gooch introduced all his well-known features into these engines. These locomotives were built by Beyer, Peacock, and Co. The "single" driving wheels were 6ft. 6in. diameter, the cylin-

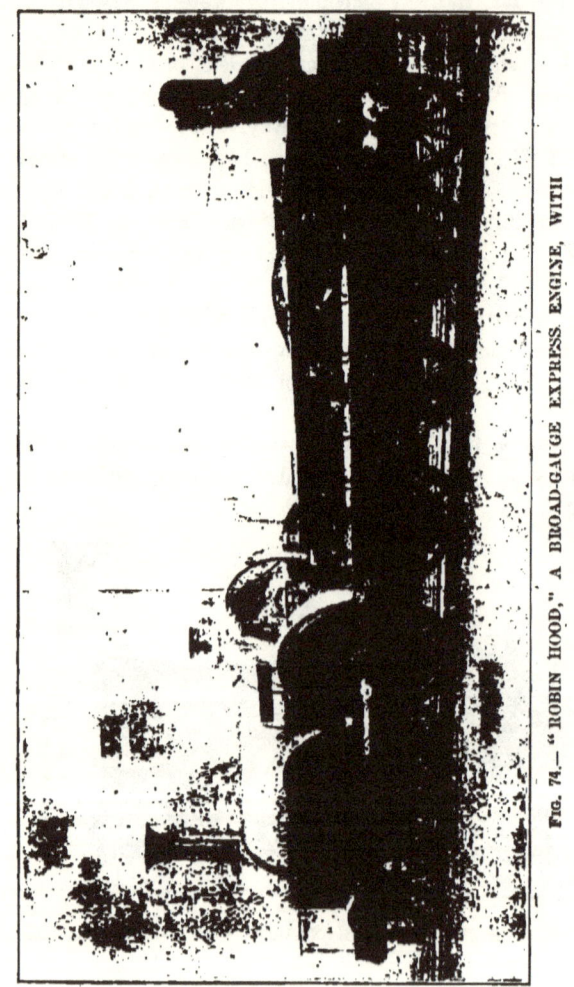

Fig. 74.—"ROBIN HOOD," A BROAD-GAUGE EXPRESS ENGINE, WITH COUPLED WHEELS 7FT. IN DIAMETER

ders being 15½in. diameter, and the stroke 22in. Compensation levers connected the leading and driving springs.

In 1855 Sir D. Gooch designed a class of coupled express boad gauge engines for the Great Western Railway. These engines had a group of four leading wheels, like the "Lord of the Isles" class. The driving and trailing wheels were coupled, and were 7ft. in diameter. At that time, no coupled wheels of so large a diameter had been constructed. The cylinders were 17in. diameter, with a 24in. stroke. R. Stephenson and Co. built the engines, of which there were 10. They were a most successful class of engine, and ran about 500,000 miles each before being "scrapped." "Robin Hood" (Fig. 74) was one of these engines. By reference to the illustration, it will be seen that the tender was fitted with the sentinel box for the "travelling porter" that formerly accompanied the G.W. broad-gauge expresses.

Fig. 75 represents the inspection or cab engine of the N.B.R., it is numbered 879, and was originally built by Messrs. Neilson and Co., in 1850, for the Edinburgh and Glasgow Railway. She is now used for inspection purposes. The cylinders are 10in. diameter by 15in. stroke. Other dimensions are: Wheels, leading and trailing, 3ft. diameter; driving, 5ft. diameter; wheel base, 15ft. 8in.; centre of leading to centre of driving, 10ft. 8in.; centre of driving to centre of trailing, 5ft. Tubes, No. 88, 1¾in. diameter outside. Heating surface: Tubes, 324 sq. ft.; fire-box, 35 sq. ft.; total, 359 sq. ft. Fire-grate, 5 sq. ft. Weight, in working order, 22 tons 1cwt. 3qrs. Tank capacity, 426 gallons.

FIG. 75.—NORTH BRITISH RAILWAY INSPECTION ENGINE, No. 879

## CHAPTER XI.

Improvements in coal-burning locomotives—Beattie's system—Trials of the "Canute"—Yorston's plan—Cudworth's successful efforts—Yarrow's apparatus—D. K. Clark's system tried on the North London and other railways—Wilson's plan fitted to engines working the O.W. & W.R.—Lee and Jacques' experiments—Frodsham's device tried on the E.C.R.—Douglas' system—The various plans reviewed—"Nunthorpe," a S. & D.R. engine—Double engine on the Turin and Genoa Railway—Crompton's engines on the E.K.R.—French locomotives on the F.C.R.—Gifford's invention of the injector—First fitted to the "Problem"—Ramsbottom's water "pick-up" apparatus—Brunel's powerful B.G. tanks for the Vale of Neath Railway—Incorporation of the Metropolitan Railway—Trial of Fowler's "hot brick" engine—Its end—Fletcher's saddle tanks—"75," T.V.R.—Second-hand locomotives on the L. & S.W.R.—The "Meteor"—Early L.C. & D.R. engines.

We have now reached an era in the "evolution of the steam locomotive" which, in its after development, amounted to a complete revolution in the character of the fuel used for locomotive purposes. The year 1855 found the locomotive, or rather those responsible for its working, on the threshold of successful experiments, which resulted in the complete substitution of the "black diamonds" in their natural state for locomotive fuel in preference to the use of coal after it had undergone the process of carbonification necessary to form coke.

It must not be forgotten that steam-users never had a preference for coke, but they were compelled to use it, because the more volatile coal produced so much smoke in the process of combustion that legislative action (which compels locomotive engines to be so constructed as "to consume their own smoke") practically prevented the use of coal until science discovered a method of consuming the smoke.

There had been various attempts to reach this desirable state, and we have from time to time in this series of articles described certain of these efforts; but none of them up to the date under review had been sufficiently successful to warrant the adoption of any one of the methods proposed as a complete smoke-consumer.

The successful efforts made by Beattie, of the London and South Western Railway, to solve the problem of smoke consumption in the locomotive so as to admit of coal being used as fuel stand out prominently. The salient points of his smoke-consuming locomotive comprised an enlarged fire-box, a combustion chamber,, the transverse division of the fire-box by means of an inclined water bridge, and the fire-box arched with fire-bricks. A perforated fire-door for the admission of air to the fire-box was another of the features of Beattie's system, as were also the use of the ashpan dampers and the employment of an auxiliary steam jet in the chimney for use when the engine was at rest and the ordinary exhaust blast consequently not available. With the addition of a feed-water heating apparatus Beattie reduced the fuel consumption to from .12 to .17lb. per ton mile.

The dimensions of the London and South Western Railway locomotive "Canute" (an engine filled with Beattie's coal burning apparatus) were:—Cylinders (outside), 15in. diameter, 21in. stroke; driving wheels, 6ft. 6in. diameter. The fire-box was 4ft. 11in. long, 3ft. 6in. wide, 5ft. 1in. deep at the back, and 4ft. 1in. in front. The combustion chamber had a flat roof, was 4ft. 2in. long, and 3ft. 6in. diameter. The tubes were 6ft. long, 1¼in. diameter, and 373 in number. Total area of fire-grate, 16 sq. ft.

The heating surface of the "Canute" was as follows:—Box, 107 sq. ft.; combustion chamber, 37 sq. ft.; tubes, 625 sq. ft.; total, 769 sq. ft; in addition to which red-hot bricks presented a surface of 80 sq. ft., not, however, for heating the water, but for the purpose of burning the smoke. Four series of trials were made with the "Canute" engine No. 135, and these are detailed in "Locomotive Engineering." The experiments are described as "1st, the engine in its usual order, with coal, bricks, and hot feed-water; 2nd, with coal, bricks, and cold water; 3rd, with coke, bricks, and hot feed-water; 4th, with coal and hot feed-water, but without the bricks." Three different kinds of coals were used for the experiments. The following is a brief summary of the experiments:—1st, a regular express train, of 10½ coaches, weighing 66 tons, or with the engine and tender, 99 tons. Average speed, exclusive of stoppages, 34 miles an hour; consumption of coal, 15lb. per train mile; water evaporated, 9.35lb. per lb. of coal consumed; average temperature of heated feed-water, 187 degrees. 2nd trial, a weighted train of 28 coaches, weighing with engine and tender 236 tons. Average speed, exclusive of stoppages, 30¾ miles an hour; coal consumed, 28¾lb. per mile, 8.87lb. of water evaporated by each pound of coal; temperature of feed-water, 212 degrees. 3rd experiment with an express train, but without the fire-bricks in the fire-box, showed that a saving of 12 per cent. was due to the use of the fire-bricks, and with coke instead of coal as fuel, the saving was 24 per cent. in favour of coal; whilst the use of the feed-water heating apparatus showed a saving of 30 per cent. of fuel. Beattie's apparatus is illustrated by Fig. 76, the "Dane," being a similar locomotive to "Canute."

FIG. 76.—THE "DANE," L. & S.W.R., FITTED WITH BEATTIE'S PATENT APPARATUS FOR BURNING COAL

As the feed-water heating apparatus was an important innovation in locomotive practice, it will be of interest if we append a description of the same. In outward appearance, the most noticeable portion of the apparatus was the condenser, a cylindrical appendage placed in a vertical position on the top of the smoke-box and in front of the chimney. From a casual glance, the condenser much resembled the steampipe of a steamship which is usually to be observed outside the smoke-stack. From the bottom of the condenser, outside the engine, a pipe conveyed the heated water and steam back to the tender. The method of working was for the exhaust steam to be discharged from the blast pipe into the condenser, which, as previously explained, was on the top of the smoke-box, and consequently right over the blast orifice. Here the exhaust steam was mixed with a jet of cold water, which was pumped into a condenser. The result of such meeting was the condensing of the steam and heating of the water, which flowed by gravitation through the pipe previously described. The supply pump for the boiler was worked off this pipe, and both the heated water and that from the tender were together pumped into the boiler. If the boiler were not being fed, the heated water from the condenser, instead of passing into the boiler, flowed through the pipe into the tender, and thus raised the temperature of the whole of the water in that vessel.

It should be mentioned that before entering the boiler the temperature of the feed-water was further increased by passing through a special heating apparatus, fixed in the smoke-box. This smoke-box chamber was heated by the exhaust steam, which passed through it after leaving the blast pipe, and before entering the external condenser placed above it. By these methods the temperature of the feed-water was raised above the boiling point before entering the boiler.

The engines of this design gave satisfaction, both as regards smoke-consuming and feed-water heating, and to Beattie, therefore, is due much of the honour of successfully overcoming the defects that previously existed in so-called "smoke-consuming" locomotives. The "Canute" can, therefore, be considered amongst the earliest of the locomotives burning coal in such a manner as to consume the smoke. It should be mentioned that in later engines built under Beattie's patent the external condenser fixed on the top of the smoke-box in front of the funnel was not used, a modified form of interior apparatus being substituted.

It must not be supposed that at this period Beattie was alone in the field of experiment relating to "smoke-consuming" locomotives. Several other engineers were engaged in the same useful research, amongst whom we mention Yorston, Cudworth, Yarrow, D. K. Clark, Wilson, Lee and Jacques (jointly), Sinclair, and Douglas. Yorston's plan was patented by Sharp, Stewart, and Co. in 1855. The fire-box was divided into two parts by a transverse mid-feather, which was perforated by a series of tubes, to allow the coal gases to escape and air to enter. The coal was fed into the portion of the fire-box next the tubes, the front part being reserved for coke; separate fire-doors were used for introducing the coke and coal into the fire-box. The air entering through the perforations in the fire-box, at the tube-plate end, was expected to force the smoke, etc., from the coal fire over the incandescent coke, where the combustion of the coal would be completed. The system, however, appears to have been better in theory than practice, as no particular steps were taken to push the invention in question.

With Cudworth's system the opposite course was adopted, and resulted in his engines taking a foremost position among those burning coal as fuel.

Mr. Cudworth, the locomotive superintendent of the South Eastern Railway appears to have made his first experiments with engine No. 142, which during July, 1857, was tried as a coke-burning locomotive; but during October and November of the same year experiments were made with this engine, fitted with Cudworth's patent grate, etc.

The principal dimensions of Cudworth's standard passenger engines were as follows:—Cylinders, 16in. by 24in. stroke; driving wheels, 6ft. diameter; wheel base, 15ft.; heating surface, 965ft.; grate area, 21 sq. ft. Total weight in working order 30½ tons, of which the leading axle supported 9 tons 9 cwt., driving 10¾ cwt., and

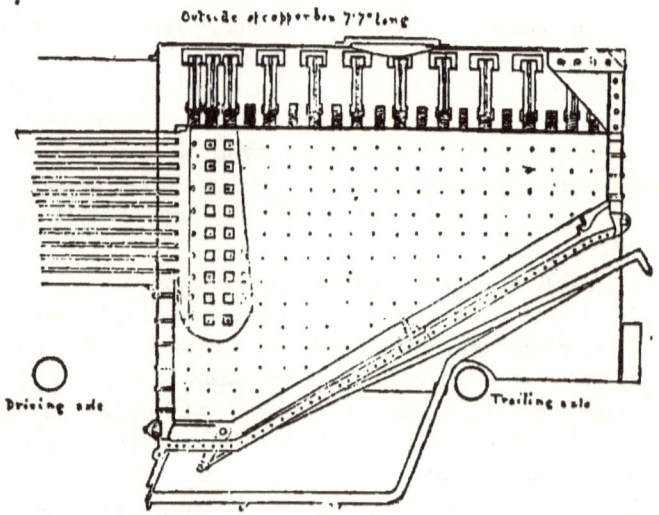

FIG. 77.— CUDWORTH'S SLOPING FIRE GRATE, FOR BURNING COAL, AS FITTED TO SOUTH EASTERN RAILWAY LOCOMOTIVES

trailing 10 tons 6 cwt. The tender was carried on six wheels, and weighed in working order 20½ tons. These engines had inside cylinders and "back-coupled" driving wheels, and for many years comprised the principal type of South Eastern passenger locomotives. Several of them are still running, but rebuilt, their former distinguishing features—viz., the large brass dome on the centre of the boiler barrel, the raised fire-box, with a brass encased Salter safety valve, the sloping fire-grate, and the peculiar chimney—all having been removed during the present locomotive *régime*.

The chief feature in Cudworth's system was the long, sloping fire-box, which was 7ft. 6in. in length, the grate being 7ft. long, illustrated by Fig. 77. The fire-box was divided into two parts by a longitudinal mid-feather, thus forming two furnaces, with separate doors; the two furnaces united at the lower end—in front of the tube-plate. The coal was introduced alternately into each furnace, being placed just within the doors; the sloping grate and the motion of the engine caused the fuel to gradually slide down the grate towards the tube-plate, and by the time the fuel had reached the lower end of the grate, the smoke had become separated from the carbon of the coal, and was consumed by the incandescent mass of fire at the lower end of the grate, as it passed over the same on its way to the tubes.

Cudworth employed neither combustion chambers nor air-bricks in his system; but air was admitted to the fire-box by means of a damper fixed in the front of the lower end of the grate. A steam-jet was fixed in the chimney to create a sufficient draught when the engine was still. Cudworth's "smoke-consuming" locomotives were as economical in coal as Beattie's, whilst the former's system was much more simple.

On March 18th, 1857, Thomas Yarrow, of Arbroath, was granted a patent for his smoke-consuming apparatus for locomotives, which was used on the Scottish North Eastern Railway. The leading characteristic of the design was a flat arch of fire-bricks constructed inside an ordinary fire-box. The lower end of the arch commenced below the bottom row of tubes, and the arch was continued upwards in a slanting direction till within 8 or 10 inches of the roof of the fire-box. Upon the top of this arch were fixed a number of tubes, through which the vapours passed before reaching the ordinary boiler tubes. Hot air was supplied to the fire by means of pipes with trumpet-shaped mouths placed in front of the ashpan. The fire-bars were fixed on a transverse rocking-shaft fitted with several short arms, upon which the ends of the fire-bars rested. To prevent the formation of clinkers, an occasional rock was given to the fire-bars by the fireman, a sector being provided for the purpose. Yarrow's system required the coal to be placed at the extreme front of the fire-box, so that the smoke was forced by the brick arch to return towards the fire-door before it could get over the arch and enter the tubes, and in the passage the denser portion of the smoke was burnt. The patent also included the use of a steam-jet in the chimney for use when the engine was not working, and the heating of the feed-water by means of the exhaust steam.

Late in 1857 D. K. Clark devised his system of smoke-consuming furnaces: the air was forced through tubes into the fire-box by the action of minute jets of steam, which acted much in the same way as the blast pipe in the smoke-box. The air-tubes were $1\frac{1}{2}$in. diameter, with the steam-jet orifice contracted to one-sixteenth inch diameter.

The first locomotive fitted with D. K. Clark's system was one of the North London Railway's tanks. This was in January, 1858, but only one side of the fire-box was fitted; four air-tubes were employed, and with a small fire the prevention of smoke was complete. In April of the same year one of the passenger engines on the Eastern Counties Railway was fitted with Clark's apparatus. Four air-tubes were fitted to one side of the fire-box, and three to the other side. In the following January a South Eastern Railway passenger locomotive was fitted with two rows of seven tubes each, through the front and back of the fire-box. In March, 1859, a Great North of Scotland Railway engine was fitted with tubes on Clark's system, with such satisfactory results that the whole locomotive stock of that railway was speedily fitted with the apparatus. No complete investigation appears to have been made as to the work performed by the jets of steam as employed by Clark. It is generally supposed that the steam had a merely mechanical effect—viz., that of drawing the air into the fire-box. It has also been suggested that the steam produced a chemical combination which facilitated the combustion of the volatile gases, besides precipitating the unconsumed carbonaceous matter on the fire. The result of the adoption of the system on the Great North of Scotland Railway's locomotives was such that the coal consumption fell to under .2lb. of coal per ton mile. A trial was also made of Clark's system on the London Brighton, and South Coast Railway, one of the old passenger engines being fitted with air-tubes and steam-jets to the front of the fire-box, with good results.

In 1858 Mr. Edward Wilson, who supplied the Oxford, Worcester, and Wolverhampton Railway with locomotive power by contract, fitted his system to several of the engines on that line. Mr. David Joy, the inventor of the celebrated Joy valve-gear, was at that time locomotive superintendent of the Oxford, Worcester, and Wolverhampton Railway, and he possesses records of many runs of the engines so fitted, and the comparisons between the fitted and unfitted engines show an immense saving of fuel by the former; indeed, the coal consumption was remarkably low considering the severe nature of the line between Oxford and Worcester. Some short time ago Mr. Joy showed the writer the tabulated results of these trials, and, if memory serves correctly, the coal consumption averaged about 21lb. per train mile. Wilson's system consisted in fixing several tubes from the bottom of the fire-box underneath the whole length of the boiler and smoke-box, so that the mouths of the air-tubes projected in front of the engine, and the resistance of the train when travelling forced the air through the tubes into the fire-box. By his method Wilson obtained a forced draught without the expenditure of the steam, which was necessary in Clark's system.

Lee and Jacques' system was introduced on the East Lancashire Railway in July, 1858. It consisted of a narrow fire-brick arch, and a deflector fixed at the top of the underhung fire-door. The deflector projected in a downward sloping direction into the fire-box. A valve for controlling the supply of air to the fire-box was fitted to the fire-door, and this valve was worked by means of a sector. The air entered the fire-box through the valve, and the deflector caused the air to be projected downwards on to the fuel, whilst the brick arch prevented the immediate escape of the gases, and kept them within the fire-box sufficiently long for the smoke to be consumed.

In December, 1858, Mr. Sinclair, the locomotive superintendent of the Eastern Counties Railway, commenced to fit some locomotives with the deflecting plate, etc., on a plan introduced by a Mr. Frodsham. The fire-door was underhung, and the baffle-plate was fixed above it, to direct the air down on to the fuel; whilst instead of a brick arch, two steam-jets were used, one on each side of the door. These also helped to force the air on to the burning fuel and to drive the liberated, but unconsumed, smoke back into the fire, when it was consumed.

Mr. Douglas's plan was adopted by the Birkenhead Railway. He combined the use of an inclined fire-grate of large area, and a baffle-plate. In January, 1858, when first introduced, the deflector was fixed to the inner side of the fire-door, but in June of the same year an underhung fire-door and movable baffle-plate were employed. These afterwards gave place to a plain inverted scoop, to project the air right on to the fire.

After reading the description of the various plans adopted for the consumption of the smoke, readers will at once observe that each and every designer had the same object in view—viz., to supply a sufficient volume of air to the fire, and mix the air with the unconsumed gases given off by the burning coal, and then to prevent the immediate escape of this gaseous mixture from the fire-box. Being retained within the heated fire-box, the temperature of the vapour was raised sufficiently, so that the vapour readily burnt when forced

FIG. 78.—"NUNTHORPE," A STOCKTON AND DARLINGTON RAILWAY PASSENGER ENGINE, BUILT IN 1856

by the steam deflector, or brick arch (according to the system adopted), back on to the incandescent fuel. As stated, the object of all the inventors was the same, but the methods adopted were different, and these latter (though some systems had advantages that others lacked) were successful in each case; but from the whole could be chosen some that certainly were more noteworthy, both as regards simplicity of application and design, and others that were more successful in attaining the object in view—viz., a consumption of the smoke given off by the coal. In these four years—1855-59—however, the problem of consuming the coal smoke, was successfully accomplished, and the era of the coal-burning locomotive definitely inaugurated.

O

## 194  EVOLUTION OF THE STEAM LOCOMOTIVE

Fig. 78 is an illustration of the "Nunthorpe," No. 117 of the Stockton and Darlington Railway. This engine shows a distinct advance in locomotive construction; indeed, it is possible at the present time to see on some lines engines somewhat similar in appearance still at work. She was built by Gilkes, Wilson and Co., in 1856, and was intended for passenger traffic. Four of the six wheels were coupled, these being 5ft. in diameter. The cylinders were inside, 16in. in diameter, and with 19in. stroke. The tender was on six wheels, and the tank capacity was 1,200 gallons. The cost of the engine was £2,550. It will be observed that the weather-board of the "Nunthorpe" afforded very

Fig. 79.—BEATTIE'S 4-COUPLED TANK ENGINE, L. & S.W.R., 1857

little protection to the driver and fireman, but its inclusion in the design of the engine was a step in the right direction.

In 1857 Beattie designed a handy class of passenger tank engines for the L. and S.W.R. Three were built at first, and named "Nelson," "Howe," and "Hood." They had four coupled wheels, 5ft. diameter, and a small pair of leading wheels. The cylinders, which were outside, were 15in. diameter, the stroke being 20in. These engines are illustrated by Fig. 79. They were good locomotives, and "Hood" and "Howe" continued in work till 1885.

Fairlie is usually given the credit of introducing double locomotives with a centre foot-plate. By reference to Chapter IX., it will be seen that the design was patented by Pearson, of the Bristol and Exeter Railway, as long ago as 1847, and in 1855 a double engine, built by R. Stephenson and Co., was at work on the Giovi incline of the Turin and Genoa Railway. The incline in question commences 7¾ miles after leaving Genoa, and is six miles long, the average gradient being 1 in 36. The double locomotive was of the tank type. The wheels were 3ft. 6in. diameter, the cylinders 14in.

diameter, and the stroke 22in. The machine actually appears to have been two engines placed fire-box to fire-box, and connected by means of a foot-plate between the two fire-boxes. The combination, with fuel and water, weighed 50 tons. In fine weather a load of 100 tons was hauled up the Giovi bank at 15 miles an hour; in bad weather the load was reduced to 70 tons.

The first portion of the East Kent Railway from Chatham to Faversham was opened in January, 1858, the original locomotives being designed by Crampton, who was one of the contractors for the construction of the line. The engines in question were "tanks," and weighed 32 tons each—at that period considered an excessive weight for an engine. They were also unsteady and generally unsatisfactory, frequently running off the metals.

Mr. Robert Sinclair was appointed locomotive superintendent of the Eastern Counties Railway in 1858, and his first design of engines was a class for working the goods traffic, of which only six were constructed, Rothwell and Co. being the builders. The engines had a pair of leading wheels, 3ft. 7in. diameter, and two pairs of coupled wheels, 5ft. diameter; the cylinders were 18in. diameter, the stroke being 22in.

During the following years another class of goods engines (Fig. 80) were built by various firms from Mr. Sinclair's improved design. Indeed, as will be seen later on, some were even constructed by the French firm of Schneider and Co. These had outside cylinders, and inside frames to all wheels. The coupled wheels (D. and T.) were 6ft. 3in. diameter, and the leading 3ft. 9in. diameter. The boiler was 10ft. 9in. long by 4ft. 2in. diameter, and contained 203 tubes, of 1¾in. diameter; heating space, 1,122 sq. ft.; weight, 35½ tons. Twenty-one of these engines, built by Neilson and Co., had Beattie's patent fire-box, which was surmounted by a large dome. These were numbered 307 to 327. When Mr. W. Adams was appointed locomotive superintendent of the Great Eastern Railway, he rebuilt several of these engines with a leading bogie in place of the pair of wheels.

In November, 1858, a design of locomotive engine was patented, four pairs of coupled wheels being employed, all of which were located under the boiler barrel. The two leading pairs of wheels had outside axle-boxes, and the two trailing pairs inside axle-boxes, the latter having a lateral motion. The cylinders

196   EVOLUTION OF THE STEAM LOCOMOTIVE

FIG. 80.—SINCLAIR'S OUTSIDE CYLINDER, 4-COUPLED GOODS ENGINE, EASTERN COUNTIES RAILWAY. (REBUILT)

were inside, under the smoke-box, but the method proposed for working the locomotive was of a curious type, being somewhat after the fashion employed in ancient steamboats, the pistons working out towards the front buffer beams, but connected to the leading wheels by outside cranks working off the cross-heads.

A design for four-wheel tank engines was patented by S. D. Davison, in February, 1859, the leading feature being plate-iron frames formed into tanks for holding a supply of water.

Attention must now be given to an invention that has proved of enormous value to the locomotive engineer, but which from its simplicity of action, yet apparent impossibility, was not at first deemed worthy of practical use. On July 23rd, 1858, a patent was granted to H. J. Giffard, a Frenchman, for his injector, or boiler feeder, which in a short period almost completely superseded feed pumps, with their attendant friction, uncertainty of action, and excessive outlay for maintenance and repair. But above these minor disadvantages of the feed pumps, the injector removed from the minds of locomotive engineers that great source of danger, a short supply of water in the boilers, as well as the additional expense and inconvenience of "exercising" the locomotives solely for the purpose of filling the boiler, or, where such a method was inconvenient, of working the engine over a "race" for the same purpose. The theory of the injector did not originate with Giffard, for as long ago as 1806 Nicholson mentioned it as applicable for forcing water, whilst other philosophers have suggested its utility; indeed, the principle was used in connection with vacuum sugar boiling pans 20 years before Giffard's patent. The story of Giffard's accidental discovery of the action of steam and water in supplying a steam boiler with additional water reads almost like an extravagant romance, but many other great inventions and scientific discoveries had beginnings that appeared quite as improbable. The action of the injector, although curious, is well known, and therefore needs no description here. It is stated that Ramsbottom's "Problem," built at Crewe in November, 1859, was the first locomotive fitted with Giffard's "injector." This engine was the prototype of the world-famous "Lady of the Lake" class. Her dimensions were, outside cylinders, 16in. by 24in.; single driving wheels, 7ft. 7½in. diameter; weight in working order, 27 tons. These engines have inside frames and bearings to all the six wheels.

An invention of Mr. Ramsbottom in connection with the improvement of the working of the locomotive deserves attention at this point. We refer to his self-filling tender apparatus, as introduced in 1860 on the London and North Western Railway system, and afterwards partially on the Lancashire and Yorkshire Railway, but which until the last year or so has not been used on other lines. The speed competition of recent years, and the expiration of the patent, has now caused the Great Western, Great Eastern, and North Eastern to adopt the water pick-up apparatus. One advantage of the system is, of course, the considerable reduction in the dead weight—a not unimportant factor in express train running. The superiority of Ramsbottom's system is easily seen by comparing the small light tenders in use on the London and North Western Railway with the gigantic ones adopted by the Great Northern, Midland, and other lines running long distances without stopping, but which systems are unsupplied with the water trough and the necessary pick-up apparatus. The first pair of water troughs appear to have been put down near Conway, on the North Wales section of the London and North Western Railway. They were of cast-iron, 441 yards long, 18in. wide, and 7in. deep, the water being 5in. deep. At each end of the main trough was an additional length of 16 yards, rising 1 in 100. It was towards the end of 1860 that the first trial of the trough system was made. Here, again, as in the case of the "injector," the arrangement requisite to produce the effect is so simple that at first blush the effect appears to be the result of some marvellous secret power rather than the operation of a simple natural law, the effect of the travelling scoop upon the water being exactly the same as if the water were forced against a stationary scoop at a velocity equal to that at which the train is travelling. The lowest speed at which the apparatus works properly is something about 22 miles an hour. This speed, however, brings it within the scope of fast goods trains, whilst express trains can scoop up the water when travelling at 50 miles an hour, and can pick up about 1,500 gallons in the length of the trough—quarter of a mile. The speed of the train would not appear to have much effect upon the water picked up in passing over a trough, as although with a slower train less water would be raised per second, yet the extra length of time spent in travelling over the trough would compensate for the smaller amount of water raised per second. The water supply-pipe is fixed inside the tender; it is slightly curved throughout its entire length, and is expanded towards its upper end to about ten times the area of the bottom, in order to reduce the speed or force of the incoming stream, which is directed downwards by the bent end or delivering mouth at the top of the pipe. To the lower end of this pipe is fitted a movable dip-pipe, which is curved forward in the direction of the motion of the tender, so as to act as a species of scoop. This dip-pipe is rendered movable and adjustable in various ways, with a view to its being drawn up clear of any impediments, such as ballast heaps lying on the way, and also to regulate the depth of immersion in the water of the feed-water trough, the dip-pipe being capable of sliding up inside the feed-pipe by a convenient arrangement of rods and levers.

In order that the dip-pipe may enter and leave the feed-trough freely at each end, the rail surface at that part of the line is lowered a few inches, a descending gradient at one end of the trough serving to allow the dip-pipe to descend gradually into the trough, whilst a rising gradient at the opposite end enables it to rise out of the trough again, the intervening length of line between the two gradients being level. To meet emergencies, Mr. Ramsbottom provided a small ice-plough, to be used occasionally during severe frost for the purpose of breaking up and removing any ice which might form in the trough. This plough consisted of a small carriage mounted on four wheels, and provided with an angular-inclined perforated top, which worked its way under the ice on being pushed along the bottom of the trough, and effectually broke it up and discharged it over each side.

A very powerful class of broad-gauge saddle tank locomotives was designed by Brunel for working the heavy coal traffic over the severe gradients of the Vale of Neath Railway. These engines were supported by six coupled wheels of 4ft. 9in. diameter, the cylinders being 18in. diameter, and the stroke 24in. The heating surface was 1,417.6 sq. ft.; the water capacity of tanks was 1,500 gallons. The engines, which were fitted with Dubs' wedge motion, were built by the Vulcan Foundry Company, and weighed 50 tons in working order. A noteworthy performance of one of these locomotives consisted in hauling a train of 25 loaded broad-gauge trucks, each weighing 15 tons, the gross weight, including the engine, amounting to 425 tons. This train travelled up a bank of 1 in 90 for a distance of $4\frac{1}{2}$ miles. Such a load on the gradient mentioned is equal to one of 1,275 tons on the level, and in a general way we do not find engines hauling trains of the latter weight upon our most level lines. The Vale of Neath performance must, therefore, be regarded as an exceptional locomotive feat. These engines were numbered 13, 14, and 15, and not being provided with compensating beams between the wheels, it is stated that one axle frequently carried 20 tons of the total weight. During 1860 these three locomotives were, under the advice of Mr. Harrison, rebuilt as tender engines, to reduce the weight on the wheels, the excessive amount of which had been very destructive to the permanent-way. The cost of the alterations to the engines and the addition of the tenders was £700 each engine. About the same time some of the other Vale of Neath six-wheels-coupled engines were converted into four-wheels-coupled bogie locomotives.

The locomotive now to be described had but a very shadowy existence; it was rather a tentative essay to produce a steam locomotive without the aid of a fire. The idea when proposed by Sir John Fowler was not new, for more or less successful essays had already been made on a small scale, with engines, the steam for propelling which was generated in the same manner as in Fowler's locomotive.

In 1853 a railway was incorporated as the North Metropolitan; the next year a new Act was obtained, and the title changed to the Metropolitan. This authorised the construction of a railway from the Great Western Railway at Paddington to the General Post Office; powers were afterwards obtained to allow the City terminus to be in Farringdon Street instead of at the Post Office. The Great Western Railway subscribed £175,000 of the capital, and for the convenience of that Company's through traffic the Metropolitan was laid out on the mixed-gauge, and when it was first opened it was worked on the broad-gauge only, by the Great Western Railway—a most sensible arrangement, and one which ought never to have been relinquished, seeing how well adapted the wider vehicles were for conveying the immense crowds that travel by every train on this line.

The Act of Incorporation specially provided that the line was to be worked without annoyance from steam or fire. At first it was proposed to convert the water into steam by means of red-hot bricks placed around the boiler, and Mr. (afterwards Sir) John Fowler designed such a locomotive, which was built by a Newcastle firm, and tried on the Metropolitan Railway between Bishop's Road and Edgware Road Stations before the line was opened. The first trial took place on Thursday, November 28th, 1861. The following is an account of the trip:—"The engine was of considerable size, and it was stated that it could run on the railway from the Great Western at Paddington to Finsbury Pavement without allowing the escape of steam from the engine or smoke from the fire. A few open trucks were provided with seats, and when the gentlemen were seated, the new engine propelled them under the covered way of the Metropolitan Railway to the first station at the eastern side of the Edgware Road, and back again to the Great Western Station, the steam and smoke being shut off. The tunnel, or covered way, was perfectly fresh and free from vapour or smoke. On the signal being given to work the engine in the ordinary way, a cloud of smoke, dust, and steam soon covered the train, and continued until it emerged from the tunnel into the open air. The experiment was perfectly successful, but it was understood that engines so constructed would be rather more expensive to work than those running in the ordinary way." To work the Metropolitan Railway on this system would have required the erection of immense boilers at both ends of the line to heat the water for the locomotive, and also furnaces for making the bricks red-hot, whilst the charging of the locomotive boilers with hot water and the fire-boxes with hot bricks would have occupied some considerable time at the end of each trip.

It is, of course, well known that the experiment was very far from being "perfectly successful." Indeed, "failure" would be a much better definition of the hot-brick engine, since the proposed method of working was not carried out. We understand the engine was sold to Mr. Isaac Watt Boulton, the well-known purchaser of second-hand locomotives, and for some time remained in his "railway museum" before being finally scrapped. The Metropolitan Railway had, consequently, upon the failure of the hot-brick engine, to fall back upon the Great Western Railway for working the underground line, until Sir John Fowler's later design of engines, constructed by Beyer, Peacock, and Co., were ready to work the traffic.

In 1862 Fletcher, Jennings, and Co., of Whitehaven, designed a handy type of saddle tank engine for shunting purposes, etc. The engine ran on four wheels, 3ft. 4in. diameter, the wheel base being 6ft. The cylinders were 10in. diameter, with 20in. stroke. Allan's straight link motion was employed, and was worked off the leading axle (it will be understood that the four wheels were coupled). This method of actuating the valves was not conducive to good working.

as, of course, if the coupling-rods worked slack the valve gear motion became disorganised.

Fig. 81 is a photograph of engine No. 75, of the Taff Vale Railway, built at the Company's Cardiff Works in 1860. The six-coupled wheels were 4ft. 8in diameter, the cylinders were 16in. diameter, and the stroke was 24in. No. 75 weighed 32 tons in working order; the steam pressure was 130 lbs. per sq. in. She was employed in the heavy mineral traffic of the Taff Vale Railway, and from her design well calculated to work over the heavy gradient of that system.

FIG. 81.—SIX-COUPLED MINERAL ENGINE, TAFF VALE RWY., BUILT 1860

In 1862 the L. and S.W. Railway purchased some second-hand engines from a contractor. They were built by Manning, Wardle, and Co., Leeds, and comprised six-wheels-coupled saddle tank engines. The wheels were 3ft. diameter; cylinders, 12in. by 18in. stroke; wheel base, 10ft. 3in.; length over buffers, 21ft. 6in.; weight, empty, 14 tons 8 cwt., loaded, 16 tons 4 cwt. The fire-box was surmounted by a safety valve enclosed within a high fluted pillar. The steam pressure was 120lb. One of these engines is leased to the Lee-on-the-Solent (Light) Railway, and may be seen working the traffic on this little line, which, by the way, spends over twopence to earn each penny of its gross income.

Before leaving the London and South Western Railway and its goods locomotives, it is as well to record the dimensions of the

"Meteor," No. 57, constructed at Nine Elms in 1863 from the designs of Mr. Beattie. The cylinders were 16½in. diameter, 22in. stroke; the leading wheels were 3ft. 3in., and the coupled (D. and T.) wheels 5ft. diameter; the wheel base was 14ft., of which 8ft. 2½in. was between the coupled wheels. The leading wheels were under the boiler, and the front buffer beam was about 6ft. in advance of the centre of this axle. An immense dome was fixed on the raised fire-box; the safety valve was within an inverted urn-shaped case on the boiler barrel. The weather-board had slight side-wings, and was curved upwards at the top, and so formed an incipient cab. The fire-box sloped from the tube-plate towards the foot-plate. The total weight, in working order, was 32 tons 18 cwt., of which 11 tons 9 cwt. was on the leading, 11½ tons on the driving, and 9 tons 18 cwt. on the trailing axle. The tender was supported on six wheels, 3ft. 9¾in. diameter, and had a tank capacity of 1,950 gallons.

By a marvellous addition of a big head and a bigger tail (to say nothing of various legs), the diminutive body of the East Kent Railway had, in August, 1859, blossomed into the London, Chatham and Dover Railway; and for this railway 24 locomotives were supplied by various firms from Crampton's designs. They were numbered 3 to 26. The design was peculiar—a leading bogie having wheels 3ft. 6in. diameter, and a base of 4ft., and four-coupled wheels 5ft. 6in. diameter. The cylinders were outside, and had a stroke of 22in., the diameter being 16in. As in the "London" and other Crampton engines, the cylinders were placed about mid-way between the smoke and fire-boxes, whilst the connecting-rods actuated the rear pair of coupled wheels, so that in describing the position of the wheels of these engines we should have to enumerate them as "leading bogie," "centre," and "driving." A compensation lever connected the centre and driving wheels. Gooch's valve gear was used. Like other engines of Crampton's design, this class was a failure, and within three or four years they were rebuilt as six-wheel engines, with inside cylinders and outside frames; some of them, as reconstructed without a bogie, are still in active service on the London, Chatham and Dover Railway.

Before the grave faults inherent in the previously described class of engines had been fully appreciated, the London, Chatham, and Dover Railway had arranged for a second batch of engines from another of Crampton's designs. These consisted of five engines constructed by R. Stephenson and Co. in 1862. The locomotives in question were worked on the principle patented by W. Bridges Adams, and previously described in an earlier chapter—viz., an intermediate driving shaft, coupled by outside rods to the driving wheels, situated behind the fire-box. The cylinders were 16in. diameter by 22in. stroke, and within the frames. The driving wheels were 6ft. 6½in. diameter, and bogie wheels 4ft. 0½in. diameter. Cudworth's sloping fire-box, fitted with a longitudinal mid-feather, was employed. The heating surface amounted to 1,200 sq. ft., made up of 130 sq. ft. fire-box and 1,070 sq. ft. tubes, which were 2in. diameter, 10ft. 10in. long, and 189 in number. The grate area was 26 sq. ft.

The engines in question were named, etc., as follows:—

| Company's No. | Name. | Builder's No. |
|---|---|---|
| 27 | "Echo" | 1981 |
| 28 | "Coquette" | 1382 |
| 29 | "Flirt" | 1383 |
| 30 | "Flora" | 1384 |
| 31 | "Sylph" | 1385 |

As remarked in describing the previous class, Crampton's engines were in this case also found to be unsuitable, so that the London, Chatham and Dover Railway rebuilt the five engines, when the intermediate driving shaft was provided with a pair of wheels, and the engines became "four-coupled bogies." The diameter of the cylinders was increased to 17in.; the Cudworth fire-box was dispensed with, and the heating surface reduced, the present dimensions being—fire-box, 100 sq. ft.; tubes, 987 sq. ft.; grate area, 16¼ sq. ft.; weight in working order: on bogie, 14 tons 12 cwt.; driving wheels, 14 tons 12 cwt.; and on trailing wheels, 10 tons; total, 38 tons 16 cwt.

## CHAPTER XII.

"Brougham," Stockton and Darlington Railway—L. & N.W.R. engines at the 1862 Exhibition—Sinclair's "Single" engines for the G.E.R.—French locomotives on the G.E.R.—L. & S.W.R. tank engines, afterward converted to tender engines—Conner's 8ft. 2in. "Single" engine on the Caledonian Railway—The liliputian "Tiny," the Crewe Works locomotive—"Dignity and Impudence"—Bridges Adams's radial axle tank engines—His spring tyres—Account of the St. Helens Railway locomotive with these innovations—Broadgauge engines for the Metropolitan Railway—Rupture between the Great Western and Metropolitan—Sturrock to the rescue—G.N. tender engines on the Metropolitan—Delivery of the Underground Company's own engines—Great Northern "condensing" locomotives—The Bissell bogie truck well advertised—End of the "hot brick" engine—Sturrock's steam-tender engines on the G.N.R.—Sinclair's tank engine with Bissell trucks—Fell's system of locomotive traction—Tried on the Cromford and High Peak line—Adopted on the Mount Cenis Railway—Spooner's locomotives for the Festiniog Railway—Fairlie's double bogie engines—The "Welsh Pony" and "Little Wonder"—Fairlie's combined trains and engines—Cudworth's trailing bogie North London engines, a model for tank locomotive constructors—Pryce's designs for the North London Railway.

Fig. 82 illustrates the "Brougham," No. 160, of the Stockton and Darlington Railway. This engine was designed for hauling passenger trains. She was a bogie engine, as will be noticed by reference to the illustration, and had four-coupled wheels 6ft. in diameter. The cylinders, placed outside, were 16in. in diameter, with a stroke of 24in. The tender was on six wheels, and the tank was capable of carrying 1,400 gallons. No. 160 was constructed in 1860, not a very long time prior to the amalgamation with the North Eastern Railway Company, by R. Stephenson and Co., of Newcastle, at a cost of £2,500.

The London and North Western Railway exhibited at the London International Exhibition of 1862 a locomotive constructed at Wolverton from the designs of Mr. McConnell; the engine was built the previous year, was numbered 373, and named "Caithness." The cylinders were 18in. by 24in.; driving wheels, 7ft. 7½in. diameter; L. and T., 4ft. 7½in.; steam pressure, 150lb.; wheel base, 18ft.; heating surface (14 tubes 1⅞in. diameter, 9ft. 4in. long), 980.319 sq. ft.; fire-box, 242.339 sq. ft.; weight in working order (engine and tender) 59 tons 14 cwt. A combustion chamber 2ft. 8in. long was provided. Two other engines of this design were built, No 372

"Delamere" and No. 272 "Maberley." Apparently these engines were not very successful, as we do not find accounts of their later performances.

In 1862 Fairbairn and Co. constructed for the Great Eastern Railway a class of "single" engines designed by Mr. R. Sinclair. These locomotives had outside cylinders, 16ft. by 24in.; driving wheels, 7ft. 3in., and leading and trailing wheels, 3ft. 9in. diameter; heating surface, tubes (203, 1¾in. diameter), 957.6 sq. ft.;

FIG. 82.— "BROUGHAM," No. 160, STOCKTON AND DARLINGTON RAILWAY

fire-box, 94.9 sq. ft.; grate area, 15.27 sq. ft,; weight, 32 tons, of which 13 tons 13 cwt. 1 qr. was on the driving axle. Gooch's link motion was employed.

The design in question was of rather attractive appearance, the open splasher being an attractive feature, as was also the cab—somewhat of an innovation 35 years ago. Mr. S. W. Johnson succeeded Mr. Sinclair at the end of 1865 as Great Eastern Railway locomotive superintendent, and under the *régime* of the former some of these engines were rebuilt with a leading bogie, and the diameter of the cylinders was increased to 18in. Another form of cab was introduced, the Salter safety valve on the dome was removed, and one of Ramsbottom design placed on the flush top fire-box, which had superseded the raised pattern as employed in this class of engine by Mr. Sinclair. One of the engines of this class (No. 0295) was in active service as recently as July, 1894. In connection with this class of engine a special circumstance needs mention—viz., that 16 of these locomotives were made—not "in Germany," but in the country of her foe; the French engineering firm with the German name of Schneider, in 1865, contracting to supply the 16 locomotives at a less price than any English maker. This event was certainly a curiosity in the economic history of this country's trade. We import many articles; let us hope, however, that foreign locomotives will not again be seen on English railways. There is some consolation to be found in the statement that all the British locomotive builders were so full of orders at the time that they practically refused to accept orders for the engines in question by tendering for them at outside prices, so that consequently the order had to be given to a foreign firm.

In 1863 Beyer, Peacock and Co. commenced to construct a class of tank engines for the London and South Western Railway from the designs of Mr. J. Beattie. The locomotives in question had outside cylinders 16½in. by 20in. stroke; four coupled wheels, 5ft. 7in. diameter; and a pair of leading wheels, 3ft. 7¾in. diameter. The boiler contained 186 tubes, 1⅝in. diameter. The heating surface was made up of tubes 715.17 sq. ft., and fire-box 80 sq. ft. The grate area was 14.2 sq. ft.

A lock-up safety valve was placed on the front ring of the boiler barrel, and two of Salter's pattern on the immense dome which surmounted the raised fire-box. The steam pressure was 130lb. The engine weighed in working order 29 tons 17 cwt., of which 10½ tons was on the driving axle. We have already stated that the engines were built as tanks, but Mr. W. Adams, who had succeeded Mr. J. Beattie as locomotive superintendent of the London and South Western Railway, added tenders to some of their engines in 1883. It is a common practice to rebuild tender engines as "tanks," but the opposite practice is somewhat of a novelty. The tenders were supported on six wheels, 3ft. 9¾in. diameter, and weighed 20¾ tons in working order, the water capacity being 1,950 gallons.

An engine that attracted considerable attention at the 1862 Exhibition was one built by Neilson and Co. from the designs of Mr. B. Conner, locomotive superintendent of the Caledonian Railway (Fig. 83). The engine in question had outside cylinders, 17¼in. diameter, with a stroke of 24in.; driving 8ft. 2in. in diameter, with inside bearings and underhung springs. The trailing and leading wheels had outside bearings. The engine had 1,172 sq. ft. of heating surface; the

## EVOLUTION OF THE STEAM LOCOMOTIVE

grate area was 13.9 sq. ft.; wheel base, 15ft. 8in.; weight, empty, 27¼ tons; in working order, 30 tons 13 cwt., of which 14 tons 11 cwt. was on the driving axle.

Colburn describes the locomotive as a "fine, well-constructed engine, standing gracefully on its wheels, large, yet compact, and qualified to run at any speed with ease and steadiness." Nor can this description be in any measure contradicted. For, until Stirling built his famous 8ft. 1in. "singles" for the Great Northern Railway, Conner's 8ft. 2in. Caledonian engines were far and away the most

FIG. 83.—CONNER'S 8FT. 2IN. "SINGLE" ENGINE, CALEDONIAN RAILWAY
(REBUILT)

graceful locomotives ever placed on the 4ft. 8½in. gauge. In general design, the engine was a modification of the old Crewe pattern engine. The dome was, however, of rather a peculiar shape: it was placed on the top of the raised fire-box. The driving axle was of cast steel, and the tyres of Krupp steel. The large number of spokes in the driving wheels was noticeable, being at only 10in. centres at the rim of the wheels. The slide-valves were provided with 1½in. lap. A great improvement was the provision of a cab, and that of not disproportionate dimensions, considering the "year of grace" in which the engine was constructed. Trains of nine carriages were hauled at an average speed of 40 miles an hour, with a coal consumption of 2¼lb. per mile; 14 loaded carriages were frequently taken up the terrible Beattock bank, 10 miles in length, at 30 miles an hour.

The late Khedive of Egypt was so taken with the appearance of this engine when it was at the Exhibition that he immediately ordered one for his own railway. He was searching for a locomotive to convey him at 70 miles an hour, and Conner's 8ft. 2in. single appeared to be the one most likely to fulfil his requirements. Nor do we hear that he was in any way disappointed with his purchase.

It is interesting to know that the Caledonian Railway has still a specimen of this notable design unscrapped—may it ever remain so. To prevent our appetite becoming vitiated with a galaxy of Brobdingnagian locomotives, we will descend to the other end of the scale, and detail the Liliputian "Tiny," as used in the Crewe locomotive works. The railway is of 18in. gauge, and was opened in May, 1862, for a length of three-eighths of a mile. In its course the engine traverses curves of 15ft. radius each, no difficulty being found in going round these curves with loads of 12 to 15 tons, or in taking 7ft. 6in. wheel forgings or tyres on edge by means of trucks specially adapted for the purpose. This engine has four wheels coupled; inside cylinders, 4¼in. diameter, and 6in. stroke; the wheels are 15in. in diameter, on a base of 3ft. The total heating surface is about 42 sq. ft. A No. 2 Giffard's injector supplied the boiler with water; this precious liquid is stored in a saddle tank, with a capacity of 28 gallons. "Tiny," when "right and tight and ready for action," weighs only 2½ tons.

The duties of the Liliputian engines consist in hauling materials to and from different parts of the works, and as the 18in. rails are in most places laid parallel with the standard gauge lines, "Tiny" is also called upon to fly shunt the trucks, etc., when necessary.

An engine of this type, the "Nipper," forms with the giant "Cornwall" that well-known photographic picture—the railway "Dignity and Impudence."

Fig. 84 represents Sharp, Stewart, and Co.'s standard design of passenger engine of this period. The "Albion" was delivered to the Cambrian Railway in May, 1863. She was an inside cylinder engine, with a pair of leading wheels, and an enclosed Salter safety valve. Altogether, the "Albion" is a fair example of locomotive practice 26 years ago.

We have on previous occasions referred to the improvements in locomotive construction introduced by Mr. W. Bridges Adams, and we now have again to record a successful employment of his design. In the first week of November, 1863, Mr. James Cross, locomotive engineer of the St. Helens Railway, completed a tank locomotive, supported on eight wheels, the leading and trailing pairs of which were fitted with the radial axle boxes patented by Mr. W. B. Adams;

P

210  EVOLUTION OF THE STEAM LOCOMOTIVE

whilst the four coupled wheels were fitted with spring tyres, which were another invention of the same engineer.

The St. Helens Railway was famous—or, from an engineer's point of view, we should say, perhaps, infamous—for the severe gradients, sharp curves, and numerous points, crossings, and junctions. The inclines were as steep as 1 in 35, 1 in 70, and 1 in 85, whilst the curves were constructed with radii of 300ft. and 500ft., and reverse or

FIG. 84.—" ALBION," CAMBRIAN RAILWAYS, 1863

S curves were also more frequent than pleasant. The St. Helens Railway was only 30 miles long, but within two miles of the St. Helens Station no less than 12 miles of sidings were located. We do not mean to suggest that the whole line of railway was so thickly covered with siding connections, but such were distributed over the remaining mileage of the railway in too plentiful profusion. Here, then, was a length of railway containing the three great hindrances to smooth and quick running, but the locomotive about to be described was so constructed as to successfully overcome these impediments.

This engine had inside cylinders, 15in. diameter and 20in. stroke. The coupled wheels were 5ft. 1in. in diameter, the rigid wheel base being 8ft., but as these wheels had spring tyres, each pair of wheels was practically as free to traverse the curves as uncoupled wheels. Other dimensions were:—Heating surface, 687 sq. ft.; grate area,

16.25 sq. ft.; total wheel base, 22ft.; weight in working order, on leading wheels, 7 tons 15 cwt.; on driving, 11¾ tons; on rear coupled, 11¼ tons; on trailing, 10 tons, including 4¼ tons water and 1¼ tons coal. Total weight, 40¾ tons.

The boiler contained 121 tubes, 10ft. 11in. long, and 1⅞in. diameter; steam pressure, 140lb.; water capacity of tank, 950 gallons. The fire-grate was 5ft. long, and sloped from the door to the tube-plate. The springs of the coupled wheels were connected by means o a compensation lever. The dome was placed on the raised fire-box and fitted with a screw-down safety valve; a second valve of the same pattern was fixed on the boiler barrel. A roomy and well-enclosed cab, fitted with side windows, thoroughly protected the enginemen.

Adams's radial axle-boxes are, of course, still in use on the Great Northern Railway, London, Chatham, and Dover Railway, and other lines, so that a detailed account here is not necessary, the salient feature being that they are made with a radius, having its centre in the centre of the adjoining axle, the axle-box guide-boxes being curved to fit. In the engine we are now describing the radius of the boxes was 7ft., and the lateral play of the boxes was 4½in. on each side. The spring-pins were not fixed on the top of the boxes, but were each fitted with a small roller to allow the boxes to freely traverse. The axle-boxes weighed 3½ cwt. each.

It will be understood that when an engine fitted with these boxes enters a right-hand curve the flanges of the leading wheels draw the boxes to the right, so that the engine itself remains a tangent to the curve, whilst, since the axle-boxes are themselves curved, the effect is that the right-hand side axles are brought nearer the rigid wheels, and consequently the radial wheels on the opposite side of the engine further from the fixed wheels, the whole effect of the radial axle-boxes being that the trailing and leading axles actually become radii of the curves being traversed, although the flanges continue parallel to the rails.

Adams's spring tyres require a more precise description, and before we describe them, readers may perhaps be reminded that Adams had strong views on the subject of railway rolling-stock wheels. He enters rather fully into the matter in his book, "Roads and Rails," especially in the chapter dealing with "the mechanical causes of accidents." In this, Adams maintains that the usual forms of wheels are in reality rollers, and not wheels.

The spring tyres had been tried on the North London Railway, Eastern Counties, and on another locomotive on the St. Helens Railway, before the engine now under review was constructed. Upon the coupled wheels of the new locomotive for the latter railway, double spring hoops were employed, the single form having been used in the three previously mentioned engines. The plan adopted was as follows:—

"The tyres chosen were constructed with a deep rib in front; this was bored out, internally, to a depth of $\frac{3}{4}$in., and to a conical section, and, of course, parallel to the tread. A flat edge, $\frac{3}{8}$in. wide, was thus left on either side.

"The springs, formed of tempered hoop steel, were placed on the inner surface of the tyres. Corresponding curves were turned across the outer circumference of the wheels. The wheels were forced into the cones containing the springs, and retained by three 1in. bolts, and a flat ring in the groove at the back of the tyre, the effort of the spring tyres being to allow of a slight lateral motion in running round curves and also to give a better grip of the rails, as the tyres, by reason of the weight upon them being transmitted through the tyre springs, slightly flattened upon the rails, and so presented a larger surface for adhesion between the tyres and rails."

The following interesting account of the working of the radial axle and spring tyre locomotive on the St. Helens Railway is extracted from a paper by Mr. J. Cross, the designer of the locomotive, and read before the Institution of Civil Engineers. Mr. Cross stated that "the engine was completed in the first week of November, 1863, and has since been running very regularly, taking its turn of duty with passenger trains or coal trains, or as a shunting engine; and about the numerous works connected by sharp curves with the St. Helens line. The motion round curves is free from all jerking, and on straight lines the speed is more than 60 miles an hour; either end of the engine being first, without any train behind to give steadiness; and the motion is so smooth that it has only been by taking the actual time that the engineers have convinced themselves of the fact of the speed exceeding 40 miles an hour. It was built to traverse curves of 200ft. radius. This it does with the greatest facility, and it has regularly worked the passenger trains round a curve of 1,000ft. radius, going directly off the straight line by a pair of facing points at a speed of more than 30 miles an hour, and it has gone round curves of 132ft. radius. It has also run a train of 12 passenger carriages, weighted up to 100 tons, exclusive of its own weight, at 60 miles an hour on the level. From the advantages it possesses over the ordinary mixed engines for weighting the trailing coupled wheel, it, without difficulty, on a wet, slippery day, started, and took this load up a gradient of 1 in 70, drawing seven of the carriages with a load weighing 72 tons 5 cwt., up a gradient of 1 in 36, round a curve of 440ft. radius; and coal trains of 250 tons are worked over long gradients of 1 in 200 with the greatest ease.

"It is evident, then, that engines on this principle, affording facilities for the use of high power in hilly countries, are peculiarly adapted for Metropolitan lines, where sharp curves are a necessity (being equally safe whichever end is foremost), and are also well suited for light lines in India and the Colonies. It may likewise be remarked that carriages and wagons on this principle would carry heavier freights, with a saving in the proportion of dead weight, while their friction round curves would be less than at present."

The improvements adopted in the construction of this locomotive for the St. Helens Railway were so successful that, as usual, other claimants, who appropriated the radial axle-boxes as their invention, were soon contending with Adams and Cross as to who was entitled to the honour of introducing the improvement.

The first portion of the Metropolitan Railway was opened on January 18th, 1863, and the line was then worked on the broad-gauge by the Great Western Railway for a percentage of the receipts. The Great Western Railway provided the stations, staff, locomotives, and rolling stock.

Mr. D. Gooch, in 1862, designed a special class of tank engines for working the Metropolitan Railway. They were six-wheel engines, the driving and trailing wheels being 6ft. diameter and coupled. The cylinders were outside. A special form of fire-box and baffle-plate was employed, and tanks were provided beneath the boiler barrel, into which the exhaust steam was discharged by means of a reversing valve fitted to the bottom of the blast pipe. When in the open air, the waste steam escaped up the chimney in the usual manner.

The first of these engines were named: Bee, Hornet, Locust, Gnat, Wasp, Mosquito, Bug, Khan, Kaiser, Mogul, Shah, and Czar. Later ones were named after flowers and Great Western Railway officers.

A dispute arose between the two companies at the beginning of August, 1865, and immediately developed into a complete rupture. The smaller *quasi vassal* railway, through the energy displayed by its chief officers, successfully overcame the apparently insurmountable obstacles that beset it, and consequently the Metropolitan Railway asserted its complete independence of the Great Western Railway, and has since maintained it.

It was indeed a nine days' wonder that the Metropolitan Railway was called upon to perform, for it had to obtain from somewhere locomotives and carriages to work the underground line, commencing on the morning of August 10th, 1863.

Mr. Sturrock, the locomotive superintendent of the Great Northern Railway, had at this time under construction a class of condensing-tank engines that he had designed to work the Great Northern Railway traffic over the Metropolitan Railway. The directors of the Metropolitan Railway in this emergency applied to Mr. Sturrock for assistance, and by working day and night he managed to fit up some Great Northern tender engines with a temporary condensing apparatus.

The difficulty was to provide some kind of condensing apparatus on the Great Northern tender engines, it being necessary to use flexible connecting pipes between the engine and tender strong enough to withstand the steam pressure, but Mr. Sturrock was successful enough to contrive the necessary flexible pipes by which the exhaust steam was conveyed from the engine to the water tank of the tender, but these pipes very frequently burst, and all concerned were far from sorry when the proper engines were delivered.

An order for eighteen had already been placed with a well-known Manchester firm of locomotive builders by the Metropolitan Railway, Beyer, Peacock, and Co. building them from the designs of the late Mr. (afterwards Sir) John Fowler.

The type is well known to London readers, the engines having side tanks, a leading bogie, the wheels of which were 3ft. diameter, with a base of 4ft. The driving and trailing wheels (coupled) were 5ft. 9in. diameter, their base being 8ft. 10in.; the total wheel base being 20ft. 9in., or to centre of bogie, 18ft. 9in. The cylinders were outside, slightly inclined from the horizontal, 17in. diameter, and 24in. stroke. The grate area was 19 sq. ft. The fire-boxes had sloping grates, which were 6in. deeper at the front than the back. The boiler barrel was 4ft. in diameter, and 10ft. 3in. long; it contained 166 tubes, 2in. diameter, the total heating surface being 1,014 sq. ft. The working pressure was nominally 130lb. per sq. in., but when working through the tunnels, condensing the steam, and with the dampers closed, a very much lower pressure resulted. The frames were inside, the dome (fitted with a Salter valve) was on the boiler barrel, close to the smoke-box, a sand-box being also fixed on the boiler barrel at the back of the dome.

The bogie truck was built of plate frames, and was on the Bissell system, turning on a centre-pin fixed to the engine frame, at a radial distance of 6ft. 8in. from the centre of the truck. "Locomotive Engineering" says that "this radial length ensures a nearly correct radiality of the bogie to curves of all radii, the proper length of the radius to ensure exact radiality of the centre of the bogie for all curves being 7ft. 2in., or 6in. more than the actual length—a difference which is, perhaps, of no great importance in practice."

For the purpose of effectually condensing the exhaust steam the side tanks were only filled with water to within 6in. of the top, and the steam was discharged upon the surface of the water, from a 7in. pipe on each side—one to each tank. Into the mouth of these 7in. pipes a 4in. pipe was projected a short distance, and the other end of the 4in. pipe was below the surface of the water, so that a portion of the steam was discharged right into the water in the tanks, and agitated the water sufficiently to prevent the surface of the water from becoming too hot, as would have been the case if the same portion of the water had always been presented to the waste steam. The tanks held 1,000 gallons, and at the end of a journey the water had become too warm to properly condense the exhaust, and it therefore became necessary to quickly empty the tanks and to take in a fresh supply of cold water.

To expeditiously perform the former operation, each tank was provided with a pipe 7in. in diameter; this led to a cast-iron valve-box being placed below the foot-plate. By means of a screw, worked from the foot-plate, a 10in. valve was operated, and the water in the tanks could be discharged into the pits below the engine in the course of some 60 seconds.

The following list gives the names and builders' numbers of the first locomotives constructed for the Metropolitan Railway:

| Engine No. | Name. | Builder's No. | Engine No. | Name. | Builder's No. |
|---|---|---|---|---|---|
| 1 | Jupiter. | 412 | 10 | Cerberus. | 421 |
| 2 | Mars. | 413 | 11 | Lutona. | 422 |
| 3 | Juno. | 414 | 12 | Cyclops. | 423 |
| 4 | Mercury. | 415 | 13 | Daphne. | 424 |
| 5 | Appollo. | 416 | 14 | Dido. | 425 |
| 6 | Medusa. | 417 | 15 | Aurora. | 426 |
| 7 | Orion. | 418 | 16 | Achilles. | 427 |
| 8 | Pluto. | 419 | 17 | Ixion. | 428 |
| 9 | Minerva. | 420 | 18 | Hercules. | 429 |

These engines were fitted with a very small coal bunker, only 18in. wide. Weight of engine in working order: on bogie, 11 tons 3½ cwt.; driving, 15 tons 9½ cwt.; and trailing, 15 tons 10 cwt. Total weight, 42 tons 3 cwt.

Mr. Sturrock's engines for working the Great Northern trains over the Metropolitan Railway were numbered 241 to 250, their leading dimensions being:—Cylinders (inside), 16½in. diameter, 22in. stroke; leading and driving wheels (coupled), 5ft. 6in.; trailing wheels, 4ft. diameter; wheel base, L. to D., 7ft. 6in.; D. to T., 11ft. 9in.; total, 19ft. 3in. Weight, empty, 32 tons 4 cwt. 1 qr.; in working order, 39 tons 12 cwt. 2 qrs.

These Great Northern Railway locomotives were fitted with Adams's radial axle-boxes to the trailing wheels, and commenced working at the end of October, 1865.

The patentee of the Bissell bogie truck did not intend to hide the light of his invention under a bushel, for he advertised the improvement in a truly American style. The following advertisement was to be found in the columns of the sober railway newspapers soon after the Metropolitan locomotives were at work:—

"Important to Railway Directors, Engineers, and the Travelling public.

"No more accidents from engines running off the line (see Queen's letter to Railway Directors copied in the railway papers January 28th, 1866).

"The Bissell bogie, or safety truck, for locomotive engines, so much prized on American and foreign railroads for the great safety and economy it affords on curved roadways, after years of probationary trial in England, has at length been adopted by

John Fowler, Esq., C.E., F.G.S., upon all the new engines, eighteen in number, now working on the Metropolitan Railway, and by Robert Sinclair, Esq., C.E., upon twenty new eight-wheeled engines on the Great Eastern Railway, which may be seen daily. The royalty for the use of the Bissell Patents has been reduced to £10 per engine, so that every engine requiring a bogie underframe should be provided with the Bissell safety truck. Apply to———."

Whilst on the subject of railway advertisements we take the opportunity to record the obituary announcement of the tentative "hot brick" engine, previously referred to, designed to work on the Metropolitan Railway. It appeared in the railway newspapers during the early months of 1865, and was to the following effect: "Metropolitan Railway. One locomotive engine for sale, either entire or in parts. For particulars apply to the Locomotive Superintendent, Bishop's Road, Paddington."

Reference must here be made to Mr. Sturrock's system of steam tenders, as adopted by him to work the heavy coal and goods trains on the Great Northern Railway. In addition to the usual engine, the pistons of a pair of cylinders, 12in. diameter, with a stroke of 17in. actuated the centre axle of the tender, and the six tender wheels were coupled by outside rods. The tender wheels were 4ft. 6in. diameter. The steam tenders weighed about 35 tons, with water and coal, and of this weight over 13 tons was on the driving wheels. After use in the tender cylinders, the exhaust steam was condensed in the tender tank. Forty-six of these steam tenders were constructed, and some are still running, but as simple tenders, the propelling apparatus having been done away with many years ago. Fig. 85 represents a Great Northern engine fitted with one of Sturrock's patent steam tenders.

Mr. Robert Sinclair, whilst locomotive superintendent of the Great Eastern Railway, only designed one type of tank engine, and Neilson and Co. constructed the first of this class in 1864. Twenty of the class were built, being originally intended to work the Enfield Town Branch, but in later years these engines were used on the North Woolwich line. The engines (Fig. 86) were supported by eight wheels, the leading and trailing being 3ft. 7in. diameter, and the driving and back coupled 5ft. 6in. diameter. The cylinders were outside, 15in. diameter, and 22in. stroke. The leading and trailing wheels were fitted with the Bissell truck, referred to in the advertisement just quoted. So that

218  EVOLUTION OF THE STEAM LOCOMOTIVE

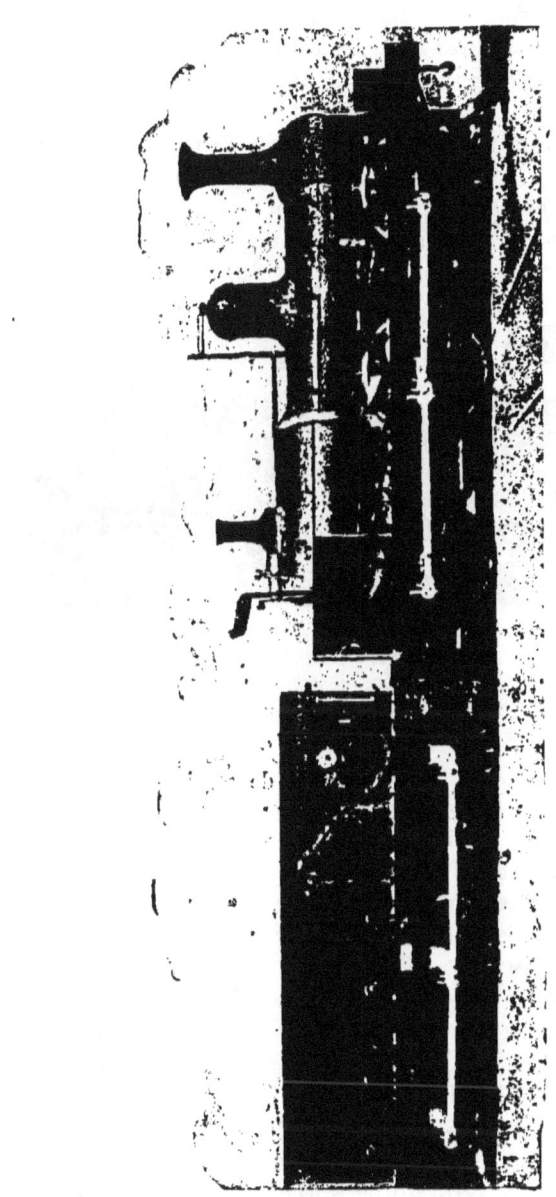

FIG. 85.—A GREAT NORTHERN RAILWAY ENGINE, FITTED WITH ONE OF STURROCK'S PATENT STEAM TENDER

although the whole wheel base was 17ft. 4in., the rigid base—that of the coupled wheels—was only 6ft. The boiler was 13ft. 6in. long, and the water was carried in the tanks beneath the boiler and between the frames. An enclosed cab with front and rear spectacle plates was provided.

This improvement so delighted the Great Eastern Railway drivers that they presented a testimonial to Mr. R. Sinclair in May, 1864, in which they described him as the "inventor" of the weather-board or "cab," as fitted to locomotives. The tank engines in question weighed

FIG. 86.—SINCLAIR'S DESIGN OF TANK ENGINE FOR THE EASTERN COUNTIES RAILWAY

38 tons 6 cwt. 3 qrs., of which weight 20 tons 5 cwt. 2 qrs. was on the coupled wheels.

In January, 1863, Mr. J. B. Fell patented a locomotive designed for working over extremely steep gradients. At that time there was a break 47 miles long in the continuity of the iron road communication between France and Italy by the Mount Cenis route. This break has in later years been abolished by the construction and working of the famous Mount Cenis tunnel. Brassey and Co. in 1863 proposed that during the construction of the tunnel a temporary mountain railway worked on Fell's system should be built over the mountain. An experimental locomotive was, therefore, constructed at the Canada Works, Birkenhead. This engine weighed $14\frac{1}{2}$ tons loaded. The boiler was 2ft. 9in. diameter, and 7ft. $9\frac{1}{4}$ in. long, and contained 100 tubes of $1\frac{1}{2}$in. external diameter. The heating surface was 420 sq. ft., and the grate area $6\frac{1}{2}$ sq. ft. The engine had two sets of machinery—one for working the vertical wheels, acting on the ordinary carrying rails, and the other actuated the special horizontal clutch wheels.

which were pressed against the centre rail. The outside cylinders which worked the four-coupled vertical wheels, of 2ft. 3in. diameter, were 11¾in. diameter, the stroke being 18in. The horizontal coupled wheels were 16in. diameter, with a base of 19in.; these were driven by inside cylinders 11in. diameter and 10in. stroke. A pressure of 12 tons, actuated by means of a screw apparatus, could be applied to the horizontal wheels.

By permission of the London and North Western Railway, an experimental railway, 800 yards long, was laid down upon the Whalley Bridge Incline of the Cromford and High Peak Railway.

The gauge was 3ft. 7⅝in., and there were 180 yards of straight line on a gradient of 1 in 13.5, and 150 yards of curves, with radii of 2½ and 3½ chains, on a gradient of 1 in 12. The third rail upon this line, to be clipped between the horizontal driving wheels of the engine, was laid on its side, 7½in. above the other rails.

In the course of a series of experiments carried on from September, 1863, to February, 1864, the engine, working up to a pressure of 120lb. to the square inch, never failed, with a maximum load of 30 tons, to take a load of 24 tons up the above inclines and round the curves. The outer cylinders working on the four vertical wheels could only draw up, besides the weight of the engine, a loaded wagon weighing seven tons; while the inside cylinders, acting upon the horizontal wheels, which pressed with 12 tons against the middle rail, enabled the engine to take up 24 tons on the same day and under the same conditions. The inside cylinders alone were able to carry up the engine itself, round the curves, and exhibited the power of taking up altogether 17 tons.

The results of the experiments on the High Peak Railway were considered so satisfactory that the line up Mount Cenis was commenced without delay. The engine was not properly adapted for working the mountain traffic, in consequence of the crowded and complicated nature of the machinery, and also because the feed-oil dropped on to the horizontal wheels and lessened the bite on the centre rail. The weight on the horizontal wheels was increased to 16 tons, and an additional pair of guide wheels acting on the centre rail was provided at the trailing end of the engine, after the High Peak experiments.

The Board of Trade was at that time so far interested in railway matters as to send out Captain Tyler, one of its inspectors, to report on the Mount Cenis Railway. We extract from his report the following account of the working of this engine on the mountain railway:—

"In the course of two days I took six trips with this engine up and down the experimental line, carrying each time a load of 16 tons, in three wagons, including the weight of the wagons, and it performed in the ascent 1,800 metres in $8\frac{1}{2}$min., with a loss of 14lb. of steam and of 5 1-3in. of water in the gauge glass, at steam pressure, varying between 92 and 125lb. to the square inch in the boiler, as the average of all those experiments.

"The speed attained was in every case greater than that which it is proposed to run with the same load with the express trains; and the average speed, as above given, was at the rate of 13 1-3 kilometres (or 8 1-3 English miles) per hour, instead of 12 kilometres (or $7\frac{1}{2}$ English miles) per hour, which is the highest running speed allowed in the programme given to the French Government for this part of the line.

"The weather was fine and calm, and the bearing rails were in first-rate order; but the middle rail, as well as the horizontal wheels, were oily, and, therefore, in a condition very unfavourable for good adhesion."

A second engine was built on Fell's system specially for working over the steep Mount Cenis Railway, and in its construction several improvements, suggested by the shortcomings of the first engine, were introduced.

The second engine was built partly of steel, and weighed 13 tons empty, and 16 tons 17 cwt. fully loaded, afterwards increased to 17 tons 2 cwt. The boiler was 8ft. $4\frac{1}{2}$in. long, and 3ft. 2in. in diameter, and contained 158 tubes of $1\frac{1}{2}$in. external diameter. Fire-box and tubes contained altogether 600 superficial feet of heating surface, and there were 10ft. of fire-grate area. There were only two cylinders, with a diameter of 15in. and stroke of 16in., which worked both the four-coupled horizontal and four-coupled vertical wheels, which were all 27in. in diameter. The wheel base of the vertical wheels was 6ft. 10in., and that of the horizontal wheels, 2ft. 4in. The maximum pressure in the boiler was 120lb., and the effective pressure on the piston was 75lb. to the square inch.

Besides possessing a greater amount of boiler power, this engine travelled more steadily than No. 1, its machinery was more easily attended to, and the pressure upon its horizontal wheels could be regulated by the engine-driver at pleasure from the foot-plate. This pressure was applied through an iron rod connected by means of right and left-handed screws, with a beam on each side of the middle rail, and these beams acted upon volute springs which pressed the horizontal wheels against that rail.

The pressure employed during the experiments was $2\frac{1}{2}$ tons on each horizontal wheel, or 10 tons altogether; but the pressure actually provided for, and which when necessary was employed, was 6 tons upon each, or 24 tons upon the four horizontal wheels.

The vertical wheels were worked indirectly by piston-rods from the front, and the horizontal wheels directly by piston-rods from the back of the cylinders.

Having already given Captain Tyler's account of his experiments with the first engine, we cannot do better than reproduce his statement concerning the second of the Fell engines, built for the Mount Cenis Railway.

Captain Tyler stated that with the new engine he "was able to take up 1,800 metres of the experimental line with the same load as before, of 16 tons in three wagons, in $6\frac{1}{4}$ minutes, or at a speed of $17\frac{3}{4}$ kilometres per hour, as against 12 kilometres per hour which it is proposed to run with the express trains. The steam pressure in the boiler fell from 112lb. to $102\frac{1}{2}$lb., and 3in. of water were lost in the gauge-glass, the feed having been turned on during the latter period only of this experiment.

"The engine exerted in this instance, omitting the extra resistance from curves, about 177 horse-power; or, adding 10 per cent. for the resistance from curves, 195 horse-power, or more than 12 horse-power to each ton of its own weight, and nearly 60 horse-power in excess of what was required to take the same load up the same gradient and curves at 12 kilometres per hour, as proposed in the programme. I observed on the following day that 40lb. of steam-pressure in the boiler, or one-third of the maximum pressure employed, was sufficient to move the engine alone up a gradient of 1 in $12\frac{1}{2}$; and the friction of carriages or wagons being proportionately much less than that of an engine, the same engine ought, *à fortiori*, to be able to move a gross load of three times its own weight, or 48 tons, at its greatest working pressure, up the same gradient."

Having now given some details of locomotives constructed for working on a foreign steep grade railway, it will not be out of place to describe the special forms of engines designed for the Welsh narrow-gauge line, usually called the Festiniog Railway. The line has been open for a great number of years, but up to June, 1863, had only been used for conveying slates from the quarries to the shipping port. Horses were employed to haul the empty trucks up to the quarries, the loaded wagons running down to Portmadoc by gravity.

The average gradient for $12\frac{1}{2}$ miles was 1 in 92, the steepest 1 in 60. The radii of the curves ranged between two and four chains. Unlike the Mount Cenis line just reviewed, the Festiniog Railway was worked with locomotives depending solely on the adhesion of the carrying wheels, no central rail being provided. The gauge was 1ft. $11\frac{1}{2}$ in.

The engines were designed by Mr. C. E. Spooner, the engineer of the railway. At first two were constructed, England and Co. being the builders. These miniature iron horses (one was more correctly called the "Welsh Pony") had two pairs of coupled wheels, with a wheel base of 5ft. The cylinders, which were outside the framing, were $8\frac{1}{2}$in. in diameter, with a length of stroke of 12in., and they were only 6in. above the rails.

The maximum working pressure of the steam was 200lb. to the square inch. Water was carried in tanks surrounding the boilers, and coal in small four-wheel tenders.

The heaviest of these engines weighed $7\frac{1}{2}$ tons in working order, and they cost £900 each. They could take up, at 10 miles an hour, about 50 tons, including the weight of the carriages and trucks, but exclusive of that of the engine and tender. They actually conveyed daily on the up journey an average of 50 tons of goods and 100 passengers, besides parcels. Two hundred and sixty tons of slates were taken down to Portmadoc daily. The engines were well adapted for convenience in starting and in working at slow speeds, but their short wheel base and the weight overhanging the trailing wheels gave them more or less of a jumping motion when running.

Safety guards, similar in form to snow ploughs, were afterwards added in front of the engines, behind the tenders, and under the platforms of the break-vans, in consequence of their being so near to the rails.

After a few years' experience of these four-wheel locomotives, the directors of the Festiniog Railway determined to experiment with an engine constructed on Fairlie's double-bogie system, and the "Little Wonder" was constructed. In February, 1870, several trials were made with this engine, when a train of 72 wagons, of a total length of 648ft., and of a gross weight, including the engine, of 206 tons 2 qrs., was drawn up an incline of 1 in 85 at a speed of five miles an hour, the steam pressure being 200lb. per square inch. The "Welsh Pony's" best performance in these trials upon the same gradient, but with a pressure of 150lb., consisted in drawing 26 wagons, the gross load of which, with engine, amounted to 73 tons 16 cwt. Tabulated, the result of these trials were as follows :—

|  | Total resistance. lbs. per ton gross. | Gravity. lbs. per ton. | Frictional resistance. lbs. per ton. |
|---|---|---|---|
| "Little Wonder" | 40 | 26·3 | 13·7 |
| "Welsh Pony" with 150lbs steam | 51·4 | 26·3 | 25·1 |
| Do. „ 130lbs steam | 44·5 | 26·3 | 18·2 |

The general arrangements of the "Little Wonder" may be described as follows. The boiler was double, having two fire-boxes united back to back with two distinct barrels and sets of flue-tubes, and consequently a chimney at each end. A bogie was placed under each barrel, and each bogie had two pairs of wheels coupled together, worked independently by a pair of steam-cylinders to each bogie. Thus a total wheel base of 19ft. 1in. in length was covered by the bogies; each bogie had a 5ft.-wheel base, and the distance between the centres of the bogies was 14ft. 1in. The four cylinders were 8 3-16th in. in diameter, and had a stroke of 13in.; the wheels were 2ft. 4in. in diameter. The combined grate area was 11 sq. ft., and the heating surface 730 sq. ft. Fairlie's system of double engines soon came into repute for working steep gradients, and many very powerful engines were and are still constructed on his system for use on foreign railways. Fairlie, in conjunction with Samuels, adapted his system to a species of combined locomotive and carriage, and, in 1869, one was constructed for working on the London, Chatham, and Dover Railway between Swanley Junction and Sevenoaks. Seven passenger compartments were provided in this vehicle, accommodation comprising seats for 16 first-class and 50 second-class passengers; its total length was 43ft., and weight, empty, 13½ tons. The leading end was supported by the engine bogie, and the trailing end by an ordinary bogie truck. Curves of only 50ft. radius were easily passed over by the combination vehicle.

Leaving Fairlie and his combinations, both of locomotives and carriages, and also of double locomotives, we now glance at a class of tank engines designed by Cudworth for working the trains between Cannon Street and Charing Cross upon the opening of the former terminus in 1866. These engines were seven in number, and were constructed at the Canada Works. They were of the "coupled in front" pattern, with a trailing bogie. The cylinders were inside, 15in. diameter and 20in. stroke. The coupled wheels were 5ft. 6in. diameter. Outside frames were employed, and also compensation beams both to the coupled and bogie wheels. The coal bunker, with water-tank under, was of exceptional length. It was always a puzzle to the writer as to how a stout driver could manage to squeeze through the narrow entrances to the foot-plate, especially as these apertures were situate at the side of the fire-box ; but evidently the "trick was done " by following the axiom, "Where there's a will there's a way," and doubtless the drivers, if asked, would have replied, "It's very easy if you only know the way." These South Eastern Railway locomotives were numbered 235 to 241.

Mr. Wm. Cowan, locomotive superintendent of the Great North of Scotland Railway, designed a class of engine, which Neilson and Co. constructed. The design was stated to be that of a "goods" locomotive, but upon examination we find the engines in question to be no other than the popular four-coupled behind, with a leading bogie and outside cylinders. The latter were arranged in a horizontal position immediately below the frames. The coupled wheels were 5ft. 6½in. diameter, with underhung springs connected by means of an equalising lever-beam. The bogie wheels were 3ft. in diameter, with a base of 6ft. Inside bearings were supplied to the bogie axles. The boiler barrel measured 10ft. 10½in. between the tube-plates, its external diameter was 4ft. 1in., and it contained 206 tubes of 1¾in. diameter. The engine was fitted with D. K. Clarke's system of smoke consuming apparatus, previously described. The fire-box was of the raised pattern, and the steam dome was placed on it. The engine weighed 39 tons 13 cwt., and the tender 27 tons, in working order.

In general appearance this "goods" engine resembled in a remarkable degree the London and South Western Railway express passenger engines as built by Mr. Adams. The tender was carried on six wheels.

Q

Fig. 85 represents Beattie's standard design of goods engine for the L. and S.W.R. in 1866, the wheels were 5ft. 1in. diameter, the cylinders being inside, and having a diameter of 17in., the stroke 24in. Beyer, Peacock and Co. were the builders. Fig. 86 represents an engine of this class as rebuilt some years later at Nine Elms Works.

In 1868. Mr. W. Adams placed upon the North London Railway the first locomotive constructed from a design which has, in its broad features and general outline, ever since been a model of simplicity, attractiveness, and utility, showing, as the design does, what engines constructed to work important local traffic should be like.

In its original form there were some points that need alteration, as they certainly spoilt the general symmetrical effect of an otherwise

FIG. 85.—BEATTIE'S STANDARD GOODS ENGINE, L. & S.W.R., 1866

artistic appearance. We may as well allude to these defects at once, and then proceed to detail the locomotive.

The first of such blots on the design was the placing of a cylindrical sand-box on the top of the boiler barrel, between the chimney and the dome. To show that such a position for this useful appendage was not necessary, we mention that only the driving wheels were supplied with sand from this unsightly excrescence, the supply of sand for the trailing wheels (for use when running bunker in front) being placed in an unobtrusive position. If the latter sand-boxes could thus be located, why was it necessary to place that for the leading wheels in so conspicuous a position? This example of awkward location of so useful an adjunct is further emphasised when we remember that these engines run just as frequently bunker first as chimney first.

Further, in consequence of the position of this sand-box, the rod for working the sand valves was carried along the top of the boiler barrel, several inches above its surface, thus still more detracting from the symmetry of the design. The other feature we wish to allude to, is the shape of the dome cover, the whole of which was of a needlessly ugly contour. Then, again, in later years an enclosed cab was added, the back and front of which, being of sheet-iron, extending to the extreme of the coal bunker, and with no return sides, has given a rather toy-like appearance to these otherwise fine locomotives. We are glad to be able to mention that when these engines were rebuilt, the objectionable sand-box was removed, and a more pleasing form of steam dome provided, but this improvement was

FIG. 86.—BEATTIE'S GOODS ENGINE, L. & S.W.R., REBUILT

in a great measure negatived by the black enamelled iron which is now used for the cover in place of the bright brass formerly employed for the purpose.

Having thus mentioned the defects in appearance, rather than utility, of the North London Railway passenger tanks (Fig. 87), we can proceed to do justice to this really fine class of engines designed by Mr. Adams.

The outside cylinders were 17in. diameter, and the stroke was 24in. The driving and trailing wheels (coupled) were 5ft. 3in. diameter, the bogie wheels being 2ft. 9in. diameter. The heating surface was 1,015 sq. feet. The boiler was 4ft. 1in. diameter, and contained 200 tubes of 1¾in. diameter. A good feature in the design was the high steam pressure employed —viz., 160lb. per sq. in.—and there can be no doubt that much of

228   EVOLUTION OF THE STEAM LOCOMOTIVE

the success of this class of engine can be traced to the use of so high a pressure of steam at a time—29 years ago—when other lines were using a much lower pressure. Indeed, to-day it is only necessary

FIG. 87.—ADAMS'S PASSENGER TANK ENGINE, N.L.R., AS REBUILT BY MR. PRYCE.

to watch a North London and any of several other railway companies' trains starting side by side, and it will be observed that the North London generally gets away first; these engines are, in fact, capital at starting, and soon attain a high rate of speed.

# EVOLUTION OF THE STEAM LOCOMOTIVE 229

The weight was as follows:—

|  | Empty. tons cwt. | Loaded. tons cwt. |
|---|---|---|
| On bogie wheels | 15 14 | 14 14½ |
| On driving wheels | 11 11 | 14 5 |
| On trailing wheels | 11 7 | 14 12½ |
| Total | 38 12 | 43 12 |

FIG. 88.—PRYCE'S 6-COUPLED TANK GOODS ENGINE, NORTH LONDON RAILWAY.

It will be observed that, when empty, the bogie axles supported 19½ cwt. more of the gross weight than when the engine was in working order.

The wheel base of the bogie was 5ft. 8in. The coupled wheels have underhung springs connected by a compensation beam. India-rubber springs are used in connection with the hanging of the springs, and also to guide the bogie, etc., and it was found that such springs answered the use to which they were put in a most admirable manner.

In all the new engines that have lately been built, and when rebuilding old engines of this type, the cylinders have been increased to $17\frac{1}{2}$in. diameter, and other things considerably modified in detail.

Mr. Pryce has also built 24 powerful six-wheel tank engines (Fig. 88) for dealing with the N.L.R. goods traffic. These engines are very

FIG. 89.—LOCOMOTIVE AND TRAVELLING CRANE, N.L.R.

efficient. They have outside cylinders 17in. diameter, 24in. stroke, and 4ft. 4in. coupled wheels. Boiler pressure, 160lb. per sq. in. Weight in working order, 45 tons 9 cwt., all available for adhesion. The total wheel base is 11ft. 4in.; consequently, they take curves easily.

The coal consumption of these engines was very satisfactory. The trains of the North London Railway consist of twelve vehicles, weighing, empty, 90 tons 14 cwt., and loaded 112 tons 6 cwt., but the coal consumption, with very frequent stoppages, only averaged 30.28lb. per mile.

Fig. 89 represents the combined saddle-tank locomotive and crane belonging to the North London Railway, as recently rebuilt by Mr. Pryce.

## CHAPTER XIII.

Beattie's express engines—Kendall's three-cylinder engine for the Blythe and Tyne Railway—Heavy engines for the Metropolitan and St. John's Wood Railway—Sold to the Taff Vale Railway—"The most powerful locomotive in the world" for sale—"Jinks's Babies"—The "Areo-steam" locomotive on the Lancashire and Yorkshire Railway—Tank engines on the Furness Railway—Patrick Stirling's world famous " 8ft. singles " for the G.W.R.—Webb's " Precedents " for the L and N.W.R.—The " John Ramsbottom "—" The Firefly," an engine that has " played many parts "—J. Stirling's 7ft. coupled engines on the G. and S.W.R.—Stirling's reversing apparatus—Watkin's express engines for the S.E.R.—Stroudley's " Grosvenor," L.B. and S.C.R.—The era of "compounds "—W. F. Webb's first compound locomotive—Bowen-Cooke's views on the subject—The " Experiment "—7ft. 1in. compounds—" Queen Empress "—" Black Prince "—Wordsell compounds—Midland coupled expresses—Stroudley's " Gladstone " class—The " General Managers " on the North Eastern—N.B.R. locomotive, " No. 592 "—Holmes's " 633 " class—Great Eastern 7ft. coupled—Holden's liquid fuel locomotives—Serve tubes in locomotives—Sacre's 7ft. 6in. " Singles."

Fig. 90 represents the "Python," one of J. Beattie's four-coupled express engines, constructed for the L. and S.W.R. The cylinders were outside, 17in. diameter by 22in. stroke. The coupled wheels were 7ft. 1in. diameter, and the leading wheels 4ft. diameter. The heating surface was 1,102 sq. ft. Weight of engine in working order, 35 tons 11 cwt. For some years this class of engine was the favourite express engine on the L. and S.W.R.

Locomotive engineers have always one great difficulty to provide for—viz., the extra power required to start locomotives, especially on steep inclines, and as such grades are particularly *en évidence* on the mineral lines, it is not surprising to find Mr. W. Kendall, of Percy Main, Northumberland, patenting a locomotive designed to overcome the defects just indicated. The patent is dated October 26th, 1867. The engine was of the three cylinder type, with one inside and two outside cylinders. When running on a level road only the inside cylinder was used, but for starting or ascending inclines the power of all three was brought into use, the whole arrangement of the power being actuated by the reversing gear apparatus. By a peculiar adaptation of the lap of the valves, a small quantity of steam was admitted to the valves of the outside cylinders when these cylinders were not working, for the purpose of lubrication. The engine in question was built at the Percy Main Works of the Blythe and Tyne Railway. She was of the "four-coupled behind" type, with a single pair of leading wheels. The inside cylinder was connected in the usual manner to the cranked axle of the centre wheels, the outside cylinders actuating the trailing pairs of wheels. Without diagrams it is rather difficult to explain the method employed to prevent the pistons, etc., of the outside cylinders from reciprocating, but shortly it may be stated that the connecting-rod was divided into two pieces, and at the joint each end fitted into an enclosed link. When disconnected, that portion of the rod coupled to the wheels which was in the link merely travelled up and down the link, whilst the part connected with the piston, etc., was

FIG. 90.—"PYTHON," A 7FT. 1IN. COUPLED EXPRESS ENGINE, LONDON AND SOUTH WESTERN RAILWAY

at rest. By means of a screw gear this latter portion of the connecting-rod was lowered in the link, and engaged with the other part of the rod, which was coupled to the wheels, and so the outside cylinders were brought into action. If required, the outside cylinders could be used independently of the one inside cylinder, so that the engine could be a one, two, or three cylinder locomotive. Separate regulators were provided for the inside and outside cylinders, but the handles were coupled together, so that, if required, one movement actuated the admission of steam to all the cylinders. To prevent too strong a blast, the driver could, by the operation of a ball valve, discharge the exhaust steam from the outside cylinders into the atmosphere by means of a pipe in front of the engine. On the other hand, the whole of the exhaust from the three cylinders could be discharged up the chimney in the usual manner if preferred.

Upon April 13th, 1868, the Metropolitan and St. John's Wood Railway was opened for traffic. The line branches from the Metropolitan Railway at Baker Street, and was worked by the Metropolitan Company. The gradients on the short line are very severe, and it was not considered advisable to attempt to work the railway by the usual type of engine employed on the underground line; so Mr Burnett, the then locomotive superintendent of the Metropolitan Railway, designed a special class of engine for the St. John's Wood Railway. These were constructed by the Worcester Engine Company, and were numbered 34 to 38. They were provided with six coupled wheels of 4ft. diameter, with outside bearings; the cylinders were 20in. diameter, with a 24in. stroke; they were placed within the frames at 2ft. 2in. centres. The wheel base of these powerful locomotives was divided as follows:—L. to D., 6ft. 10in.; D. to T., 7ft. 2in. The boiler was 11ft. long, and 4ft. 3in. diameter, and contained 176 tubes of 2in. diameter.

The fire-boxes were exceptionally large, the measurements being: Length, outside 7ft. 1in., inside 6ft. 6in.; width, outside 4ft., inside 3ft. 6in. The depth was 5ft. 5in. in front, sloping to 3ft. 11in. at back. The steam pressure was 140lb.; heating surface, 1,165 sq. ft.; grate area, 22½ sq. ft. The water capacity of the tanks was 1,000 gallons.

These mammoth engines weighed 46 tons in working order, and it was soon discovered that they were far too powerful for working the light traffic over the St John's Wood line, the ordinary type of Metropolitan locomotives being quite capable of successfully working the trains over these inclines. So, in 1873, when the Taff Vale Railway was in urgent need of some powerful engines for hauling the heavy coal trains over the Penarth Dock lines, the Metropolitan Railway succeeded in disposing of these five engines to the South Wales Company, and they can still be seen employed on work more adapted to their construction than was that of hauling light passenger trains on the St. John's Wood Railway.

It is evident that both the patentee and builders of the "double bogie" locomotives had a very exalted opinion of the capabilities of these peculiar engines. In December, 1870, G. England and Co.

were advertising for sale by private tender to the best bidder "the most powerful locomotive at present known upon any railway in the United Kingdom, irrespective of gauge."

This "most powerful" locomotive was constructed for the 4ft. 8½in. gauge on Fairlie's double bogie system. She had four cylinders, 15in. diameter and 22in. stroke, eight wheels, all drivers of 4ft. 6in. diameter, and with steel tyres.

Amongst other useful features claimed for this "most powerful" locomotive, we read that she "would take a load up an incline at a speed exceeding that of any other engine at present known, and would round the sharpest curves with ease."

"Jinks's Babies" consisted of a batch of ten engines constructed towards the end of 1871, and early in 1872. They had outside cylinders, 17in. diameter and 30in. stroke, with a leading bogie and four coupled wheels of 7ft. diameter; they had, perhaps, as good a right to the title "most powerful" as the Fairlie engine just mentioned.

Be this as it may, however, "Jinks's Babies" were not successful. They were built at the Stockton and Darlington Locomotive Works, at Darlington, and originally numbered 238 to 240, etc., and upon the consolidation of the North Eastern Railway were renumbered 1238 to 1240, etc. They were rebuilt by Mr. Fletcher as six-wheel engines, the bogie giving place to a single pair of leading wheels, and the stroke of the pistons was reduced from 30in. to 26in. Even after this metamorphosis, "Jinks's Babies" could not be truthfully described as successful locomotives. Amongst other peculiarities the circular valves should be enumerated. The steam pressure was 140lb. per sq. in.

In 1871 the Lancashire and Yorkshire Railway fitted up an engine with an apparatus said to have been invented by Mr. Richard Eaton, but called "Warsop's Areo-Steam system," by means of which a continuous supply of heated air was forced into the bottom of the boiler, so causing the water to be continually agitated, and thereby preventing incrustation of the metal, as well as more quickly generating steam, and last—but far from least—economising the fuel. The engine experimented upon was a six-coupled goods, No. 369, with cylinders 15in. by 24in., 5ft. wheels, and working at a pressure of 130lb. per sq. in. An air pump, single acting, 6in. diameter by 2ft. stroke, with piston and metallic rings, driven from one of the main cross-heads, was secured to the framework of the engine in the place originally occupied by the feed pump. The compressed air passed along a pipe 1½in. in diameter, 6ft. long, to a coil of 1½in. lap-welded iron pipe, within the smoke-box, 61ft. in length, so arranged as to avoid contact with the blast pipe or the ashes deposited in the smoke-box by the action of the blast. After traversing the coil, the expanded air became heated to a temperature nearly as high as that of the waste gases, and thus ranging between 500 degrees and 800 degrees, or 850 degrees Fahr., lifted the self-acting valve, and entered the perforated distributing pipe within the boiler, and was constantly passing in jets through the water to the steam space, whence the combined powers of steam and air proceeded to the cylinders to carry out their duty. A very simple apparatus was used when desirable to stop compression, by keeping the inlet valve open when steam was shut off; otherwise an undue proportion of air would enter the boiler, and impede the feed-water injectors.

At the same time, occasions arose where a judicious use of the air injection was made with great advantage, even with steam shut off. It is stated that "on March 21st, 1872, there was a heavy fall of snow, and the driver of No. 369 had to make the most of his resources. In coming down Rainford bank he had but 100lb. of steam at Balcarres siding, with steam shut off. He allowed the air pump to continue work, and in 400 yards his gauge rose to 140, when he opened his regulator again to mount the incline with his heavy load, and so successfully gained the summit." The annexed table shows the working of engine No. 369, with and without the apparatus, and also of an exactly similar engine, No. 38, employed on the same length of line, and hauling the same trains. No. 38 was not fitted with the apparatus.

| Engine. | Miles run. | Coal consumed. | Average lbs. per mile. |
|---|---|---|---|
|  |  | Tons. cwt. |  |
| 369 (without apparatus) ... ... | 21,948 | 403   6 | 42·92 |
| 369 (with apparatus) ... ... | 27,934 | 472   10 | 37·89 |
| 38 (without apparatus) ... ... | 28,053 | 550   10 | 43·95 |

Although the above glowing statement is made about this invention, which was fitted to no less than six engines, and tried for a period of about five years, it was not found to be commercially successful, the power consumed in working the pump, and the cost of repairs running away with the economy supposed to have been gained in the original experiment.

About this time the Furness Railway introduced a powerful design of six-coupled tank engines. The cylinders were: Inside, 18in. diameter and 24in. stroke; heating surface tubes, 1,048 sq. ft.; fire-box, 96 sq. ft.; grate area, 15 sq. ft. The frames were "inside." The side-tanks were capable of containing 1,000 gallons of water. Weight in working order: L., 13 tons 13 cwt.; D., 16 tons 6 cwt.; T., 14 tons 15 cwt.; total, 44 tons 14 cwt. On the level this class of engine hauled 372 tons at 20 miles an hour, and up an incline of 1 in 80 a load of 367 tons was drawn at $11\frac{3}{4}$ miles an hour. The steam pressure was 145lb., and the coal consumption 40.16lb. per mile.

The name of Patrick Stirling, the late locomotive superintendent of the Great Northern Railway, will long be remembered and held in high honour amongst those of his *confrères*, consequent upon his successful design of 1870, in which year he built the first of his now world-famous 8ft. 1in. singles, a type of locomotive which immediately leaped into public favour, which for elegance and simplicity of design it is not saying too much in stating that no modern engine has surpassed or is likely to surpass. These engines soon showed the travelling public that really express speed could be safely indulged in for continuous runs of great length without fear of accident or failure. Indeed, modern express speed can date its foundation from the introduction of these engines. The Great Northern Railway undoubtedly owes its popularity and fame as the "express" route to the successful running of Patrick Stirling's 8ft. 1in. outside cylinder "single" engines.

The following may be accepted as a correct description of the earlier type of this locomotive design. Later engines of the same class have, in common with the development of locomotive design, increased in weight, grate, tube, and cylinder area, and steam pressure; but the general outline to-day, as seen in Fig. 91, is the same as that of 27 years ago, and we do not think this compliment can be paid to the design of any other locomotive built at the present time. The cylinders were 18in. diameter, with a length of stroke of 28in. The small ends of the connecting-rods were furnished with solid bushes of gun metal. The inner and the outer fire-boxes were connected together by stays,

EVOLUTION OF THE STEAM LOCOMOTIVE 237

screwed into each of the plates, without the intervention of iron girder bars. By this arrangement, which had been in use for some time in Belgium, the large amount of deposit generally existing upon girder-boxes was prevented, the facility for cleansing was much greater, and the liability of the tube holes in the copper-plate to become oval had been got rid of.

The heating surface in this engine was, in the tubes, 1,043, and in the fire-box 122 sq. ft. The fire-grate had an area of 17.6 sq. ft.

FIG 91.—8FT. 1IN. "SINGLE" EXPRESS ENGINE, GREAT NORTHERN RAILWAY

When the engine was in working order, the weights upon the driving, trailing, and bogie wheels were 15, 8, and 15 tons respectively. The distance from the centre of the trailing wheels to the centre of the bogie pin was 19ft. 5in. These engines were capable of drawing a weight of 356 tons on a level at a speed of 45 miles an hour, with a working pressure of 140lb. to the sq. in. The consumption of coal, with trains averaging sixteen carriages of 10 tons weight each, had been 27lb. per mile, including getting up steam and piloting. The cost of maintaining and renewing passenger engines on the Great Northern Railway was in 1873 estimated to amount to 2½d. per mile.

The contemporary type of engines on the "West Coast"

route was the celebrated "Precedent" class, illustrated by "John Ramsbottom" (Fig. 92).

These London and North Western Railway locomotives were constructed at the Crewe Works, from the designs of Mr. F. W. Webb,

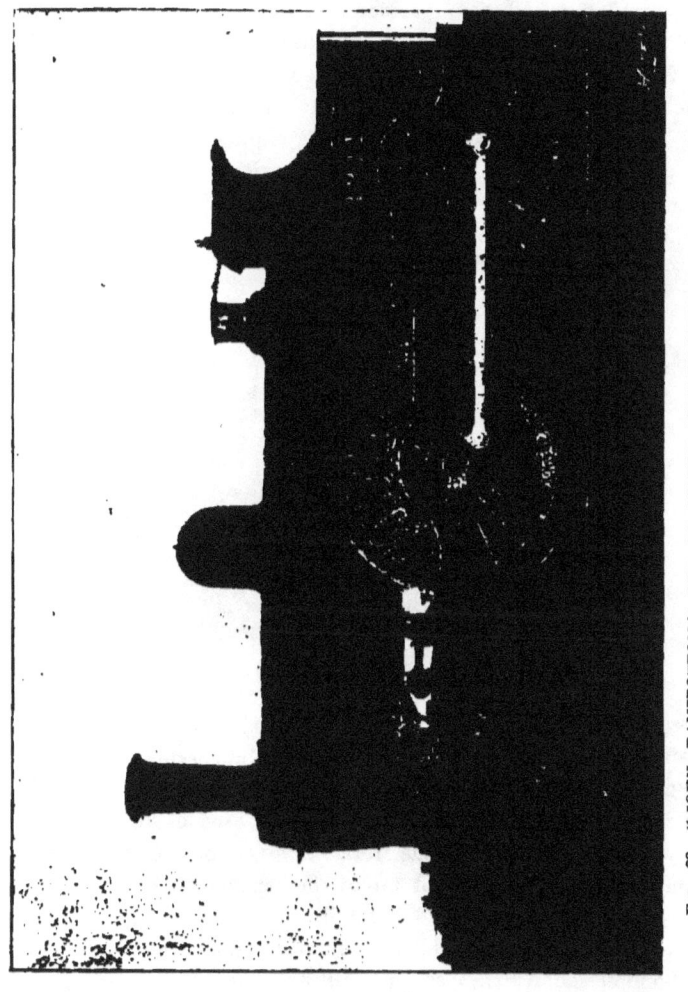

FIG. 92.—"JOHN RAMSBOTTOM," ONE OF WEBB'S "PRECEDENT" CLASS, L. & N.W. RWY.

locomotive superintendent, the first of them being constructed in December, 1874. The engines, as our readers well know, have four-coupled wheels, 6ft. 6in. diameter, and a leading pair of wheels 3ft. 9in. diameter. The principal dimensions originally were—they may vary a little in some details in certain engines—inside cylinders, 17in. diameter, with a stroke of 24in. Heating surafce, 980 sq. ft. in tubes, and 103.5 sq. ft. in fire-box; grate area, 17.1 sq. ft.; weight in working order, L., 10 tons 5 cwt.; D., 11 tons 10 cwt.; T., 11 tons; total, 32 tons 15 cwt. Steam pressure, 120lb.; wheel base, 15ft. 8in.

The most famous engine of this class is the "Charles Dickens," No. 955, built at Crewe in 1882; the "Inimitable" is shedded at Manchester, and the daily journey to and from Euston consists of 366½ miles; the trains worked by this engine are the 8.30 a.m. up, and the 4 p.m. down. As long ago as September 21st, 1891, the "Charles Dickens" had obtained the premier position in engine mileage. On that day "she"—if the shades of

FIG. 93.—"FIREFLY," A L. & S.W.R. OUTSIDE CYLINDER TANK ENGINE

"Boz" will allow the bull—completed her millionth mile, consisting of 2,651 trips between Manchester and London, in addition to 92 other journeys. During this period of 9 years 219 days the engine had burned 12,515 tons of coal. Up to the end of February, 1893, the total mileage of "Charles Dickens" amounted to 1,138,557, and up to the present time it has exceeded the enormous total of 1,600,000 miles!

In April, 1874, Mr. Webb introduced another type of locomotive for the London and North Western Railway: the "Precursor," No. 2145, gives its title to the design in question.

The cylinders were 17in. by 24in. stroke. The leading wheels were 3ft. 6in. diameter, whilst the driving and trailing wheels (coupled) were 5ft. 6in. diameter. The tubes contributed 980 sq. ft., and the fire-box 94.6 sq. ft., of the heating surface. The weight in working order was 31 tons 8 cwt.

"Firefly" (Fig. 93) is one of the numerous six-wheel outside cylinder tank engines built from the designs of J. Beattie by Beyer, Peacock and Co. for the L. and S.W.R. between 1863 and 1875. The cylinders were 15½in. by 20in. stroke, the leading wheels 3ft. 7¾in. diameter, and the coupled wheels 5ft. 7in. diameter. The heating surface was 795.17 sq. ft.; the weight, in working order, 34 tons 12 cwt. A number of these engines had the cylinder diameter increased to 16½in, and a tender added by W. Adams in 1883. "Firefly" was built in 1871.

FIG. 94.—" KENSINGTON," A 4-COUPLED PASSENGER ENGINE, L.B. & S.C R.

"Kensington" (Fig. 94), a L., B. and S.C. locomotive, was, in December, 1872, rebuilt by Mr. Stroudley in the form illustrated. The cylinders were 17in. by 24in. stroke; coupled wheels, 6ft. 6in. diameter; leading wheels, 4ft. 3in. diameter. In 1872, "Kensington" was domeless, that appendage being added later. This engine was originally a single engine, built by R. Stephenson and Co. in 1864. Altogether, this engine, like many individuals, has "played many parts."

We have now reached a period in locomotive history when the engines to be described are of comparatively modern construction, a very large proportion of them being still in work on the various lines of railway, and readers interested in such matters are probably acquainted with the particulars of the locomotives. Under such circumstances, a detailed and particular account of each design would be rather wearisome, therefore the general features of modern engines will be less fully described. At the same time any uncommon points in their design or construction will be mentioned.

The standard type of express passenger engines now used on the South Eastern Railway has developed from a class introduced by Mr. J. Stirling, when locomotive superintendent of the Glasgow and South Western Railway.

In 1873 he constructed at the Kilmarnock Works an engine with a leading bogie and four coupled wheels of 7ft. diameter. The cylinders were inside, 18in. diameter and 26in. stroke. In this design, as in the later type on the South Eastern Railway, the boiler was unprovided with a dome, but in the latter the duplex safety valve is placed about the centre of the boiler barrel, whilst on the Glasgow and South Western Railway engines it surmounts a flush-top fire-box.

Mr. Stirling's reversing apparatus is a very useful contrivance; it enables the driver to reverse his engine without the expenditure of any muscular power. At first the new reversing gear was frequently mistaken for the Westinghouse air-brake pumps. It consists of two vertical cylinders placed tandem fashion at the side of the boiler barrel. One piston-rod passes through both cylinders, and the pistons are attached to it; this rod is connected with the reversing apparatus. One cylinder contains steam, the other oil. The duty of the latter is to prevent the movement of the piston or rod. It will be understood that, since the cylinder is quite full of oil, it is impossible for the piston and connections to move unless the oil can pass from one side of the piston to the other.

This is accomplished by a handle, which also actuates the valve of the steam reversing cylinder so that when the steam is admitted into one cylinder to move the piston, the oil is at the same time permitted to flow through a valve to the other piston, and the reversing apparatus is worked.

The oil keeps the piston in any desired position. As soon as the oil cannot pass from one side of the piston face to the other, the gear is firmly locked.

Mr. A. M. Watkin became locomotive superintendent of the South Eastern Railway in 1876, and he introduced a very pretty design of express passenger engines. Twenty engines of the type were constructed: Nos. 259 to 268 by Sharp Stewart, and Co., and Nos. 269 to 278 by the Avonside Engine Company. Inside frames were provided; the leading wheels were 4ft. and the four-coupled wheels

R

6ft. 6in. diameter. The cylinders were 17in. diameter and 24in. stroke. The weight in working order was 34½ tons; the total heating surface, 1,103½ sq. ft. The splashers to the coupled wheels were of open-work design. The chimney was of the rimless South Eastern pattern; a dome was provided on the centre of the boiler barrel, and a duplex safety-valve on the fire-box top. A cab very much resembling the standard London and North Western Railway pattern was fitted to the engines.

Several of these engines, as rebuilt by Mr. Stirling, remain in work at the present time; they are principally employed on the Mid-Kent services.

In 1874 Mr. Stroudley, the then locomotive superintendent of the London, Brighton and South Coast Railway, built the "Grosvenor" with 6ft. 9in. single driving wheels, inside cylinders 17in. by 24in., and a total heating surface of 1,132 sq. ft. The "Stroudley" speed indicator was fitted to this engine. On August 13th, 1875, the "Grosvenor" conveyed a train from Victoria to Portsmouth (87 miles) without a stop. This was the first occasion on which such a trip had been performed; the time taken was 110 minutes.

No other engine exactly similar to the "Grosvenor" was constructed; but in 1877 the "Abergavenny"—with 6ft. 6in. single drivers and cylinders 16in. by 22in.—was built, and in 1880 the first of the "G" class of singles was turned out at the Brighton Works. These also have single driving wheels 6ft. 6in. in diameter, but the cylinders are 17in. diameter, the stroke being 24in. The weight on the driving wheel is 13 tons.

An interesting era in the evolution of the steam locomotive is at this point arrived at—viz., the first really practical trial of compound engines, or the use of steam twice over for the purpose of propelling a locomotive.

To Mr. Webb, the chief mechanical engineer of the London and North Western Railway, is due the honour of introducing the compound system on an extended scale in railway practice. Although 21 years have now passed since the premier attempt of giving the system a fair trial on an English railway was made, it does not seem to have gained much favour with English locomotive engineers. Indeed, at the present time, excepting a few minor trials elsewhere, the London and North Western Railway is the only company that constructs and uses compound locomotives.

Mr. Webb employs the three-cylinder type of engine, which is an adaptation of the system introduced by M. Mallet on the Bayonne and Biarritz Railway. Three engines were built from Mr. Mallet's design by Schneider and Co., Creusot, and were brought into use in July, 1876. In these locomotives Mallet employed two outside cylinders, one being 15¾in. and the other 9½in. diameter.

Mr. Webb uses three cylinders: an inside cylinder for the l.p. steam and two outside cylinders for the high-pressure steam. But at first one of Trevithick's old "single" engines was fitted up on Mallet's two-cylinder plan. This was in 1878. The engine worked successfully for five years on the Ashley and Nuneaton branch of the London and North Western Railway, and thereupon Mr. Webb decided to construct compound engines on his three-cylinder system.

The first of such engines was the "Experiment." Her outside h.p. cylinders were 11½in. diameter, the inside l.p. being 26in. diameter. Joy's celebrated valve gear was employed to regulate the admission of steam to the cylinders.

"Webb" compounds have two pairs of driving wheels, but these are uncoupled, so that practically the engines are "singles." Whether the four driving wheels work well together, or whether, on the other hand, there exists a considerable amount of either slip or skidding is another matter. The trailing pair of wheels is driven from the h.p. cylinders, and the middle pair from the inside or l.p. cylinder.

Mr. Bowen-Cooke, an authority on London and North Western Railway locomotive practice, sums up the advantages of the "Webb" compound system under the five following heads:—

1. Greater power.
2. Economy in the consumption of fuel.
3. The whole of the available power of the steam used.
4. A more even distribution of the strains upon the working parts, and larger bearing surfaces for the axles.
5. The same freedom of running as with a single engine, with the same adhesion to the rails as a coupled engine.

The 6ft. wheel type of London and North Western Railway compound was introduced in 1884. The outside cylinders are 14in. and the inside 30in. diameter, stroke 24in. Joy's gear is used for all the valves; the valves to the outside cylinders are below, and the valve of the l.p. cylinder is above the cylinder. The boiler steam-pressure is 175lb. per square in., but it is reduced to 80lb. when enter-

R 2

ing the low-pressure cylinder. The weight of the engine in working order is 42 tons 10 cwt. Heating surface: Tubes, 1,242 sq. ft.; firebox, 159.1 sq. ft.; total, 1,401.5 sq. ft. Grate area, 20.5 sq. ft.

FIG. 95.—"TEUTONIC," A L. & N.W.R. "COMPOUND" LOCOMOTIVE ON WEBB'S SYSTEM

An engine built to this design—the "Marchioness of Stafford"—was exhibited at the London Inventions Exhibition of 1885, and gained the gold medal.

In 1890 the first of the "Teutonic" (Fig. 95) class of 7ft. 1in. compounds was constructed at Crewe Works. The leading wheels of this

## EVOLUTION OF THE STEAM LOCOMOTIVE 245

type are 4ft. 1½in. diameter. Total weight in working order, 45 tons 10 cwt. In these engines Mr. Webb's loose eccentric motion is used for the low-pressure inside cylinder, but Joy's gear is retained for the h.p. outside cylinders.

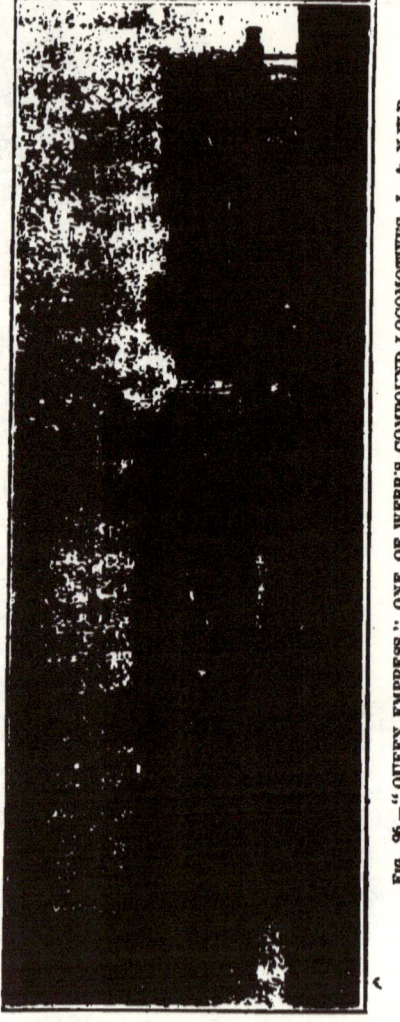

FIG. 96.—"QUEEN EMPRESS," ONE OF WEBB'S COMPOUND LOCOMOTIVES, L. & N.W.R.

Another type of compound is the "Greater Britain." During 1897 the "Greater Britain" and other engines of the class were coloured red, white, and blue, and employed to haul the Royal train when travelling over the London and North Western Railway system. They were then nicknamed the "Diamond Jubilees."

The special feature of this class is the length of the boiler, which is divided into two portions by means of a central combustion chamber. The heating surface is: fire-box, 120.6 sq. ft.; combustion chamber, 39.1 sq. ft.; front set of tubes, 875 sq. ft.; back set of tubes, 506.2 sq. ft.; total, 1540.9 sq. ft. The two pairs of driving wheels are located in front of the fire-box, and in addition there are a pair of leading and a pair of trailing wheels.

An engine of this class—the "Queen Empress" (Fig. 96)—was exhibited at the World's Fair held at Chicago in 1893. Her leading dimensions are: Two high-pressure cylinders, 15in. diameter by 24in. stroke; one low-pressure cylinder, 30in. diameter by 24in. stroke; wheels—driving, 7ft. 1in. diameter (four in number); leading, 4ft. 1½in. diameter; trailing, 4ft. 1½in. diameter. Weight on each pair of driving wheels, 16 tons. Total weight of engine in working order, 52 tons 15 cwt. Total wheel base, 23ft. 8in. Centre to centre of driving wheels, 8ft. 3in.

The most recent type of compound goods locomotives constructed by Mr. Webb has eight-coupled wheels, three pairs of which are located under the fire-box, the trailing pair being close to the back of the fire-box. The outside cylinders are below the top of the frame-plate, and incline towards the rear. This type of engine was designed by Mr. F. W. Webb, chief mechanical engineer of the London and North Western Railway, principally for working the heavy mineral traffic over that Company's South Wales district, the first engine being built in 1893. The wheels (all coupled) are 4ft. 5½in. in diameter, with tyres 3in. thick. The distance between the centres of each pair is 5ft. 9in., the total wheel base being 17ft. 3in. All the cylinders drive on to one axle—the second from the front of the engine; the two high-pressure cylinders are connected to crank pins in the wheels set at right angles to each other, the low-pressure cylinder being connected to a centre crank-pin set at an angle of 135 degree with the high-pressure cranks; the high-pressure cylinders are 15in. diameter by 24in. stroke, and the low-pressure cylinders are 30in. diameter by 24in. stroke. All the cylinders are bolted together and in line, the low-pressure being placed immediately under the smoke-box and the high-pressure cylinders on each side outside the frames, the steam chests being within the frames.

The engine weighs, in working order, 53 tons 18 cwt.
The empty weights are as follows:—

|  | Tons. | Cwts. |
|---|---|---|
| Leading wheels | 12 | 10 |
| Driving wheels | 14 | 8 |
| Intermediate wheels | 12 | 14 |
| Trailing wheels | 9 | 15 |
| Total (Empty) |  | 7 |

The latest type of a passenger compound locomotive built by Mr Webb is the 7ft. four wheels coupled engine "Black Prince" (Fig. 97), which was built at the Crewe Works in July, 1897.

The engine has two high-pressure and two low-pressure cylinders, all being in line, and driving on to one axle, the high and low pressure cranks being directly opposite each other.

One of the features of this engine is the method adopted for working the valves, two sets of gear only being used for working the four valves.

Joy's valve motion is used for the low pressure valves, the valve spindles being prolonged through the front of the steam chest, and on the end of the spindle a crosshead is fixed which engages with a lever of the first order, carried on a pivot firmly secured to the engine frame. The other end of this lever engages with a crosshead fixed on the end of the high pressure valve spindle, and by this means the requisite motion is given to the high pressure valve.

The leading end of the engine is carried on a double radial truck, the centre of which is fitted with Mr. Webb's radial axle box and central controlling spring. This arrangement permits of 1in. side play, and gives greater freedom to the truck when passing round curves than is possible in the ordinary type of bogie with a rigid centre pin.

One important object aimed at in the construction of this engine has been to get all the bearing surfaces throughout as large as possible. Each of the four journals in the radial truck is $6\frac{1}{4}$in. diameter and 10in. long. The driving axle, in addition to the two ordinary bearings, which are each $7\frac{1}{2}$in. diameter and 9in. long, has a central bearing between the two cranks, 7in. diameter and $5\frac{1}{2}$in. long. In the trailing axle the journals are $7\frac{1}{2}$in. diameter, by $13\frac{1}{2}$in. long.

This engine made its first trip on August 2nd, 1897, and since then has been principally engaged in working the "up" dining saloon express, which leaves Crewe at 5.2 p.m., running through to Willesden without a stop—a distance of $152\frac{1}{2}$ miles—and returning the same night with the Scotch sleeping saloon express, which leaves Euston at 11.50 p.m., running through to Crewe without a stop, 158 miles.

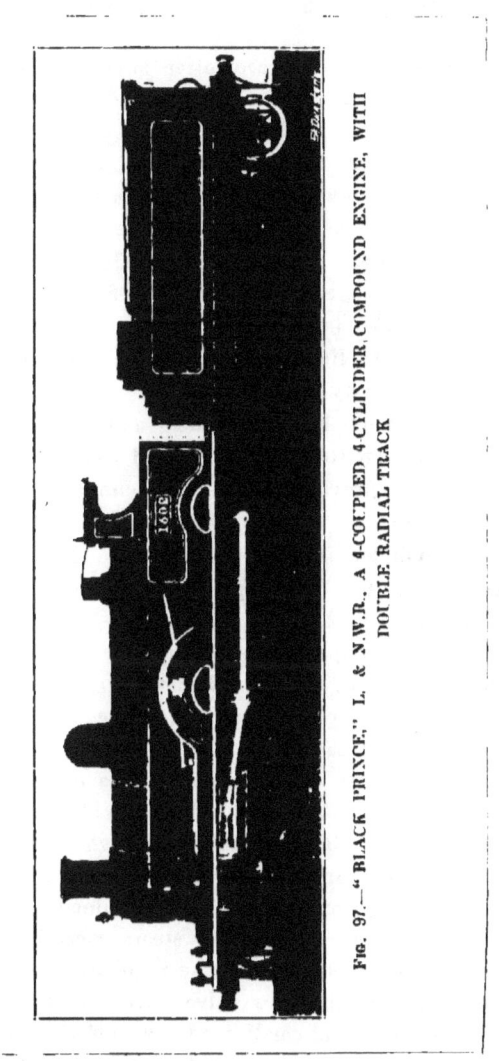

FIG. 97.—"BLACK PRINCE," L. & N.W.R. A 4-COUPLED 4-CYLINDER COMPOUND ENGINE, WITH DOUBLE RADIAL TRACK

The total distance run by this engine up to June 30th, 1898, was 52,034 miles.

The high-pressure cylinders are 15in. diameter by 24in. stroke, and the low-pressure cylinders are $20\frac{1}{2}$in. diameter by 24in. stroke.

The heating surface is: Tubes, 1,241.3 sq. ft.; firebox, 159.1 sq. ft.; total 1,409.1 sq. ft.   Grate area, 20.5 sq. ft.

A concise survey of other compound locomotives will be of interest at this juncture.

Mr. Wordsell, the then locomotive superintendent of the Great Eastern Railway, in 1882 built a compound engine, with two inside cylinders, the h.p. 18in. and the l.p. 26in. diameter; the stroke was 24in.; steam pressure, 160lb. per sq. in. The coupled wheels were 7ft. diameter. The engine was fitted with a leading bogie, the wheels of which were 3ft. 1in. diameter; with her tender she weighed 77 tons in working order; her number was 230. A similar engine, No. 702, with Joy's valve gear, was built in 1885.

Mr. Wordsell also built a two-cylinder, six-coupled goods engine for the Great Eastern Railway, on the compound principle. This was fitted with the ordinary link motion.

Mr. Wordsell, upon his appointment as locomotive superintendent of the North Eastern Railway, introduced compound engines on that line. These were provided with two cylinders, both inside, with the valves on top.

The h.p. cylinder is 18in. and the l.p. 26in. in diameter, the stroke being 24in. Mr. Worthington thus describes the North Eastern Railway standard compound goods engine:—"In outside appearance this engine is neat, simple, and substantial. It weighs 40 tons 7 cwt., and has six coupled 5ft. 1¼in. in diameter.

"The cylinders are placed, as in the passenger compound engines, beneath the slide valves and inside the frames.

"The chief features of this goods engine to be observed are the starting and intercepting valves, which enable the engine-driver to start the engine by admitting sufficient high-pressure steam to the large cylinder without interfering with the small cylinder, in case the latter is not in a position to start the train alone.

"The two valves are operated by steam controlled by one handle. If the engine does not start when the regulator is opened, which will occur when the high-pressure valve covers both steam ports, the driver pulls the additional small handle, which closes the passage from the receiver to the low-pressure cylinder, and also admits a small amount of steam to the low-pressure steam-chest, so that the two cylinders together develop additional starting power.

"After one or two strokes of the engine the exhaust steam from the high-pressure cylinder automatically forces the two valves back to their normal position, and the engine proceeds working compound."

The North Eastern Railway has other compound engines constructed on the Wordsell and Van Borries system, a 6ft. 8¼in. four-coupled, with a leading bogie of locomotive, being turned out in 1886. Engines of this type have a heating surface of 1,323.3 sq. ft., a grate area of 17.33 sq. ft., and a working pressure of 175lb. per sq. in.

Another North Eastern Railway type of compound has 7ft. 6in. single driving wheels and a leading bogie. The h.p. cylinder is 20in., and the l.p. 26in. diameter, compared with 18in. and 26in. in the four-coupled class, the stroke being the same in each design—viz., 24in.

The first of the 7ft. 6in. compound class of locomotives was constructed at the Gateshead Works in 1890. The engines of this design appear capable of doing very heavy work with a low coal consumption, the average being 28lb. per mile, which, considering the heavy traffic and speed maintained, is low, being, in fact, 2lb. per mile below that of any other class of engine engaged on the same traffic.

With a train of 18 coaches, weighing 310 tons (including 87 tons, the weight of the engine and tender), a speed of 86 miles an hour was attained on a level portion of the road, the horse-power indicated being 1,068. These engines have a commodious cab, and the tenders carrying 3,900 gallons of water, thus making it possible for the run of 125 miles, from Newcastle to Edinburgh, to be performed without a stop. There is also a class of six-coupled tank engines, with a trailing radial axle. The stroke is 24in., and the diameter of cylinders h.p. 18in., and l.p. 26in. Compound engines have also been tried on the Glasgow and South Western Railway and on the London and South Western Railway.

The advantages of express locomotives being fitted with leading bogies were speedily recognised by most of the locomotive superintendents. Mr. S. W. Johnson, the Midland chief, introduced a design of such engines in 1876. The steam pressure of the early engines of this class was 140lb., but in later years this was increased to 160lb., whilst in the recent engines the pressure is still further augmented.

The same progress is to be noticed in the diameter of the cylinders of the Midland engines, the diameter having increased from 17½in. in 1876 to 19½in. at the present time. The size of the coupled wheels has also increased from 6ft. 6in. to 7ft. 9in. The length of stroke has been the same in all engines of this design—viz., 26in.

# EVOLUTION OF THE STEAM LOCOMOTIVE 251

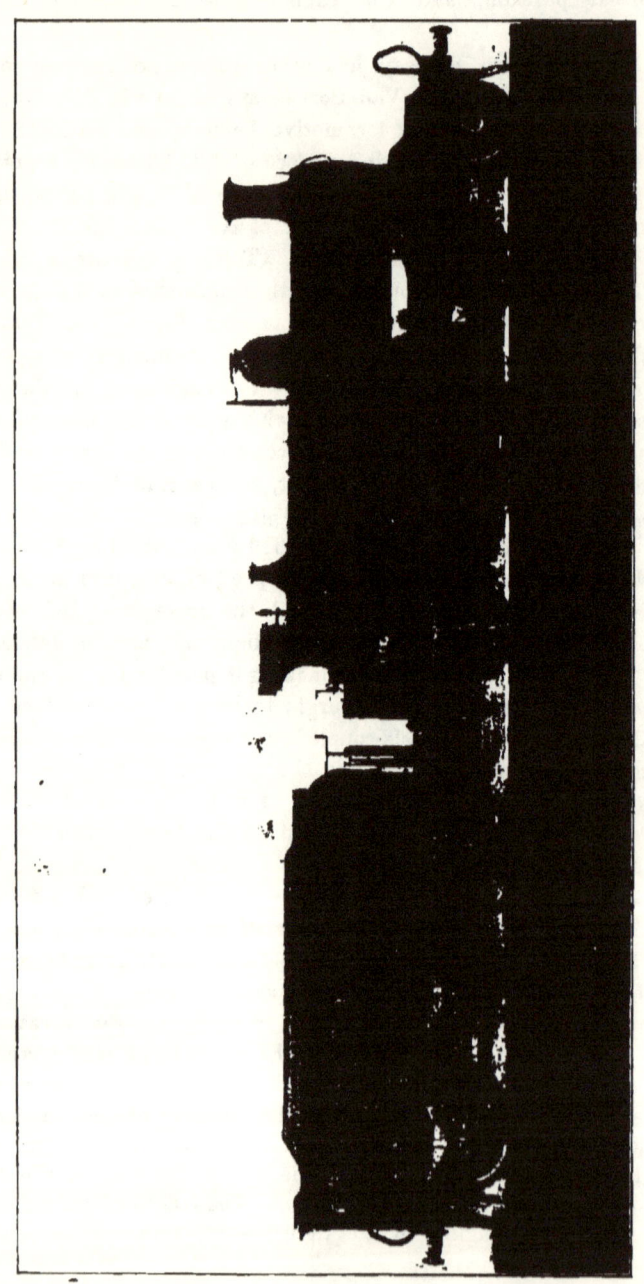

FIG. 93.—JOHNSON'S 7FT. 9IN. "SINGLE" ENGINE, MIDLAND RAILWAY

The new Midland single express engines are illustrated by Fig. 98. These locomotives have inside cylinders 19½in. diameter, with a stroke of 26in. The driving wheels are 7ft. 9in. in diameter. By standing on a railway station platform alongside one of these engines, one gets a good idea of their immense proportions, the abnormally high pitch of the boiler being especially noticeable.

Mr. Stroudley, in his "Gladstone" class of engines for the London, Brighton and South Coast Railway, adopted an entirely opposite practice. His engines had the leading and driving wheels coupled, and a pair of smaller trailing wheels. The coupled wheels are 6ft. 6in. diameter, and the trailing 4ft. 6in. diameter. The cylinders are inside, and measure 18¼in. diameter, the stroke being 26in.

The reversing apparatus is actuated by means of compressed air, supplied by the Westinghouse brake pump; whilst part of the exhaust steam is projected against the flanges of the leading wheels, and upon condensation upon the flanges forms a lubricant to the flange surface, when pressing against the inner sides of the rails. Fig. 99 is from a photograph of "George A. Wallis," an engine of the "Gladstone" class.

FIG. 99.—"GEORGE A. WALLIS," AN ENGINE OF THE "GLADSTONE" CLASS, L.B. & S.C.R.

The "Tennant" (Fig. 100) class of express engines, on the North Eastern Railway, deserves mention, being the design of a general manager during the North-Eastern locomotive interregnum of 1885.

The engines have four-coupled wheels, 7ft. diameter, and a leading pair of small wheels, cylinders being 18in. diameter, and 24in. stroke.

The cab is somewhat similar to the Stirling pattern on the Great Northern Railway.

The North British Railway engine, No. 592, was exhibited at the Edinburgh Exhibition of 1886, and Mr. Holmes, her designer, was awarded the gold medal.

The driving and trailing wheels are coupled, their diameter being 7ft. The fore part of the machine is supported on a four-wheeled bogie. The symmetrical appearance of this and other North British Railway locomotives is spoilt by having the safety valve located above the dome casing. The cylinders are 18in. diameter, and 26in. stroke. No. 602, another engine of this design, is notorious as being the first engine to cross the Forth Bridge, when formally opened by the Prince of Wales on March 4th, 1890, the Marchioness of Tweeddale driving the engine upon the occasion.

Mr. Holmes, in 1890, introduced another very similar design of North British Railway engines, but with coupled wheels only 6ft. 6in. diameter. These are known as the "633" class, illustrated by Fig. 101.

Turning to the Great Eastern Railway, we have to chronicle some types of locomotives designed by Mr. Holden. The express passenger engines have a pair of small leading wheels and four coupled wheels of 7ft. diameter, with cylinders 18in. by 24in. The valves are below the cylinders, which, by the way, are both cast in one piece.

FIG. 100. "1463," N.E.R., ONE OF THE "TENNANT" LOCOMOTIVES

In connection with this design of locomotive, the triumph of skilled mechanism, combined with the application of scientific research, deserves record, seeing that a troublesome waste product has been turned into a valuable calorific agent. We refer to the introduction of liquid fuel for locomotive purposes, as carried out under Mr. Holden's patent.

254   EVOLUTION OF THE STEAM LOCOMOTIVE

Now, sanitary authorities have large powers, and they are very fond of abusing these powers, and pushing matters to extreme issues —although at times, as we know from personal experience, they some-

FIG. 101.—HOLMES'S TYPE OF EXPRESS ENGINES FOR THE NORTH BRITISH RAILWAY

times exceed their statutory powers, and find themselves in a tight place from which they can only retreat by payment of compensation and heavy law costs.

In connection with the pollution of streams the authorities have very wide powers, and when they found the waters of the never clear or limpid Channelsea and Lea were further polluted by some oily, iridescent matter, with a pungent odour, the sanitary inspectors were soon ferreting out the offenders. The waste products from the Great Eastern Railway oil-gasworks at Stratford were found to be responsible for the nuisance, and the service of a notice requiring immediate abatement of the same was the result of the discovery.

Mr. Holden, remembering the good old proverb, "Necessity is the mother of invention," soon commenced to experiment with the matter which the sanitary authorities refused to allow to be emptied into the already impure waterways under their jurisdiction. The result of a series of trials on, first a six-coupled tank engine, and then on a single express, was a four-coupled express engine on the G.E.R., No 760, named "Petrolea."

This locomotive was constructed in 1886, and in general appearance is similar to the four-coupled express engine just described. The heating arrangements are, however, supplemented by the liquid fuel burning apparatus, which may be briefly described as follows: The oil fuel is carried on this engine in a rectangular tank of 500 gallons capacity, but in later examples occupies two cylindrical reservoirs, which contain 650 gallons, placed on the top of the tender water-tanks, one on each side.

The liquid fuel is supplied to these reservoirs through man-holes at the footplate end. The feed pipes from these tanks unite on the tender footplate at the centre, and from this junction the oil is conveyed by the flexible hose pipe to the engine, where the supply is again divided to feed the two burners situated on the fire-box front just under the footboard.

Both the liquid in the tanks and the injected air are heated before use, the former by means of steam coils in the tanks, and the latter by coiled pipes in the smoke-box. The heated liquid fuel and air are injected into the fire-box, through two nozzles in the form of fine spray, steam being injected at the same moment through an outer ring of the same nozzles. The steam divides the mixture of air and liquid into such fine particles that it immediately ignites when in contact with the incandescent coal and chalk fire already provided in the fire-box. The fire-box is fitted with a brick arch deflector.

The whole of the apparatus is controlled by a four-way cock fitted on the fire-box case, near the position usually occupied by the regulator. The positions of the cock in question are: (1) steam to warm coils in liquid fuel tank; (2) steam to ring-blowers on injectors; (3)

FIG. 102.—7FT. SINGLE ENGINE, G.E.R., FITTED WITH HOLDEN'S LIQUID FUEL APPARATUS

steam to centre jets of ejectors; and (4) steam to clear out the liquid fuel pipes and ejectors. The success of "Petrolea" was so apparent and unquestionable that Mr. Holden's patent system of burning liquid fuel was immediately fitted to other Great Eastern Railway locomotives, with the result that at the present time a number are fitted with his patent apparatus.

The following Great Eastern Railway locomotives have been fitted to burn liquid fuel:—

9 four-wheel coupled express engines.
6 single express engines. (Fig. 102)
1 six-wheel coupled goods engine.
1 six-wheel coupled tank, and
20 four-wheel coupled bogie tanks.

and the 10 engines of the new class of "single" bogie expresses. (Fig. 103.)

The application of the "Serve" corrugated tube has also been introduced on the Great Eastern Railway in connection with the liquid fuel. The goods engine and also two of the express passenger engines have the "Serve" tubes. The experiment of burning liquid fuel has been very successful, only 16lb. of oil having been consumed per mile run, against an average of 35lb. of coal per mile, with coal-fired engines.

Some very handsome Bogie Single Express Locomotives have recently been built at the Stratford Works of the G.E.R. Company to the designs of Mr. James Holden. They have been specially constructed for running the fast Cromer traffic. The boiler has a telescopic barrel 11ft. long, in two plates, and is 4ft. 3in. diameter outside the smaller ring. It contains 227 tubes 1¾in. external diameter, and the height of its centre line above the rail level is 7f. 9in.

The fire-box is 7ft. long, and 4ft. ½in. wide outside, and has a grate area of 21.37 sq. ft., and is fired with oil fuel. The total heating surface is 1,292.7 sq. ft., the tubes giving 1,178.5 sq. ft., and the fire-box 114.2 sq. ft. The working pressure is 160 lbs. per sq. in.

The driving wheels are 7ft., the bogie wheels are 3ft. 9in., and the trailing wheels 4ft. diameter. The total wheel base is 22ft. 9in., the bogie wheels centres being 6ft. 6in. apart, from centre of bogie pin to centre of driving wheel is 10ft. 6in., and from centre of driving wheel to centre of trailing is 9ft. The total length of engine and tender, over buffers, is 53ft. 3in.

The cylinders are 18in. diameter by 26in. stroke, the distance between centres being 24in. The slide valves are arranged underneath,

s

258   EVOLUTION OF THE STEAM LOCOMOTIVE

and are fitted with a small valve, which allows any water that may collect in the slide valve to drain off.

FIG. 103.—"No. 10," THE LATEST TYPE OF G.E.R. EXPRESS ENGINE, FIRED WITH LIQUID FUEL.

Steam sanding apparatus is fitted at front and back of the driving wheels.

Macallan's variable blast pipe is used, the diameter of the pipe being 5¼in., and of the cap 4⅜in.

This variable pipe is being adopted on all the Company's engines. The pipe has a hinged top, operated from the footplate. When the hinged top is on the pipe, the area is that of a suitable ordinary pipe, and when the top is moved off the area is about 30 per cent. larger.

It is found that a large proportion of the work of the engines can be done with the larger exhaust outlet, the result being a reduced back pressure in the cylinders and also a reduced vacuum in the smoke-box, and less disturbance of the fire and consequent saving of fuel.

The tender is capable of carrying 2,790 gallons of water, 715 gallons of oil fuel, and 1½ tons of coal. It is provided with a water scoop for replenishing the tank whilst running.

The weights of the engine and tender in working order are: engine, 48½ tons; tender, 36 tons; total, 84½ tons.

The oil-firing arrangements embody a number of ingenious details, among them the supply of hot air for combustion from a series of cast-iron heaters placed around the inside of the smoke-box, the air being drawn from the front through the heaters to the burners for the exhausting action of the steam jets used for injecting the oil fuel. The latter is warmed before leaving the tender in a cylindrical heating chamber, through which the exhaust steam from the air-brake pump circulates.

The regulation of the oil supply is effected by a neatly designed gear attached to the cover and hood of the ordinary fire-door, and finally the burners or injectors are so constructed that should one require cleaning, inspection, or renewal, the internal cones can all be removed from the casing by simply unscrewing one large nut. These engines have polished copper chimney tops, and are painted and lined in the standard G.E.R. style, and fitted with the Westinghouse automatic brake.

## CHAPTER XIV.

Modern L.B. and S.C.R. locomotives—Four-coupled in front passenger tank—Six-coupled tank with radial trailing wheels—Goods engines—"Bessemer," four-coupled bogie express—"Inspector"—Standard L.C. and D.R. passenger engines—Goods locomotives—Three classes of tanks—Cambrian locomotives, passenger, goods, and tank—S.E. engines—A "Prize Medal" locomotive—Stirling's goods and tank engine—His latest type of express engines—Adams's locomotives on the L. & S.W.R.—Mixed traffic engines—Passenger and six-coupled tanks—Drummond's "Windcutter" smoke-box—His four-cylinder express engine—North British passenger locomotives—Engines for the West Highland Railway—Holme's goods and tank engines—His latest express type of engine—Classification of N.B.R. locomotives—N.B.R. inspection or cab engine—L. and Y. locomotives—Aspinall's water "pick-up" apparatus—Severe gradients on the L. and Y. system—7ft. 3in. coupled expresses—"A" class of goods engines—Standard tank engines—L. and Y. oil-burning tank locomotives—Caledonian Railway engines—Drummond's famous "Dunalastairs"—Excelled by his "Dunalastairs 2"—Six-coupled "condensing" tender engines—"Carbrooke" class—Dimensions of 44 types of Caledonian locomotives

The modern engines on the London, Brighton, and South Coast Railway are designed by Mr. Billinton, and comprise—

The four-coupled in front tank with a trailing bogie, of which "Havant," No. 363, is an example. This engine was built at Brighton Works, 1897. Inside cylinders, 18in. by 26in.; diameter Coupled wheels, 5ft. 6in. diameter. Heating surface, 1,189 sq. ft. Steam pressure, 160lb. Weight in working order, 47 tons.

"Watersfield," No. 457, built at Brighton in 1895, is a specimen of the six-coupled goods tank engines, with radial trailing wheels. This class have inside cylinders, 18in. by 26in.; heating surface. 1,200 sq. ft.; diameter of wheels, 4ft. 6in.; steam pressure 160lb.; weight in working order, 51 tons.

No. 449 represents the six-coupled goods tender engines, built by Vulcan Foundry Co. in 1894, from Mr. Billinton's designs. Inside cylinders, 18in. by 26in.; wheels, 5ft. diameter; heating surface, 1,212 sq. ft.; steam pressure, 160lb.; weight in working order: engine, 38 tons; tender, 25 tons.

"Bessemer," No. 213, is one of the new type of four-coupled express passenger engines, with leading bogie, and was built at ton Works, 1897. Inside cylinders, 18in. by 26in.; diameter of coupled wheels, 6ft. 9in.; heating surface, 1,342 sq. ft.; working pressure, 170lb.; weight in working order: engine, 44 tons 14 cwt.; tender, 25 tons. Fig. 104 is from a photograph of "Goldsmith," an engine of this class.

Before closing this short description of the London, Brighton, and South Coast Railway locomotives, attention must be called to the combined engine and carriage named "Inspector," No. 481 (Fig. 105). This engine was constructed in 1869 by Sharp, Stewart, and Co., as an ordinary four-coupled passenger tank, and rebuilt in its present form some 11 years or so ago.

The cylinders are inside, $10\frac{1}{2}$in. diameter, 16in. stroke; coupled wheels, 4ft. diameter; weight in working order, about 20 tons; steam pressure, 120lb. In addition to the coupled wheels there are also a pair of leading and a pair of trailing wheels. There is no steam dome, and the side tanks are as long as the boiler barrel, being

FIG. 104.—"GOLDSMITH," ONE OF THE NEW L.B. & S.C.R. EXPRESS PASSENGER ENGINES

extended on each side to the smoke-box. The inspection car is fixed on to the back of the coal bunker, its floor is some distance below the level of the engine frames, and the car is entered from a platform at the end, which is in turn entered from the outside by steps on either side, as in a tram-car. The back of the platform is quite open, whilst the partition dividing the platform from the enclosed portion of the car is glazed, so that anyone sitting with his back to the coal bunker can see the permanent-way, etc., over which "Inspector" has just passed without leaving his seat if necessary. There is a speaking tube, to enable those in the saloon to communicate with the driver.

A special form of indicator board, not used for any other train, is carried by "Inspector"—viz., a white board with black horizontal stripes.

The modern locomotives of the London, Chatham and Dover Railway are built from designs prepared by Mr. William Kirtley, the Company's locomotive superintendent. The main line passenger engines (Fig. 106) are of the M3 class, and have the following dimensions:—

Cylinders, 18in. diameter, 26in. stroke;
Coupled wheels, 6ft. 6in. in chamber;
Bogie     „     3ft. 6in.     „
Heating surface: tubes, 1,000.2 sq. ft.; fire-box, 110 sq. ft.
Grate area, 17 sq. ft.; working pressure, 150lb.
Weight, in working order, 42 tons 9 cwt., of which the driving and trailing coupled wheels support 28 tons 18 cwt.

The standard tender, for both goods and passenger engines, is

FIG. 105.—"INSPECTOR," LONDON, BRIGHTON, AND SOUTH COAST RY.

carried on six wheels, and, loaded, weighs 34 tons; accommodation is provided for 4¾ tons of coal, and 2,600 gallons of water.

The standard goods engines have six coupled wheels, 5ft. in diameter.

Cylinders, 18in. by 26in.;
Heating surface: tubes, 1,000.4 sq. ft.; fire-box, 102 sq. ft.; working pressure, 150lb. per sq. in.;
Weight, in working order: leading 13 tons 2 cwt.; driving, 15 tons 4½ cwt.; trailing, 10 tons 19½ ct.; total, 39 tons 6 cwt.

These engines are known as "Class B2."

The tank engines consist of three classes.

The dimensions of those for working the main line and suburban services are as follows:—

Inside cylinders, with an incline of 1 in 10, 17in. diameter; 24in. stroke. Wheels, four coupled in front, 5ft. 6in. diameter. A trailing bogie with 3ft. wheels. Heating surface: tubes, 971.7 sq. ft.; firebox, 99.3 sq. ft.; grate area, 16¼ sq. ft. Tank capacity, 1,1(0 gallons of water, 2 tons of coal. Weight, in working order, 49 tons 15 cwt. Steam pressure, 150lb. per sq. in.

These engines are officially described as Class R.

FIG. 106.—"No. 192," ONE OF THE STANDARD EXPRESS PASSENGER LOCOMOTIVES, LONDON, CHATHAM AND DOVER RAILWAY

The A class of bogie tank engines were specially designed for working through tunnels. The inside cylinders are 17½in. diameter, and 26in. stroke. The coupled (leading and driving) wheels are 5ft. 6in. diameter, the wheels of the trailing bogie being 3ft. in diameter. The heating surface is made up as follows: Tubes, 995 sq. ft.; firebox, 100 sq. ft.; grate area, 16½ sq. ft.; working pressure, 150lb.; water capacity of tanks. 970 gallons; fuel space, 80 cubic ft.; weight, in working order, 51 tons.

All of these engines are fitted with steam condensing apparatus to allow of working over the Metropolitan Railway between Snow Hill and King's Cross and Snow Hill and Moorgate Street.

Class T comprises the goods or shunting tanks. These have six coupled wheels of 4ft. 6in. diameter, with a wheel bore of 15ft. The cylinders are inside, with a 17in. diameter and 24in. stroke. The heating surface is as follows: Tubes, 799.3 sq. ft.; fire-box, 88.7

FIG. 137.—STANDARD EXPRESS PASSENGER ENGINE, CAMBRIAN RAILWAYS

sq. ft.; grate area, 15 sq. ft.; steam pressure, 150lb.; tank capacity, 830 gallons; coal bunker, 48 cubic ft.; weight, in working order, 40¾ tons.

In 1889 these shunting engines were fitted with the Westinghouse Automatic Brake, which is the continuous brake adopted by the London, Chatham and Dover Railway.

In general outline the modern locomotives on the Cambrian Railways are similar to those of the London, Chatham, and Dover Railway.

The express passenger engines (Fig. 107) on the Cambrian Railway have a leading bogie, with wheels 3ft. 6in. diameter, and four coupled wheels of 6ft. diameter. The inside cylinders are inclined 1 in 15, and are 18in. diameter, and 24in. stroke. The heating surface is: Tubes, 1,057 sq. ft.; fire-box, 99½ sq. ft.; grate area, 17 sq. ft. There are 230 tubes, 10ft. 5-16th in. long, and 1¾in. diameter. The wheel base is: Centre to centre of bogie, 5ft. 6in.; leading to trailing, 7ft.; centre of bogie to driving, 9ft. 3½in.; and driving to trailing, 8ft. 3in. Boiler pressure, 16lb. per sq. inch. These engines have underhung springs to the driving and trailing wheels, are fitted with a steam sanding apparatus, the vacuum brake, screw reversing gear, and other improvements. They were built by Sharp, Stewart, and Co.,

Atlas Works, Glasgow, the particular one we have described having been turned out in 1893.

The modern goods engines numbered 73 to 77 were built by Neilson and Co., Glasgow, in 1894, the maker's numbers being 4,691 to 4,695. The six-coupled driving wheels are 5ft. 1½in.; diameter; the wheel base being: leading to driving, 7ft. 5in.; driving to trailing, 7ft. 10in. The springs to all the wheels are underhung, the driving wheel springs being of Timmis' patent design. Steam sanding apparatus is provided in front of the leading wheels. The cylinders are inside, and are inclined 1 in 10. The boiler barrel is 10ft. 3in. long, and contains 204 tubes of 1¾in. diameter; the heating surface being: tubes, 986.2 sq. ft.; fire-box, 98.3 sq. ft.; fire-grate area, 16½ sq. ft.; working pressure, 160lb. per sq. in.

The tenders have six wheels. 3ft. 10in. diameter, with a wheel base of 12ft., equally divided. Water capacity, 2,500 gallons; coal space, 200 cubic ft.

FIG. 108.—STANDARD PASSENGER TANK ENGINE, CAMBRIAN RAILWAYS

The above dimensions are those of the Cambrian Railways modern standard tender, and apply both to the passenger and goods engines.

The bogie passenger tank engines (Fig. 108) have inside cylinders, 17in. diameter, 24in. stroke, inclined 1 in 9. The coupled wheels (leading and driving) are 5ft. 3in. diameter, the bogie wheels being 3ft. 1½in. diameter. The boiler barrel is 10ft. 2¾in. long, and contains 134 tubes of 2in. diameter, and 38 tubes of 1¼in. diameter Boiler pressure, 160lb. per sq, inch. Heating surface:

Tubes, 920.1 sq. ft.; fire-box, 90 sq. ft.; grate area, 13.3 sq. ft. The tanks contain 1,200 gallons of water, and the bunkers 2 tons of coal. The total wheel base is 20ft. 1in., the coupling side-rods being 7ft. 8in. long. Weight, in working order, 45 tons 9 cwt. 3 qrs.

These engines were built by Nasmyth, Wilson, and Co., Ltd., Bridgewater Foundry, near Manchester.

Mr. James Stirling, the present locomotive superintendent of the South Eastern Railway, soon after his appointment, took steps to thoroughly renovate and classify the various types of locomotives on the system.

He has now succeeded in doing so; indeed, save for a few 6ft. D. and T. coupled, of Cudworth's design, now rebuilt without a dome, and the six-wheel four-coupled express engines built during the short Watkin locomotive régime, and now rebuilt by Mr. Stirling, nearly every engine on the South Eastern Railway is from Mr. Stirling's own designs.

It should be mentioned that Mr. James Stirling, like his brother, the late Patrick Stirling, of Great Northern Railway fame, does not believe in a steam dome. Another feature of resemblance in their designs is discovered in the style of cab. Patrick favoured a brass encased safety-valve, located on the top of the fire-box; whilst James chooses the boiler barrel for the position of that useful feature in a locomotive, which he, however, constructs after the Ramsbottom type.

The modern South Eastern Railway engines all have inside cylinders, and Mr. Stirling's excellent reversing gear previously described. They may be divided into the following classes:—

    Four-wheels coupled bogie express engine—of two sets of dimensions.

    Four-wheels coupled bogie passenger engine.

    Four-wheels coupled bogie tank engine.

    Six-wheels coupled, goods engine.

    Six-wheels coupled shunting tank engine.

The standard express class of engines was introduced about 15 years ago, and the locomotives were then painted black, but fortunately for their appearance, Mr. Stirling has recently reverted to the South Eastern Railway colour obtaining before his appointment as locomotive superintendent, and the newer engines are now painted a pleasing tint of olive green. "No. 240" (Fig. 109), an engine of this class, was exhibited at the Paris Exhibition of 1889, and obtained the Gold Medal.

The leading dimensions are: Cylinders, 19in. diameter, 26in. stroke (incline 1 in 30); leading bogie wheels, 3ft. 9in. diameter; wheel base of bogie, 5ft. 4in.; driving and trailing wheels (coupled), 7ft. diameter; wheel base of coupled wheels, 8ft. 6in. The driving wheels have Timmis's springs; the trailing wheels underhung laminated springs.

The tender is carried on six wheels of 4ft. diameter, with a wheel base of 12ft., equally divided. The tender tank holds 2,650 gallons of water, and the coal capacity is 4 tons.

FIG. 109.—"No. 240," THE SOUTH EASTERN RAILWAY ENGINE THAT OBTAINED THE GOLD MEDAL AT THE PARIS EXHIBITION, 1889

Weight, in working order: on bogie, 13 tons 12 cwt.; driving wheels, 15 tons 18 cwt.; on trailing wheels, 13 tons; tender. L., 10 tons 6 cwt.; centre, 10 tons 1 cwt.; T., 10 tons 3 cwt.; total weight of engine and tender, 73 tons.

From the above description it will be seen that these locomotives are finely proportioned and should be capable of doing excellent service. They are good for hauling heavy loads, and the "direct" line viâ Sevenoaks has some severe gradients, which these engines negotiate in fine style.

Another point in their favour is the coal consumption, the average being low, although the fuel is of inferior quality.

The speed, however, of the fast trains is disappointing. Probably it is not right to blame the engines for this, but rather the timing of the trains.

Whilst other railways are accelerating their services, the South Eastern Railway retrogrades in the matter of speed.

Yet there is not a finer length of line in the kingdom for showing what an engine can do than that between Redhill and Folkestone, or leaving the main line at Ashford and on to Ramsgate. For many miles these tracks are practically straight and level; but no advantage is taken of the circumstances so far as speed is concerned; hence travellers are apt to blame the locomotives. These probably have never had a chance to show what speed they are capable of.

Fig. 110.—STANDARD GOODS ENGINE, SOUTH EASTERN RAILWAY

Mr. Stirling's other class of bogie tender engines is very similar in appearance to the one just described, but of smaller dimensions. The engines now to be described were first constructed some years before the 7ft. coupled expresses; indeed, soon after Mr. Stirling was appointed locomotive superintendent. They are principally used for working the passenger trains on the North Kent line (London to Maidstone).

Cylinders, 18in. by 26in. (incline 1 in 15); bogie wheels, 3 ft. 8in.

## EVOLUTION OF THE STEAM LOCOMOTIVE

diameter; wheel base, 5ft. 4in.; driving and trailing wheels (coupled), 6ft. 0½in. diameter (wheel base, 8ft. 2in.); springs and framing similar to the 7ft. class: tender wheels, 3ft. 8in. diameter; wheel base, 12ft., equally divided; water capacity, 2,000 gallons; coal, 3 tons. Weights in working order: on bogie, 12 tons 12 cwt.; driving axle, 14 tons 2 cwt.; trailing, 11 tons 5 cwt.; tender, L., 8 tons 12 cwt.; centre, 8 tons 2 cwt.; T., 9 tons; total weight (engine and tender), 63 tons 13 cwt.

The tender goods engines (Fig. 110) have six wheels (coupled) of 5ft. 2in. diameter; cylinders, 18in. by 26in. (incline 1 in 9); wheel base, L. to D., 7ft. 4in.; D. to T., 8ft. 2in. The tenders are of similar dimensions to the 6ft. passenger engines, with 100 gallons additional water capacity. Weights in working order: engine, L., 12 tons 2 cwt.; D., 15 tons 3 cwt.; T., 11 tons. Tender, L., 9 tons 5 cwt.; C., 9 tons 1 cwt.; T., 9 tons 17 cwt.; total (engine and tender), 64 tons 18 cwt.

The four wheels coupled bogie tanks (Fig. 111) have the leading

FIG. 111.—STANDARD PASSENGER TANK LOCOMOTIVE, S.E. RWY.

and driving wheels coupled; these are 5ft. 6in. diameter; cylinders, 18in. by 26in. (incline 1 in 9); trailing bogie, with wheels, 3ft. 9in. diameter; side water tanks, capacity, 1,050 gallons; coal bunker capacity, 30 cwt.; wheel base, L. to D., 7ft. 5in.; D. to bogie centre, 11ft. 11in.; bogie wheel base, 5ft. 4in. Weight in working order: L., 13 tons 17 cwt.; D., 16 tons; bogie, 18 tons 16 cwt.; total, 48 tons 13 cwt.

The above is a capital type of passenger tank engine, of which the South Eastern Railway possess a large and increasing number. They are mostly constructed by Glasgow firms, whilst the tender engines are built at Ashford Works.

There is a similar type of bogie tanks, fitted with condensing apparatus, and used for working the through South Eastern trains over the Metropolitan Railway to the Great Northern Railway. Some of these engines were also used for hauling the South Eastern trains through the Thames Tunnel, when the through service between Croydon (Addiscombe Road) and Liverpool Street was in operation. For this purpose they were fitted with a short funnel, to enable them to clear the Thames Tunnel.

The illustration (Fig. 112) shows Mr. Stirling's latest type of express engine for the South Eastern Railway, the first of which commenced to work at the end of July, 1898. Several differences of detail compared with Mr. Stirling's previous South Eastern Express engines are introduced. The more noticeable are the large bright brass stand upon which the safety-valves are mounted, the improvement in the shape of the cab on the engine, whilst the sides of the tender are painted in two panels, with the Company's coat of arms between (Mr. Stirling, it will be observed, has not slavishly copied other practice in lettering the tenders S.E.R.); the springs are below the frames, and steps at the back are provided on either side of the tender.

The diameter of the wheels and cylinders, the stroke, and wheelbase remain the same. The tender is a trifle longer, making the total length over buffers 52ft. 8in., instead of 52ft. 4in. The working pressure is now 170lb. per sq. in., there being 215 tubes of 1⅜in. external diameter, 10ft. 4½in. long. The other differences in the dimensions are tabulated below:—

|  | "440" Class, illustrated by FIG. 112. | "240" Class, illustrated by FIG. 109. |
|---|---|---|
| Rail level to centre of Boiler | 7ft. 10in | 7ft. 5in. |
| Total Heating Surface | 1,100 sq. ft. | 1,020½ sq. ft. |
| To top of Chimney is 13ft. 4in. in both classes, the new Engines having Funnels 2in. shorter. | | |

Weight loaded—

| ENGINE. | | | ENGINE. | | |
|---|---|---|---|---|---|
| Bogie. | D. | T. | Bogie. | D. | T. |
| 15 tons. | 16 tons 8 cwt. | 14 tons 13 cwt. | 13 tons 12 cwt. | 15 tons 0 cwt. | 13 tons. |

| TENDER. | | | | TENDER. | | | |
|---|---|---|---|---|---|---|---|
| L. | C. | T. | | L. | C. | T. | |
| 10 tons 15 cwt. | 10 tons 18 cwt. | 12 tons 9 cwt | | 10 tons 6 cwt. | 10 tons 1 cwt. | 10 tons 3 cwt. | |
| TOTAL | | 80 tons 3 cwt. | | TOTAL | | 73 tons. | |
| Water Capacity of Tender | | 3,000 galls | | | | 2,650 galls. | |
| Coal | ,, | ,, | 3 tons. | | | | 4 tons. |

With the increased weight, boiler pressure, and heating surface of these engines, coupled with a compromise towards a steam dome, such fine locomotives ought to be quite equal to hauling the heavy trains

# EVOLUTION OF THE STEAM LOCOMOTIVE 271

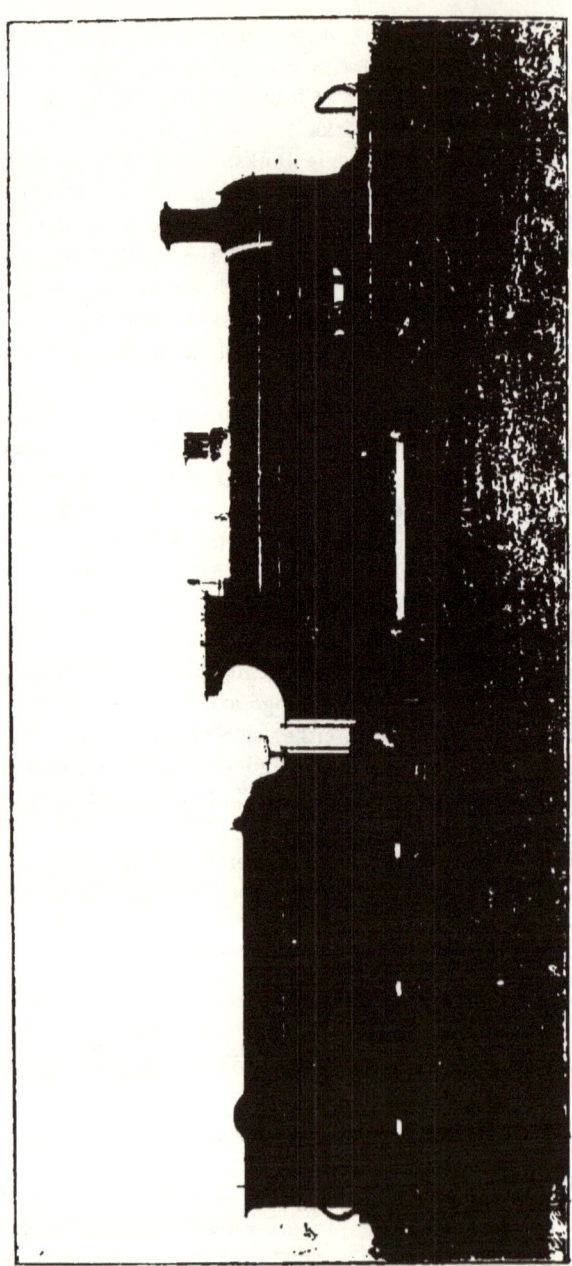

Fig. 112.—LATEST TYPE OF EXPRESS PASSENGER ENGINE, SOUTH EASTERN RAILWAY

run by the South Eastern Railway at high speeds. Mr. Stirling is to be congratulated upon the appearance of the machines.

The standard engines now in work on the London and South Western Railway were constructed from the designs of Mr. Adams, the late locomotive superintendent, who resigned about three years ago. Mr. D. Drummond, who succeeded Mr. Adams, has built several new types of engines, viz., large bogie tank engines, six-wheels-coupled goods engines, four-wheels-coupled bogie express engines, as well as a "four-cylinder" engine, which latter is decidedly a new departure in London and South Western Railway practice. The most important of Mr. Adams' designs can be classified thus:

Four-coupled bogie express engine and tender.
Four-wheels-coupled in front, mixed engine and tender.
Six-wheels-coupled goods engine and tender.
Four-wheels-coupled bogie tank engine; and
Six-wheels-coupled bogie shunting tank engine.

There are two classes of four-wheels-coupled bogie passenger engines, both of the same design, but of different dimensions.

The appended table will show the variations in the two classes:

| Design No. | Cylinders. | Boiler Pressure. | Length of Boiler Barrel. | Length of Firebox. | Diameter of Bogie Wheels. | Diameter of Coupled Wheels. | Heating Surface Tubes. | Fire Box. | Grate Area. | Number of Tubes. | Tractive Force on Rails. | Water Capacity of Tender. |
|---|---|---|---|---|---|---|---|---|---|---|---|---|
| | in. | lbs. | ft. | ft. in. | ft. in. | ft. in. | Square feet. | | | | lbs. | gals. |
| 26 | 17¼ x 26 | 160 | 11 | 6 0 | 3 3 | 6 7 | 1121 | 110 | 17 | 216 | 10,079 | 3,000 |
| 43 | 19 x 26 | 175 | 11 | 6 10 | 3 7 | 6 7 | 1193.7 | 112·12 | 19·75 | 230 | 13,069 | 3,300 |

These engines (Fig. 113) have outside cylinders, underhung springs to the coupled wheels, the springs being connected by means of a compensation beam; a dome on the boiler barrel, and a Ramsbottom safety valve on the fire-box. A notable feature in the design is the distance the frames project in front of the smoke-box. The style of cab is also very neat.

A great feature in Mr. Adams's later engines was his patent vortex blast pipe, the introduction of which very considerably reduced the coal consumption of the locomotives fitted with the invention.

The mixed traffic engines have inside cylinders, 18in. diameter, 26in. stroke, leading and driving wheels (coupled) 6ft. diameter, and trailing wheels 4ft. diameter, underhung springs and compensation beams to the coupled wheels; steam pressure, 160lb. The heating surface and grate area are similar to the "26" design. Tractive force on rails, 11,700lb. Tender capacity the same as "43" design.

The six-coupled goods engines have inside cylinders, 17½in. diameter, 26in. stroke; wheels, 5ft. 1in. diameter; steam pressure, 140lb.; underhung springs; boiler barrel, 10ft. 6in. long, 4ft. 4in. diameter; fire-box, 5ft. 10in. long, 5ft. high. The smoke-box front

FIG. 113.—ADAMS'S STANDARD EXPRESS ENGINE, LONDON AND SOUTH WESTERN RAILWAY

inclines, so that the box is wider at the base than at the top, as is the case with the London and North Western goods engines. There are 218 tubes of 1¾in. external diameter; the heating surface being: tubes, 1,079 sq. ft.; fire-box, 108 sq. ft.; grate area, 17.8 sq. ft. Tractive force on rails, 10,442lb.; water capacity of tender, 2,500 gallons.

The suburban and other short distance passenger traffic is performed by tank engines, having the leading and driving wheels coupled, and a trailing bogie. The cylinders are inside, 18in. diameter, 26in. stroke; coupled wheels, 5ft. 7in. diameter; bogie wheels,

T

3ft. diameter; heating surface and grate area the same as in the "26" class and the mixed traffic engines already described. Steam pressure, 160lb. per sq. in.; fuel space of bunkers, 80 cubic ft.; water capacity of tanks, 1,200 gallons.   Tractive force on rails, 12,573lb.

The six-wheels-coupled shunting tanks are altogether of smaller dimensions, the cylinders being 17½in. diameter, and having a 24in. stroke wheels, 4ft. 10in. diameter; boiler barrel, 9ft. 5in. long and 4ft. 2in. diameter, containing 201 tubes of 1¾in. external diameter. The heating surface is: tubes, 897.76 sq. ft.; fire-box, 89.75 sq. ft.; the fire-box is 5ft. long and 4ft. 9in. high, the grate area being 13.83 sq. ft.; the steam pressure, 160lb.  Tractive force on rails,

Photo by] [F. Moore

FIG. 114.—A "WINDCUTTER" LOCOMOTIVE ENGINE, "No. 136," L. & S.W.R., FITTED WITH A CONVEX SMOKE-BOX DOOR

12,672lb.; fuel capacity of bunker, 77½ cubic ft.; capacity of water tanks, 1,000 gallons.

The London and South Western Railway at one time had an extraordinarily large number of different designs of locomotives, and at the present time the number of designs in use probably exceeds that on any other British railway, despite the fact that the older classes are being rapidly "scrapped," although some of the very ancient

# EVOLUTION OF THE STEAM LOCOMOTIVE 275

FIG. 115.—DRUMMOND'S 4-CYLINDER ENGINE, LONDON AND SOUTH WESTERN RAILWAY

types have in recent years been rebuilt with new boilers. The older engines of Battie design mostly have names, but this practice, unfortunately, has been disregarded by recent London and South Western Railway locomotive superintendents, save in the case of one tank engine, named "Alexandra," under special circumstances, and even this name has lately been removed.

Since Mr. Drummond has become chief at Nine Elms, two at least of his innovations deserve notice. One is an experiment with a windcutter smoke box door (Fig. 114), constructed in the belief that the wind resistance is thereby decreased. In addition to the tender engine 136 being so fitted, this form of convex smoke-box door is fitted to a L. and S.W.R. tank engine, and also to some of the tender goods engines.

Another type of engine, designed by Mr. Drummond, that has attracted considerable attention, is the four-cylinder engine (Fig. 115), built at Nine Elms at the end of 1897. This engine is supported on four driving wheels (uncoupled) of 6ft. 7in. diameter, and a leading bogie. Joy's valve-gear is used for the outside cylinders; all the cylinders are 15in. diameter; the stroke is 26in. A very large heating surface, including the water tubes in the fire-box, amounting to 1,700 sq. ft., is provided. The steam pressure is 175lb. per sq. in. The tender is carried on two four-wheel bogies, and carries 4,300 gallons of water. The motion is reversed by means of a steam apparatus. A portion of the exhaust steam is discharged at the back of the tender.

The locomotive works of the North British Railway are situate at Cowlairs, Glasgow, and Mr. M. Holmes is the present locomotive superintendent.

Originally the North British Railway works were located at St. Margaret's, near Edinburgh, but when the Edinburgh and Glasgow Railway was amalgamated with the North British, in 1865, the Cowlairs works of the former were chosen as the locomotive headquarters of the Company.

Considerable power is required to work the trains over the North British system, as not only are the trains heavy, but many are run at a good speed, whilst steep gradients are not unknown.

It is not, therefore, surprising that "single" engines should be absent from the locomotive stock.

## EVOLUTION OF THE STEAM LOCOMOTIVE

The passenger engines are mostly of the four-coupled leading bogie type. (Fig. 116).

The principal passenger engines have the coupled wheels of 6ft. 6in. and 7ft. diameter, both with cylinders 18in. by 26in. The steam pressure is 140lb. usually, but some of the engines are credited with an additional 10lb. per sq. in.

The other dimensions are:

|  | 7 FT. WHEELS. | 6 FT. 6 IN. WHEELS. |
|---|---|---|
| Heating Surface Tubes | 1,007 sq. ft. | 1,148 sq. ft. |
| Fire-box | 119 ,, | 118 ,, |
| Grate Area | 22 ,, | 20 ,, |
| Weight in Working order | 45 tons 5 cwt. | 46 tons 10 cwt. |

The driving wheels of both sizes have a weight of 15 tons 12 cwt. upon them. The tenders weigh 32 tons, and hold 5 tons of coal and 2,500 gallons of water.

Engines of these classes work the East Coast expresses between

FIG. 116.—FOUR-COUPLED PASSENGER ENGINE WITH LEADING BOGIE, N.B.R.

Edinburgh and Berwick, 57 miles 42 chains. The booked time is 72 minutes, but the runs are frequently performed under the hour; indeed, a train has been timed from start to stop in 57 minutes 21 seconds, on the journey up from Edinburgh to Berwick.

For working the West Highland Railway Mr. Holmes designed a class of four-coupled bogie engines of exceptional power. The coupled wheels are only 5ft. 7in. in diameter; cylinders, 18in. diameter, 24in. stroke; heating surface tubes, 1,130.41 sq. ft.; fire-box, 104.72 sq. ft.;

grate area, 17 sq. ft.; steam pressure, 150lb. per sq. in.; weight of engine in working order, 43 tons 6 cwt., of which $14\frac{1}{2}$ tons rest on the driving axle.

The tender is similar to that previously described.

Goods engines are very numerous on the North British Railway, the most modern ones being known as the "18in. standard" type. These have six coupled wheels of 5ft. diameter; cylinders, 18in. by 26in. stroke; heating surface tubes, 1,139.96 sq. ft.; fire-box, 107.74 sq. ft.; grate area, 17 sq. ft.; weight in working order, 40 tons 13 cwt., of which 15 tons 8 cwt. are supported by the driving wheels. The tender is of the usual type. Other goods engines have cylinders 17in. diameter, with 26in. stroke.

The short distance passenger traffic is worked by four classes of tank engines, one type of which is very similar to the London, Brighton, and South Coast "terriers," though of larger dimensions. These have cylinders 15in. by 22in., coupled wheels, 4ft. 6in. diameter, tanks to hold 600 gallons of water, and weigh $33\frac{1}{2}$ tons in working order. Another class of bogie tank has coupled wheels 5ft. in diameter, a leading bogie with solid wheels 2ft. 6in. diameter, cylinders 16in. by 22in. stroke. These engines originally condensed the exhaust steam, but the usual practice is now followed, and the exhaust is used as a blast for increasing the draught.

The two other classes of tank engines have the following dimensions:

494 Class: Cylinders, 17in. by 26in.; diameter of driving wheels, 6ft.; water capacity, 950 gallons; coal, 30 cwt.; weight, 47 tons 4 cwt.

586 Class: Cylinders, 17in. by 24in.; diameter of driving wheels, 5ft. 9in.; water capacity, 1,281 gallons; coal, 50 cwt.; weight, 50 tons 7 cwt.

There is a handy little type of saddle tanks, known as "shunting pugs." These run on four (coupled) wheels of 3ft. 8in. diameter; they have outside cylinders, 14in. diameter and 20in. stroke. The wheel base is 7ft.; weight in working order, 28 tons 15 cwt.; water capacity of saddle tank, 720 gallons.

Mr. Holmes' latest type of express engines for the N.B.R. (Fig. 117) has a working pressure of 175lb. per sq. in. The principal dimensions being: Cylinders, $18\frac{1}{4}$in. diameter by 26in. stroke. Wheels: Bogie, 3ft. 6in. diameter; driving and trailing, 6ft. 6in. diameter;

wheel base, 22ft. 1in.; centre of bogie to centre of driving wheels, 9ft. 10in.; centre of driving to centre of trailing wheels, 9ft. Tubes No. 254, 1¾in. diameter outside. Heating surface: Tubes, 1,224 sq. ft.; fire-box, 126 sq. ft.; total, 1,350 sq. ft. Fire-grate, 20 sq. ft. Weight of engine in working order, 47 tons. Weight of tender in working order, 38 tons. Tank capacity, 3,500 gallons.

The North British Railway locomotive stock comprises about 800 engines, but many of these are in the A or duplicate list, and are not, therefore, included in the statutory returns.

FIG. 117.—HOLMES'S LATEST TYPE OF EXPRESS ENGINE, NORTH BRITISH RAILWAY

The North British Railway tender locomotives are classified under seven headings—four goods and three passenger.

By a recent return the number of engines under each head was:

```
                              GOODS.
18in. cylinder, 6 wheels coupled main line  ...   ...   ...   144
1st class, 6 wheels coupled   ...   ...   ...   ...   ...   267
2nd class, 6 wheels coupled   ...   ...   ...   ...   ...     8
      (Of which 1 (No. 17a) is on the duplicate list.)
3rd class, 6 wheels coupled   ...   ...   ...   ...   ...    75
      (Of which 2 (18a and 250a) are on the duplicate list.)
                            PASSENGER.
1st class, 4 wheels coupled   ...   ...   ...   ...   ...   121
2nd class, 4 wheels coupled   ...   ...   ...   ...   ...    22
      (Of which 5 (268a, 269a, 394a, 395a, and 404a) are on the duplicate list.)
3rd class, 4 wheels coupled   ...   ...   ...   ...   ...    29
      (One (247a) is on the duplicate list.)
```

The locomotive works of the Lancashire and Yorkshire Railway are situate at Horwich, in the vicinity of Bolton, and are the newest of the immense assemblages of workshops and factories designated by the various railways as their "works," which have been erected by the principal railway companies. It is not, therefore, surprising to find that the Horwich works are quite equal to all, and exceed many other, of the railway establishments in the matter of modern machine tools, and in the general completeness of the undertaking.

Mr. J. A. F. Aspinall is chief mechanical engineer of the Lancashire and Yorkshire Railway, and under his supervision the locomotive stock of the railway has been raised to a degree of excellence seldom equalled and never exceeded.

This position has been attained because Mr. Aspinall has always shown a determination to introduce the best features of all kinds into his locomotive designs. The Joy valve gear is very extensively employed in the construction of Lancashire and Yorkshire locomotives, and has always given excellent results on that line.

For many years past the Lancashire and Yorkshire Railway has adopted the Ramsbottom system of water tanks, while the pick-up apparatus is actuated by a vacuum arrangement patented by Mr. Aspinall. The water troughs are situate at nine different places on the system—viz:—

| | Register No. |
|---|---|
| Horbury Junction: East end of Horbury Junction Station | 3 |
| Hoscar Moss: Between Hoscar Moss and Burscough Bridge | 7 |
| Kirkby: Between Kirkby and Fazakerly | 5 |
| Lea Road: Between Lea Road and Salwick | 6 |
| Rufford: Between Rufford and Burscough North Junction | 8 |
| Smithy Bridge: West end of Smithy Bridge Station | 1 |
| Sowerby Bridge: West end of Sowerby Bridge Tunnel | 2 |
| Walkden: Between Moorside and Wardley and Walkden | 4 |
| Whittey Bridge: West end of Whittey Bridge Station | 9 |

Very severe gradients are to be found on the Lancashire and Yorkshire Railway, many stretches of 1 in 50, between which rate of inclination and that of 1 in 100 very many banks exist, some of which are of considerable length; whilst from Baxenden to Accrington the line falls 1 in 40 for two miles at a stretch, and at the same rate for $1\frac{1}{4}$ miles, from Padiham Junction to Padiham Station, and also for $1\frac{3}{4}$ miles at 1 in 40 from Hoddlesden Junction to Hoddlesden. From Britannia to Bacup the gradients are as follows:

Fall 286 yards, 1 in 61.
Fall 550 yards, 1 in 35.
Fall 154 yards, 1 in 70.
Fall 1,056 yards, 1 in 34.

EVOLUTION OF THE STEAM LOCOMOTIVE    281

But this bank is even eclipsed in severity by the Oldham incline of 1 in 27 for three-quarters of a mile. All these stiff banks are worked by locomotive engines without the help of stationary engines.

Every train which leaves Victoria Station, Manchester, in an eastward direction, has to start off by ascending a serious incline of 1 in 77, followed by another of 1 in 65, round a sharp S curve, on its way to Newton Heath, or else to ascend gradients towards Miles Platting of 1 in 59 and 1 in 49.

The locomotive stock consists of 1,333 engines. Of this number, 590 are of the standard types described below as being of the three leading types designed by Mr. Aspinall The balance is made up

FIG. 118.—FOUR-WHEELS-COUPLED, SADDLE TANK ENGINE, LONDON AND NORTH WESTERN RAILWAY

mainly of engines of older forms, which are gradually being replaced with engines of the standard type, though a large number of these engines have been altered so as to require a boiler of one type only.

The locomotive sheds of the Lancashire and Yorkshire Railway are situate and numbered as below:—

Newton Heath, No. 1; Low Moor, No. 2; Sowerby Bridge, No. 3; Leeds, No. 4; Mirfield, No. 5; Wakefield, No. 6; Normanton, No. 7; Barnsley, No. 8; Knottingley, No. 9; Goole, No. 10; Doncaster, No. 11; ———, No. 12; Agecroft, No. 13; Bolton, No. 14; Horwich, No. 15; Wigan, No. 16; Southport, No. 17; Sandhills, No. 18; Aintree Sidings, No. 19; Bury, No. 20; Bacup, No. 21; Accrington, No. 22; Burnley, No. 23; Skipton, No. 24; Lower

Darwen, No. 25; Hellifield, No. 26; Lostock Hall, No. 27; Chorley, No. 28; Ormskirk, No. 29; Fleetwood, No. 30; Blackpool (Talbot Road), No. 31; and Blackpool (Central), No. 32.

FIG. 118A.—STANDARD EXPRESS PASSENGER LOCOMOTIVE, LANCASHIRE AND YORKSHIRE RAILWAY

We will now proceed to describe some of the types of Lancashire and Yorkshire locomotives.

The "H." or Standard class of four-wheels coupled passenger engines is illustrated by engine No. 1,093 (Fig. 118A). The cylinders are inside, and the axles also have the bearings inside. The principal dimensions are:

```
Cylinders, 18in. diameter by 26in. stroke.
Bogie wheels, 3ft. 0¼in. diameter
Coupled wheels (driving and trailing), 7ft. 3in. diameter
Wheel base, 21ft. 6¼in.
Centre of bogie to centre of driving wheel, 10ft. 2¼in.
Centre of driving to centre of trailing wheel, 8ft. 7in.
Weight loaded (bogie), 13 tons 16 cwt.
    "      "   (driving), 16 tons 10 cwt.
    "      "   (trailing), 14 tons 10 cwt.
Total, 44 tons 16 cwt.
Boiler, 4ft. 2in. diameter, 10ft. 7⅜in. long.
Firebox, 6 ft. long, 4ft. 1in. wide, 5ft. 10in. high.
Number of tubes, 220.
Tubes (outside diameter), 1⅜in.
Heating surface, tubes, 1,108·73sq.-ft.
    "        "    firebox, 107·68 sq.-ft.
Total, 1,216·41sq.-ft.
Firegrate area, 18·75sq.-ft.
Pressure of steam per sq. inch, 160lbs.
Weight of 6-wheel tender, loaded, 26 tons 2 cwt. 2 qrs.
Capacity of water tank of tender, 1,800 gallons.
Fuel capacity of tender, 3 tons.
```

The "A" or standard class goods engines have cylinders, boilers, heating surface, steam pressure, etc., the same as the "H" class of passenger engines just described; whilst a similar pattern of tender is employed, the six-coupled wheels are 5ft. 1in. diameter; the wheel base is: L. to D., 7ft. 9in.; D. to T., 8ft. 7in.; total, 16ft. 4in.

FIG. 119.— STANDARD 8-WHEEL PASSENGER TANK ENGINE, L. & Y.R.

Weight in working order: L., 13 tons 16 cwt. 2 qr.; D., 15 tons; T., 13 tons 6 cwt. 2 qr.; total, 42 tons 3 cwt.

Tank engines are employed to work the trains between Manchester and Blackburn, a distance of 24½ miles, of which 13 miles are on rising gradients, and six on falling gradients, most of them being steeper than 1 in 100. The most serious gradients affecting the working of this line are those from Bolton up towards Entwistle, where, for a mile and a quarter, the gradient is 1 in 72, and for the following 4½ miles is 1 in 74; a more serious incline than the celebrated one over Shap Fell. These tank engines are fitted with the water pick-up apparatus which can be used when running either chimney or bunker in front. The trains each consist of thirteen coaches, which including the engine weigh about 250 tons.

The engines (Fig. 119) have eight wheels—viz., a pair of leading radial, two pairs of coupled, and a pair of trailing radial. The cylinders are inside, and have 26in. stroke, the diameter being 18in. The diameters of the wheels are:—

Radial, 3ft. 7¼in.
Coupled (driving and trailing), 5ft. 8in.
Wheel base, 24ft. 4in., divided as follows:—Front radial wheel to centre of driving, 7ft. 10½in.; driving to rear coupled, 8ft. 7in.; rear coupled to trailing radial, 7ft. 10¼in.
Weight loaded (leading radial wheel), 13 tons 10 cwt.
  "     "    (driving), 16 tons 12 cwt.
  "     "    (rear coupled), 15 tons 2 cwt.
  "     "    (trailing radial), 10 tons 15 cwt.
Total, 55 tons 19 cwt.

FIG. 120.—OIL-FIRED SADDLE TANK SHUNTING ENGINE, LANCASHIRE AND YORKSHIRE RAILWAY

The boiler, fire-box, etc., dimensions are the same as the "H" class. The tanks of these locomotives hold 1,340 gallons of water, and the bunkers two tons of coal.

EVOLUTION OF THE STEAM LOCOMOTIVE 285

The above three classes form the leading types of locomotives of Mr. Aspinall's designing. Fig. 120 illustrates a four-wheel-coupled saddle tank locomotive designed by Mr. Aspinall, and fired with oil, on Holden's system. It is used for shunting at Liverpool.

Fig. 121.—"DUNALASTAIR," CALEDONIAN RAILWAY

At the present time the locomotives of the Caledonian Railway hold first place in the popular mind for speed and hauling capacity. This result has been attained through the remarkable performances of the engines of the "Dunalastair" class, constructed at St. Rollox

Works from the designs of Mr. J. F. McIntosh, the present locomotive superintendent of the Caledonian Railway.

These engines (Fig. 121) have been frequently described, but it is as well to recapitulate the leading dimensions. The cylinders are inside, 18¼in. diameter and 26in. stroke. The engine is supported by a leading bogie, and by four-coupled wheels of 6ft. 6in. diameter. The bogie wheel base is 6ft. 6in.; centre of bogie to driving wheel, 9ft. 11in.; D. to T., 9ft.; total length over buffers (engine and tender), 53ft. 9¾in. The weight in working order is: Engine—bogie, 15 tons 14 cwt. 3 qr.; D., 16 tons; T., 15 tons 5 cwt.; tender—L., 12 tons 13 cwt.; M., 13 tons 4 cwt.; and T., 13 tons 4 cwt. 2 qr.; total, 86 tons 1 cwt. 1 qr.

The tractive force is 14,400lb. Water capacity of tender is 3,570 gallons. The working pressure is 160lb. The leading feature of the engine consists of the large heating surface—viz., tubes, 1,284.45 sq. ft., and fire-box, 118.78 sq. ft. To obtain this result the boiler has been "high pitched," giving the engine a rather squat appearance, and causing the driving wheels to appear to be of smaller diameter than is actually the case.

An extended cab is provided for the protection of the driver and fireman. The splendid work performed by these machines has frequently been chronicled, the principal feature being the daily run from Carlisle to Stirling, 118 miles, in 123 minutes, without a stop: this trip includes the tremendous pull up the Beattock Bank, with a rise of 650ft. in ten miles. Yet Sir James Thompson, the general manager of the Caledonian Railway, said of this class of engine, " But, effective as it is, we are already improving upon it, and it will undoubtedly be superseded by our next type of engines."

As Sir J. Thompson intimated, Mr. McIntosh improved upon the above type, the result being the excellent "Dunalastair 2" (Fig. 123). These fine engines also are employed to haul the heavy West Coast corridor trains between Carlisle and Glasgow, and Edinburgh and the North.

From Glasgow to Carlisle one of the engines hauls the 2.0 p.m. corridor train without a pilot throughout the journey, the weight of the train, excluding passengers, luggage, and tender of engine, is upwards of 350 tons. The dimensions are: wheels, 6ft. 6in., D. and T. coupled, with leading bogie; cylinders 19in. by 26in. Tender runs on two four-wheel bogies; water capacity, 4,125 gallons. The weights on

FIG. 122.—ONE OF McINTOSH'S FAMOUS "DUNALASTAIR 2," CALEDONIAN RAILWAY EXPRESS LOCOMOTIVES

wheels are as follows: engine—bogie, 16 tons 6 cwt.; driving wheels, 16 tons 17 cwt.; trailing, 15 tons 17 cwt.—total, 49 tons. Tender: front bogie, 22 tons 11¾ cwt.; hind bogie, 22 tons 6¼ cwt.—total, 45 tons. Total weight of engine and tender in working order, 94 tons. Total length over buffers (engine and tender), 57ft. 3¾in.; tractive force, 16,840lb.; working pressure, 175lb. per sq. in. Heating surface: tubes, 1,381.22 sq. ft.; fire-box, 118.78 sq. ft.—total, 1,500 sq. ft.

Bogie wheel base, 6ft. 6in.; centre of bogie to driving, 10ft. 11in.; driving to trailing, 9ft.; distance between bogie centres of tender, 11ft. 3in.; total tender wheel base, 16ft. 9in.

Another new type of engine introduced by Mr. McIntosh has 5ft. 6in. coupled wheels. It is a passenger-goods, or mixed traffic engine (Fig. 123), for working goods, mineral and heavy passenger and

Fig. 123.—SIX-WHEELS-COUPLED CONDENSING ENGINE, CALEDONIAN RAILWAY

excursion trains through the Glasgow Central Underground Railway. Wheel base, L. to D., 7ft. 6in.; D. to T., 8ft. 9in.; cylinders, 18in. by 26in.; six-wheeled tender; water capacity, 2,800 gallons.

Another good design of Caledonian Railway engines is the "Carbrook" (Fig. 124) class, constructed from Mr. D. Drummond's specification with a leading bogie, and four-coupled wheels of 6ft. 6in. diameter. The weight of these engines is: bogie, 14 tons 15 cwt.; D., 15 tons 4 cwt.; T., 15 tons; L., 10 tons 16 cwt. 2 qr.; M., 14 tons 6 cwt. 3 qr. Wheel base: bogie, 6ft. 6in.; centre of bogie to D., 9ft. 10in.; D. to T., 9ft.; total length over buffers (engine and tender), 54ft. 6in. Water capacity of tender, 3,560 gallons. The safety valve is located on top of the dome, an unsymmetrical practice which spoils the outline. There is also another type of Mr. Drummond's engines, with cylinders, 18in. by 26in. stroke. Wheel base and water capacity as in the "Dunalastair" class; but the weight and tractive force are dissimilar. The former, on bogie, is 14 tons 13 cwt. 2 qr.; D., 15 tons 7 cwt. 3 qr.; tender, L., 12 tons 13 cwt.; M., 13 tons 4 cwt.; T., 13 tons 4 cwt. 2 qr.; total, 84 tons 6 cwt. 3 qr. The tractive force is 12,900lb.

FIG. 124.—"CARBROOK," ONE OF DRUMMOND'S EXPRESS ENGINES FOR THE CALEDONIAN RAILWAY

To give full details of all the 44 types of Caledonian Railway engines would be rather wearisome to the reader, so of the remaining classes, particulars only are appended:—

PASSENGER ENGINES WITH TENDERS.

Diameter of Driving Wheels:
- 5ft. 9in. four-coupled, with leading bogie. Cylinders, 18in. by 26in. stroke.
- 7ft. single, with leading bogie and pair of trailing wheels. Cylinders, 18in. by 26in. stroke.
- 7ft. four-coupled, with leading bogie. Cylinders, 18in. by 24in. stroke. (This is a rebuilt type of engine.) Tender only holds 1,880 gallons.
- 7ft. four-coupled, with a small pair of leading wheels. Cylinders, 17in. diameter by 24in. stroke. No dome on boiler.
- 6ft. 6in. four-coupled (D. and T.), with a small pair of leading wheels. Cylinders, 17in. by 24in. stroke. No dome to engine, and only four wheels to tender, with a water capacity of 1,428 gallons.
- 6ft. D. and T. coupled, small leading wheels. No dome. Cylinders, 17in. diameter by 22in. stroke. Six-wheel tender.

8ft. 2in. single, small leading and trailing wheels. No dome. Cylinders, 17in. diameter by 24in. stroke. Six-wheel tender.

6ft. D. and T. coupled, small leading wheels, rebuilt by Drummond, with safety valve on dome. Cylinders, 18in. by 24in.

5ft. D. and T. coupled, with leading bogie. Cylinders, 18in. by 24in. Four-wheel tender; water capacity, 1,550 gallons.

7ft. single. Cylinders, 17¾in. by 22in. No dome. Four-wheel tender, 1,384 gallons.

FIG 125.—McINTOSH'S 5FT. 9IN. CONDENSING TANK ENGINE, CALEDONIAN RAILWAY

PASSENGER TANKS.

.5ft. L. and D. coupled, trailing bogie; cylinders, 16in. by 22in.; Drummond valve; water capacity of tanks, 830 gallons.

5ft. single (for use of officials): cylinders, 9½in. diameter by 15in. stroke; well tank holds 520 gallons; bunker, 30 cwt. of coal; wheel base: L. to D., 6ft. 6in.; D. to T., 7ft. 6in. Weight: L., 7 tons 10 cwt. 3 qr.; D., 11 tons 6 cwt. 2 qr.; T., 7 tons 16 cwt. 1 qr.; tractive force, 2,489lb.

5ft. D. and T. coupled; cylinders, 17½in. by 22in. Water, 820 gallons.

5ft. 6in. L. and D. coupled; cylinders, 16in. by 20in. Water, 450 gallons.

4ft. 6in. L. and D. coupled with trailing bogie; cylinders, 18in. by 22in. Water capacity, 950 gallons.

3ft. 8in. L. and D. coupled, and pair of trailing wheels; cylinders, 14in. by 20in. stroke. The saddle tank holds 800 gallons.

5ft. 8in. radial L. and T. wheels, and 4 coupled wheels (eight wheels in all); cylinders, 17½in. by 22in. Water in side tanks, 1,200 gallons. Coal in bunker, 3 tons.

5ft. D. and T. coupled with leading bogie; cylinders, 17in. by 24in.

5ft. 9in. L. and D. coupled, with trailing bogie; cylinders, 18in. by 26in. This class is fitted with condensing apparatus. (Fig. 125.)

## GOODS ENGINES WITH TENDERS.

6ft. D. and T. coupled, with pair of leading wheels. Cylinders, 18in. diameter, by 24in. stroke. Six-wheel tender; water capacity, 1,840 gallons.

5ft. six-wheels coupled. Cylinders 18in. by 26in. 6-wheel tender; water capacity, 2,500 gallons.

The following engines have no domes:—

5ft. 6-wheels coupled. Inside cylinders, 17in. by 24in. 6-wheel tender; water capacity, 1,800 gallons.

5ft. (mineral engine) L. and D. coupled, small trailing wheels, no dome. Cylinders, 17in. by 24in. 4-wheel tender; 1,542 gallons.

5ft. 6-wheels-coupled mineral engine. Wheel base: L. to D., 5ft. 6in.; D. to T., 5ft. 6in.; all wheels under boiler barrel. Cylinders, 18in. by 24in. 6-wheel tender; water capacity, 1,840 gallons.

5ft. 6-wheels coupled mineral engine. Cylinders 17in. by 24in. 4-wheel tender; water capacity, 1,383 gallons.

5ft. 6in. L. and D. coupled and small pair trailing wheels; inside cylinder, 16in. by 20in. 4-wheel tender.

4ft. 8in. L. and D. coupled, mineral engine. Cylinders, 17in. by 20in. 4-wheel tender; water capacity, 1,000 gallons.

(A similar class of engines has cylinders 17in. diameter by 18in. stroke.)

5ft. D. and T. coupled, with pair of small leading wheels. Cylinders, 17in. by 24in. 4-wheel tender; 1,545 gallons.

5ft. D. and T. coupled, with small leading wheels. Cylinders, 17in. by 20in. 6-wheel tender; water capacity, 1,700 gallons.

## MINERAL TANK ENGINES.

4ft. 6in. 6-wheels coupled, saddle tank, holding 1,000 gallons of water; safety valves on dome; cylinders 18in. by 26in.

4ft. 6in. 6-wheels coupled, side tanks, with condensing apparatus; cylinders, 18in. by 26in.

4ft 6in. 6-wheels coupled; saddle tank; cylinders, 18in. by 26in.

4ft. 6-wheels coupled; saddle tank, 1,000 gallons; cylinders, 18in. by 22in. stroke.

4ft. 6-wheels coupled; saddle tank, 940 gallons. Cylinders, 17in. by 20in. No dome.

3ft. 8in. 4-wheels-coupled; wheel base, 7ft.; saddle tank, 800 gallons. Cylinders, 14in. by 20in.

There is a similar class of engine built by Neilsons, the difference being in the weight. That of the former is, on leading axle, 13 tons 14 cwt. 1 qr.; on driving axle, 13 tons 13 cwt. 1 qr.

Weight of Neilson's class: L., 13 tons 10 cwt. 3 qr.; D., 13 tons 9 cwt. 1 qr.

3ft. 8in. 6-wheels-coupled, saddle tank; water capacity, 900 gallons. Drummond's safety valves. Cylinders, 14in. by 20in. stroke.

Lastly, a class of 4-wheel engines, with coupled wheels, 3ft. 6in. diameter; side tanks hold 500 gallons. No dome, cab, or weather board; wheel base, 6ft. 3in. Cylinders, 14in. diameter, 22in. stroke.

FIG. 126.—"No. 143," TAFF VALE RAILWAY INCLINE TANK LOCOMOTIVE

Engine "No. 143" (Fig. 126) is one of three peculiar locomotives, specially constructed for working on the Pwllyrhebog Incline, of 1 in 13, on the Taff Vale Railway. The fire-box and roof slopes backwards, so that when the engine works bunker first up the incline, the water is level over the top of the fire-box. She is fitted with two drawbars for attaching a wire rope. This rope is coupled to a low draw-bar under the drag-plate, so as to keep the rope below the axles of the wagons, which follow the engine down the incline, or are pushed up before the engine. "143" has cast-iron "Sleigh" brakes acting on the rails, in addition to the usual steam brakes on the wheels. The dome is placed on the fire-box, and the regulator is within it, so as to ensure dry steam when working on the incline. Wheels, 5ft. 3in. diameter. Cylinders, 17½in. by 26in. Weight, 44 tons 15 cwts.

A FAVOURITE LOCOMOTIVE OF THE ISLE OF WIGHT CENTRAL RAILWAY

# CHAPTER XV.

Great Western "convertible" locomotives—The value of names in locomotive practice—Water troughs on the G.W.R.—Dean's 7ft. 8in. singles—His "Armstrong" class—An extension smoke-box on the G.W.R.; the "Devonshire" class—7ft. "singles"—"2202" and "3225," four-coupled G.W. engines—The "Barrington"—Great Western passenger tanks—"Bull Dog" design—"No. 36," Great Western Railway—A six-wheel coupled goods engine with a leading bogie—Ivatt's advent on the Great Northern, and his innovations —"Domes" to the fore—New goods and tank engines—Rebuilt "Stirlings" —Ivatt's inside cylinder four-coupled bogie engines—His chef d'œuvre "990"—A ten-wheel tank on the G.N.R.—"266," the latest Great Northern engine—Possibilities of the future—Great North of Scotland locomotives—Manson's designs—James Johnson's tank and tender engines—Furness engines, passenger and goods—The 1896 "express" design—Pettigrew's new goods engines—Highland Railway engines—A Great Central Railway locomotive—Some Irish locomotives—Belfast and Northern Counties Compounds—The "Restrevor" class, G.N. (I.)—Great Southern and Western standard passenger design—A locomotive for an Irish "light" railway.

The broad gauge having been finally abandoned on the G.W.R. in May, 1892, it became necessary to re-arrange the locomotive power. Previous to that date Mr. W. Dean, the G.W.R. Locomotive Superintendent, had constructed at Swindon several six-wheeled express locomotives (Fig. 127), with "single" driving wheels, 7ft. 8in. in diameter, inside cylinders 20in. in diameter, and a stroke of 24in., and weighing 44 tons 4 cwt., of which 13 tons 4 cwt. was on the leading axle.

This class of engine was designed to work the West of England expresses between London and Newton Abbot, consequent upon the conversion of the gauge, and the locomotives were therefore built upon strictly narrow-gauge dimensions, but some few of them were worked on the West of England expresses whilst the gauge was yet broad, and for this purpose the wheels were fixed outside the framing. In this condition they had a very curious and ungainly appearance, intensified by the squat chimney, large dome, and bulged fire-box covering.

After the alteration of the gauge had been effected, and the wheels of the engines of this class had been fixed in their normal position, their appearance was considerably improved, but there still remained about the locomotives a somewhat indescribable want of symmetry and unison of outline. However, it was decided to substitute a bogie for the pair of leading wheels, whilst the diameter of the cylinders was reduced to 19 inches. These alterations, coupled with other minor improvements, added to the admittedly good qualities of the engines as locomotive machines, soon caused the class, thus improved, to gain a high place in the estimation of both experts and the railway public. The amount of bright brass about the engines and the names carried by them—mostly those of famous broad-gauge engines, or popular broad and narrow-gauge, Great Western Railway officials—have also added to the prestige of the design. Let cynics say what they will, one feels more interest for, say, the "Rover" than he can ever expect to for plain "No. 999."

The adoption of water troughs on the Great Western Railway, and the addition of the "pick-up" apparatus to the tenders of these engines, enables the Great Western Railway to perform many daily runs for length and speed that, a few years back, would rightly have been considered quite phenomenal. Happily, we improve with giant strides in matters locomotive at the tail end of the 19th century.

[Photo]  [F. Moore
FIG. 127.—7FT. 8IN. "SINGLE" CONVERTIBLE ENGINE, GREAT WESTERN RAILWAY

With the adoption of the normal gauge over the whole of the Great Western Railway system, engines of this class are now used on the expresses on all sections where the character of the gradients allows such engines to be run with proper economy. Under these circumstances, it is not surprising to learn that additional batches of engines of Mr. Dean's 7ft. 8in. "single" design (Fig. 128) are being added to the Great Western Railway locomotive stock at not infrequent intervals. At the present time, there are 71 of these engines at work, and nine others under construction—probably a larger number of one class of modern express locomotives than can be found elsewhere.

Fig. 128.—"Empress of India," a standard Great Western 7ft. 8in. single express locomotive

The huge pipe for delivering the feed-water to the boilers of these engines, formerly placed in a conspicuous position, has been removed, an alteration that has added much to the beauty of outline of these fine-looking locomotives.

Mr. Dean has constructed a class of four-coupled engines, with a leading bogie, known as the "Armstrong" class. In its salient features, the design is a modification of the 7ft. 8in. single class described above, but naturally several of the dimensions are dissimilar in the two classes. "Armstrong" is No. 7, "Gooch" (Fig. 129), No. 8, "Charles Saunders," No. 14, and "Brunel," No. 16.

FIG. 129.—"GOOCH," A 4-COUPLED EXPRESS ENGINE, GREAT WESTERN RAILWAY

Immediately subsequent to the change of gauge in May, 1892, a class of tank engines, with wheels four-coupled in front and a trailing bogie, was built for working the fast passenger traffic west of Newton Abbot. The bogies of these engines were fitted with Mansel wheels—quite an exceptional practice in locomotive building.

Mr. Dean has since designed another class of locomotive to work the fast train traffic over the severe gradients and curves so common to the Great Western Railway main line west of Newton Abbot.

These engines are popularly called the "Devonshire" or "Pendennis Castle" class (Fig. 130), after the name given to the first engine

298  EVOLUTION OF THE STEAM LOCOMOTIVE

FIG. 130.—"PENDENNIS CASTLE," ONE OF THE GREAT WESTERN "HILL CLIMBERS"

constructed on the plan. A prominent feature of the design is the "extension" smoke-box—a feature copied from modern American practice. Before constructing the "Pendennis Castle," Mr. Dean had fitted another engine—No. 426—with an extended smoke-box, and the result of the trials made with this locomotive satisfied the Great Western Railway locomotive superintendent as to the advantages of the arrangement.

The cylinders of this class are 18in. diameter, the stroke being 26in. The coupled wheels (D. and T.) are 5ft. 7½in. diameter, that of the (leading) bogie being 3ft. 7½in. The use of Mansel wheels has also been adopted both for the bogies and the tenders of the locomotives of this class. The frames are double, and are specially contracted at the smoke-box end to allow sufficient play to the bogie wheels. Both inside and outside bearings are provided for the driving axle. The boiler is of steel, the heating surface being: Tubes, 1,285.58 sq. ft.; fire-box, 112.60 sq. ft.; steam pressure, 160lb.; grate area, 19 sq. ft.; weight of engine, 46 tons, of which 15 tons 7 cwt. is on the driving axle, 17½ tons on the bogie, and 13 tons 3 cwt. on the trailing (coupled) axle. The tender holds 2,000 gallons of water, and weighs, loaded, 24 tons. Ten engines of this design were originally constructed at Swindon—viz. :—

| | | |
|---|---|---|
| 3252 Duke of Cornwall. | 3255 Cornubia. | 3259 Lizard. |
| 3253 Pendennis Castle. | 3256 Excalibur. | 3260 Merlin. |
| 3254 Boscawen. | 3257 Guinevere. | 3261 Mount Edgcumbe. |
| | 3258 King Arthur. | |

These proved so satisfactory in performing the peculiar duties required from passenger engines on the West of England main line of the Great Western Railway that a second batch of twenty was put in hand. These commenced running in the early months of 1898. They are named and numbered as follow :—

| | | |
|---|---|---|
| 3262 Powderham. | 3272 Amyas. | 3282 Ma Istowe. |
| 3263 Sir Lancelot. | 3273 Armorel. | 3283 Mounts Bay. |
| 3264 St. Anthony. | 3274 C rnishman. | 3284 Newquay. |
| 3265 St. German. | 3275 Chough. | 3285 St. Erth. |
| 3266 St. Ives. | 3276 Dartmoor. | 3286 St Just. |
| 3267 St. Michael. | 3277 Earl of Devon. | 3287 St. Agnes. |
| 3268 Tamar. | 3278 Eddystone. | 3288 Tresco. |
| 3269 Tintagel. | 3279 Exmoor. | 3289 Trefusis. |
| 3270 Trevithick | 3280 Falmouth. | 3290 Torbay. |
| 3271 Tre Pol and Pen. | 3281 Fowey. | 3291 Tregenna. |

Several of these engines have the tenders fitted with the water pick-up apparatus.

The names, it will be observed, should specially please the patrons of the Great Western Railway residing in Devon and Cornwall, and help to palliate the keen regret with which the abolition of the broad-gauge was felt in those counties.

Among types of Great Western locomotives, one may be mentioned

FIG. 131.—SINGLE EXPRESS ENGINE, 6-WHEEL TYPE, GREAT WESTERN RAILWAY

—the 7ft. "singles" (Fig. 131), largely used for hauling the express trains on the Birmingham and Northern lines. The cylinders are 18in. diameter, the stroke being 24in. Heating surface, 1,250.31 square feet.

Many of the passenger trains on the Gloucester and Weymouth sections are worked by the 6ft. 6in. four-coupled engines, illustrated by engine 2,202 (Fig. 132). The leading dimensions of this class are:

FIG. 132.—6FT. 6IN. 4-COUPLED PASSENGER LOCOMOTIVE, GREAT WESTERN RAILWAY

Cylinders, 17in. diameter; stroke, 24in.; heating surface, 1,363.5 sq. ft. Weight of engine and tender, in working order, 59 tons 8 cwt.

North of Wolverhampton, for working the West to North expresses, and for other fast trains in the North Western district of the G.W.R., the engines represented by 3,225 (Fig. 133) are largely used.

FIG. 133. - 6FT. 4-COUPLED PASSENGER ENGINE, GREAT WESTERN RAILWAY

This class has cylinders 18in. by 24in. stroke; leading wheels 4ft. diameter, and coupled, driving, and trailing wheels, 6ft. diameter. The heating surface totals to 1,468.82 sq. ft.; and the weight of engine

FIG. 134.—"BARRINGTON," NEW TYPE OF 4-COUPLED ENGINE, GREAT WESTERN RAILWAY

and tender, including the load of 4 tons of coal and 3,000 gallons of water, amounts to $74\frac{1}{2}$ tons.

"Barrington" (Fig. 134) is one of Mr. Dean's latest type of express passenger engine. These powerful locomotives are somewhat of the "Devonshire" type, having an extended smoke-box, whilst the "Belpaire" fire-box is also introduced. In the framing, it will be noticed, early G.W. practice is reverted to. The cylinders are 18in. by 26in. stroke. The bogie wheels are 4ft., and the coupled wheels 6ft. 8in. in diameter. The engine weighs 51 tons 13 cwt.; the tender, with the same amount of water and coal as "3,225" class, 32½ tons.

A good deal of the G.W. passenger trains are hauled by smart little six-wheel (four-coupled) tank engines, which are specially noted for getting away quickly, and immediately attained high speeds. "No. 576" (Fig. 135) represents a coupled-in-front engine of this description, but the more generally-known Great Western Railway pas-

FIG. 135.— 4-COUPLED-IN-FRONT PASSENGER TANK ENGINE, G.W.R.

senger tank engines have the driving and trailing wheels coupled; these are 5ft. diameter, the cylinders being 16in. diameter by 24ft. stroke.

Mr. Dean's latest creation for the Great Western Railway is named "Bull Dog," No. 3,312, and the design will be known as the "Bull Dog" class. Except that the bogie wheels have spokes, the wheels, framing, and motion are similar to the "Devonshire" class (Fig. 130). The boiler is of gigantic proportions; the fire-box is of the Belpaire type, and projects over the top and sides of the boiler barrel. The smoke-box is extended, and steaming reversing gear is employed, whilst another improvement, Davies and Metcalfe's patent exhaust steam injector, is fitted to the engine; and is being extensively adopted on Great Western Railway locomotives. The name-plates are on the sides of the fire-box; the clack valves are below the boiler barrel, behind the smoke-box. The cab of the "Bull Dog" extends to the edge of the footplate, with a door in the front on the fireman's side.

Before closing these remarks on modern Great Western Railway locomotives, some description of No. 36 is necessary. Here again we have an adaptation of American practice—a six-wheels-coupled engine, with a leading bogie, and an extension smoke-box. The cylinders are inside, 20in. diameter by 24in. stroke, with the steam chests below them. The driving wheels are 4ft. 6in. diameter, the bogie wheels only 2ft. 8in. diameter. All the wheels have outside bearings, and the driving wheels have inside bearings in addition. The boiler contains 150 "Serve" tubes of $2\frac{1}{2}$in. diameter. The total heating surface is 2,385 sq. ft.; steam pressure, 165lb.; grate area, 35 sq. ft. The weight is as follows: On bogie, 12 tons 6 cwt.; leading coupled wheels, 15 tons 12 cwt.; driving wheels, 16 tons 11 cwt.; and trailing wheels, 15 tons 1 cwt. Total weight of engine, $59\frac{1}{2}$ tons; of tender, 32 tons; together, $91\frac{1}{2}$ tons. The tender is fitted with a water pick-up apparatus. This locomotive has been employed in hauling goods trains for many months past, and it is stated to have hauled a train weighing 450 tons through the Severn Tunnel—despite the severe gradients and length—in ten minutes, although for such a load two goods engines of the usual Great Western design would be required, and they would take 18 minutes to perform the trip.

Consequent upon the death of the late Mr. Patrick Stirling—one of the best locomotive superintendents of his time—the directors of the Great Northern Railway appointed Mr. H. A. Ivatt to the supreme command at Doncaster. Mr. Ivatt received his early training in the science of locomotive construction at Crewe, and left the Great Southern and Western Railway (Ireland), where he was locomotive superintendent, to succeed Mr. P. Stirling on the Great Northern Railway.

Mr. Ivatt, having decided opinions of his own relative to locomotive design, soon set to work to introduce his ideas on the Great Northern system; so that after many years—more than two decades—of domeless locomotives, Doncaster awoke one morning to find a Stirling 8ft. "single" fitted with a steam dome encased in a green-painted cover. It was certainly a great surprise—the colour especially, for many had hoped to see bright brass—but those interested survived the shock, and waited to see some engines of Mr. Ivatt's design on the Great Northern Railway.

Several engines, with pronounced Ivatt features, were soon running, but the main designs of all of them are cast after distinctly Stirling models, as they were already under construction at the time of Mr. Ivatt's appointment.

In the 1070 class (four-coupled, six-wheeled engines) we find that the dome and cab, amongst external signs, are the work of the new chief

FIG. 136.—"No. 1312," ONE OF MR. IVATT'S (1073) SMALLER CLASS OF 4-COUPLED BOGIE ENGINES FOR THE GREAT NORTHERN RAILWAY

at Doncaster; whilst those of the 1073 design have his leading bogie, splasher over the coupled-wheels, dome, and cab.

Coming to "No. 34," a rebuilt 8ft. "single," Mr. Ivatt is responsible for the dome, cab, and safety-valve casing, whilst in the 1206, six-coupled saddle-tanks, we again find the dome and new pattern valve casing.

Readers will notice that we have only referred to the apparent details that are attributed to Mr. Ivatt, but, by reference to the appended tables of dimensions, they will find that several alterations that do not so readily meet with notice have been made in other matters connected with the Great Northern locomotives.

### 6FT. 6IN. FOUR-WHEELS COUPLED ENGINE, No. 1070.

#### CYLINDERS.
| | |
|---|---|
| Diameter | 17¼in. |
| Stroke | 26in. |

#### WHEELS.
| | |
|---|---|
| Driving | 6ft. 6in. diameter. |
| Trailing | 6ft. 6in. diameter. |
| Leading | 4ft. 0in. diameter. |

#### WHEEL CENTRES.
| | |
|---|---|
| From centre of trailing to centre of driving wheels | 8ft. 3in. |
| From centre of driving to centre of leading wheels | 9ft. 8in. |
| Total wheel base | 17ft. 11in |

EVOLUTION OF THE STEAM LOCOMOTIVE 305

### BOILER.

| | |
|---|---|
| Length of barrel | 10ft. 1in. |
| Diameter of barrel | 4ft. 5in. |
| Length of firebox casing | 5ft. 6in. |

### HEATING SURFACE.

| | |
|---|---|
| Tubes | 1,020·7sq.-ft. |
| Firebox | 103·1sq.-ft. |
| Total | 1,123·8sq.-ft. |
| Grate area | 17·8sq.-ft. |
| Tubes | 215—1¼in. diameter outside |

### 6FT. 6IN. FOUR-WHEELS COUPLED BOGIE ENGINE, No. 1073.
(Illustrated by Fig. 136.)

### CYLINDERS.

| | |
|---|---|
| Diameter | 17¼in. |
| Stroke | 26in. |

### WHEELS.

| | |
|---|---|
| Driving | 6ft. 6in. diameter. |
| Trailing | 6ft. 6in. diameter. |
| Bogie | 3ft. 6in. diameter. |

### WHEEL CENTRES.

| | |
|---|---|
| From centre of trailing to centre of driving wheels | 8ft. 3in. |
| From centre of driving to centre of bogie pin | 9ft. 9in. |
| Centres of bogie wheels | 6ft. 3in. |
| Total wheel base | 21ft. 3in. |

### BOILER.

| | |
|---|---|
| Length of barrel | 10ft. 1in. |
| Diameter of barrel | 4ft. 5in. |
| Length of firebox casing | 5ft. 6in. |

### HEATING SURFACE.

| | |
|---|---|
| Tubes | 1,020·7sq.-ft. |
| Firebox | 103·1sq.-ft. |
| Total | 1,123·8sq.-ft. |
| Grate area | 17·8sq.-ft. |
| Tubes | 215—1¼in. diameter outside. |

FIG. 137.— THE LATEST TYPE OF 6FT. 6IN. COUPLED ENGINE,
GREAT NORTHERN RAILWAY

Fig. 137 represents the larger and later type (just out) of the 6ft. 6in. four-coupled engine, with a leading bogie, on the Great Northern Railway. In these engines the boiler diameter has been augmented by 3in., so that it bulges out over the splashers; the heating surface is increased to 1,250 sq. ft., while the fire-box is greatly enlarged, having 120 sq ft. This enlargement of the fire-

X

box has involved a lengthening of the side rods and coupled-wheel base by 9in. The fire-grate area is 20.8in., instead of 17.8in., in the smaller engines. The chimney, which is much shorter owing to the height of the boiler, is built up in three pieces.

### 8FT. SINGLE PASSENGER ENGINE No. 34.

**CYLINDERS.**

| | |
|---|---|
| Diameter | 18in |
| Stroke | 26in. |

**WHEELS**

| | |
|---|---|
| Driving | 8ft. 0in. diameter. |
| Trailing | 4ft. 6in. diameter. |
| Bogie | 3ft. 10in. diameter. |

**WHEEL CENTRES.**

| | |
|---|---|
| From centre of trailing wheel to centre of driving | 9ft 0in. |
| From centre of driving wheel to centre of bogie pin | 10ft. 9in. |
| Centres of bogie wheels | 6ft. 6in. |
| Total wheel base | 23ft. 3in. |

**BOILER.**

| | |
|---|---|
| Length of barrel | 11ft. 2in. |
| Diameter of barrel | 4ft. 9in. |
| Length of firebox casing next to barrel | 6ft. 9in. |
| Length of firebox casing at bottom | 7ft. 2in. |

**HEATING SURFACE.**

| | |
|---|---|
| Tubes | 980sq.-ft. |
| Firebox | 114·2sq.-ft. |
| Total | 1,094·2sq.-ft. |
| Grate area | 23· 6sq -ft. |
| Tubes | 184—1¾in. diameter outside. |

### 4FT. 6IN. SIX-WHEELS COUPLED SADDLE TANK ENGINE, No. 1200.

**CYLINDERS.**

| | |
|---|---|
| Diameter | 18in. |
| Stroke | 26in. |

**WHEELS.**

| | |
|---|---|
| Driving | 4ft. 6in. diameter. |
| Trailing | 4ft 6in. diameter. |
| Leading | 4ft 6in. diameter. |

**WHEEL CENTRES.**

| | |
|---|---|
| From centre of trailing to centre of driving wheels | 8ft. 3in. |
| From centre of driving to centre of leading wheels | 7ft. 3in. |
| Total wheel base | 15ft. 6in. |

**BOILER.**

| | |
|---|---|
| Length of barrel | 10ft. 6in. |
| Diameter of barrel | 4ft. 5in. |
| Length of casing | 5ft. 6in. |

**HEATING SURFACE.**

| | |
|---|---|
| Tubes | 1,061·13sq.-ft. |
| Firebox | 103·1sq.-ft. |
| Total | 1,164·23sq -ft. |
| Grate area | 17·8sq -ft |
| Tubes | 215—1⅞in. diameter outside. |

Mr. Ivatt's express passenger engine No. 990 (illustrated as a frontispiece to this volume) is quite a new departure in British locomotive practice, having a leading bogie, four-coupled wheels in front of the fire-box, and a pair of trailing wheels under the foot-plate. The dimensions are:—

### CYLINDERS.
| | |
|---|---|
| Diameter | 19in. |
| Stroke | 24in. |

### WHEELS.
| | |
|---|---|
| Trailing | 3ft. 6in. |
| Coupled | 6ft. 6in. |
| Bogie | 3ft. 6in. |

### WHEEL CENTRES.
| | |
|---|---|
| From centre of trailing to centre of driving wheels | 8ft. 0in. |
| Centres of coupled wheels | 6ft. 10in. |
| Centre of leading coupled to centre of trailing bogie wheel | 5ft. 3in. |
| Centres of bogie wheels | 6ft. 3in. |
| Total wheel base | 26ft. 4in. |

### BOILER.
| | |
|---|---|
| Length of barrel between tube plates | 13ft. 0in. |
| Diameter of barrel | 4ft. 8in. |
| Length outside firebox casing | 8ft. 0in. |

### HEATING SURFACE.
| | |
|---|---|
| Tubes | 1,302sq.-ft. |
| Firebox | 140sq.-ft. |
| Total | 1,442sq.-ft |
| Grate Area | 26.7sq.-ft. |
| Tubes | 191—2in. external diameter. |

Mr. Ivatt has also designed a new class of 10-wheel tank engines for the G.N.R., the leading dimensions being—

### CYLINDERS.
| | |
|---|---|
| Diameter | 17½in. |
| Stroke | 26in. |

### WHEELS.
| | |
|---|---|
| Coupled | 5ft. 6in. |
| Trailing | 3ft. 6in. |
| Bogie | 3ft. 6in. |

### WHEEL CENTRES.
| | |
|---|---|
| From centre of trailing to centre of back coupled | 6ft. 0in |
| From centre of back coupled to centre of driving | 8ft. 3in. |
| From centre of driving to centre of trailing bogie | 6ft. 9in. |
| Centres of bogie wheels | 6ft. 3in. |
| Total wheel base | 27ft. 3in. |

### BOILER.
| | |
|---|---|
| Length of barrel | 10ft. 1in. |
| Diameter of barrel | 4ft. 5in. |
| Length of firebox casing | 5ft. 6in. |

### HEATING SURFACE.
| | |
|---|---|
| Tubes | 1,020.7sq.-ft. |
| Firebox | 103sq.-ft. |
| Total | 1,123.7sq.-ft. |
| Grate Area | 17.8sq.-ft. |
| Tubes | 215—1¾in. external diameter |

There now remains to be described Mr. Ivatt's newest engine, a 7ft. 6in. single-wheeler with leading bogie, a large boiler, 11ft. 4in.

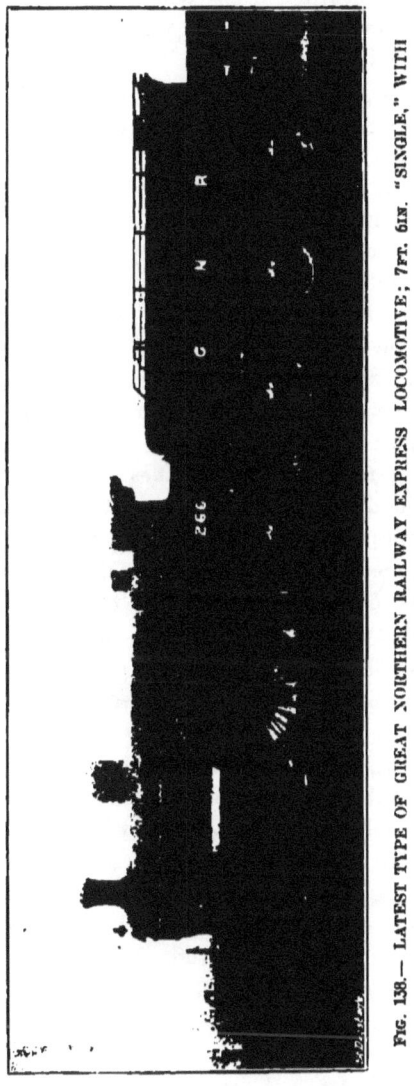

FIG. 138.— LATEST TYPE OF GREAT NORTHERN RAILWAY EXPRESS LOCOMOTIVE; 7FT. 6IN. "SINGLE," WITH INSIDE CYLINDERS AND A LEADING BOGIE

long and 4ft. 5in. diameter, giving 1,268 sq. ft. of heating surface, 175lb. steam pressure, and a fire-box 7ft. long. The boiler centre stands 8ft. 3in. above the level of the rails. The cylinders are 18in.

diameter by 26in. stroke. The grate area is 23 sq ft. As this engine is only just out of the Doncaster shops none of her performances have as yet been recorded, but if she prove as good as she looks the Great Northern Railway will have a valuable addition to its already numerous "single" locomotives.

And now, perhaps, may be ventured an opinion on Mr. Ivatt's innovations in Great Northern locomotive practice. In the first place, from an aesthetic point, there can be no two opinions that a dome greatly improves the appearance of a locomotive, but one of bright brass is infinitely superior to one covered with green paint. "To win the eye is to win all," and plenty of bright brass about a locomotive is certainly an attraction; a large amount of the popularity of the Great Western engines is due to the fine display of brass. The same reason that causes us to prefer a brass dome makes us sorry to see the Stirling brass casing of the safety valve give place to Mr. Ivatt's design. Green paint undoubtedly is a good thing, but then you can have "too much of a good thing." Again, a curved splasher for coupled wheels, following the outlines of both wheels, looks much neater than the design used with Class 1073. The bogie is decidedly an improvement; so is an extended cab, but graceful outlines might be used in connection with the latter. Mr. Ivatt has certainly introduced some decided improvements into the composition of the Great Northern Railway locomotives, but the *tout ensemble* might be more pleasing; a few alterations in matters of detail would give observers a more appreciative opinion of modern Great Northern Railway engines.

Now water-troughs are so much in fashion, it should not be difficult to find suitable locations for them on the Great Northern system, and with a double-bogie tank engine, with outside cylinders, a 9ft. or larger driving wheel, York ought to be reached in less than three hours from King's Cross, and without an intermediate stop. Will the 19th century see such an achievement? We hope so, but fear to prophesy; its sands are almost run.

The Manson engines of the Great North of Scotland Railway deserve notice. As long ago as 1878 and 1879 it was decided to place heavier and more powerful engines on that railway. The engines weighed 41 tons 5 cwt. each, and the tender 28 tons 5 cwt. in working order. The working pressure was 150lb. per square inch.

In 1884 Mr. Manson, who succeeded Mr. Cowan, got some six

wheel coupled inside cylinder tank engines from Kitson and Co., of Leeds. The following are the principal dimensions, viz.:—

| | |
|---|---|
| Cylinders | 16in. by 24in. |
| Coupled wheels | 4ft. 6in. diameter. |
| Wheel base | 13ft. 8in. |
| Tubes | 140—1¾in. external diameter |
| Heating surface—Tubes | 690sq.-ft. |
| Heating surface—Firebox | 66sq.-ft. |
| Total | 756sq.-ft. |
| Working steam pressure | 140lb. per sq.-in. |
| Weight in working order | 37 tons 7 cwts. |

In the same year Messrs. Kitson and Co. also supplied some four coupled passenger engines, with leading bogie and a six-wheeled tender. The cylinders are "inside," and the bogie is Kitson's swing link type, which this Company has used since 1884. These engines were delivered with a brick arch in the fire-box, but this was afterwards taken out, and air tubes put into the front and rear of the fire-box, so as to consume the smoke. The principal dimensions are:—

| | |
|---|---|
| Cylinders | 17½in. by 26in |
| Coupled wheels | 6ft. 0in. diameter. |
| Bogie wheels | 3ft. 0in. diameter. |
| Tender wheels | (6), 3ft. 9in diameter. |
| Wheel base of engine | 20ft. 8in. |
| Wheel base of tender | 11ft. 0in. |
| Total wheel base of engine and tender | 40ft. 3⅜in. |
| Tubes | 189—1¾in. external diameter. |
| Heating surface—Tubes | 946sq.-ft. |
| Heating surface—Firebox | 90sq.-ft. |
| Total | 1,036sq.-ft. |
| Tank capacity | 2,000 gallons. |
| Working steam pressure | 140lb. per sq.-in. |
| Weight in working order—Engine | 37 tons 2 cwts. |
| " " " Tender | 29 tons 0 cwts. |
| Total | 66 tons 2 cwts. |

In 1888 Mr. Manson brought out his engine with inside cylinders, having the valves placed on the top, which were of the balanced type introduced by Mr. Cowan. The valves were driven by the ordinary Stephenson link motion working on a rocking shaft. In other respects the engine very much resembled those just described, except that the engine and tender were coupled by a central bar and one solid central rolling block in place of side spring buffers.

The cylinders were 18in. by 26in. and the coupled wheels 6ft. 0½in. diameter.
The engine weighed .. .. .. .. 41 tons 9 cwts.
The tender weighed .. .. .. .. 29 tons 0 cwts
In working order.
They were built by Messrs. Kitson and Co.

In 1890 Mr. Manson increased the capacity of the tender to 3,000 gallons, and in doing this introduced a bogie tender. The tender was carried on eight wheels 3ft. 9½in. diameter. The four trailing wheels were fixed, and the four leading carried a bogie similar to that on the engine.

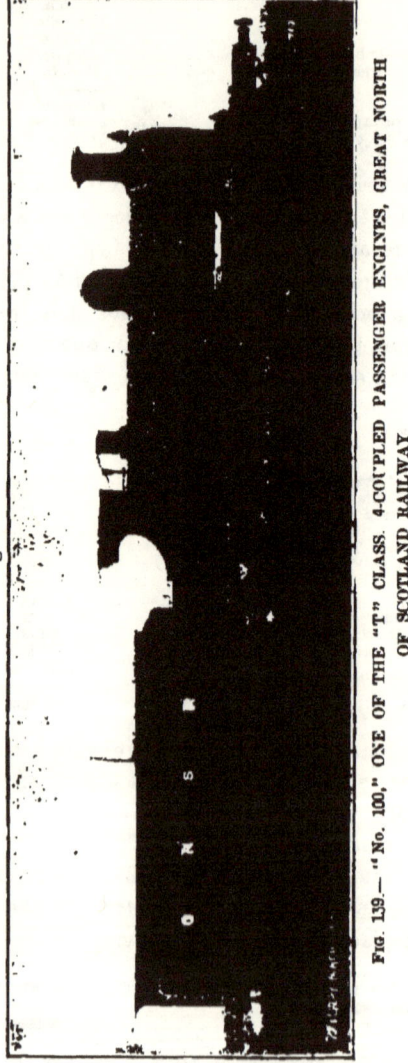

FIG. 139.—"No. 100," ONE OF THE "T" CLASS. 4-COUPLED PASSENGER ENGINES, GREAT NORTH OF SCOTLAND RAILWAY

The wheel base of the tender was 16ft. 6in., and the weight in working order 38 tons. The engine for these tenders was the same as that just described. These were built by Stephenson and Co.

In 1893 Mr. James Johnson, who succeeded Mr. Manson, designed some heavy bogie tank engines. They were four wheels coupled in front, with trailing four-wheeled bogie. The valves were of the ordinary type, placed between the cylinders, which were "inside."

These engines were fitted with the brick arch, and since that time all the Company's engines have had the air tubes removed, and brick arches fitted.

The following are the principal dimensions:—

| | |
|---|---|
| Cylinders | 17½in. by 26in. |
| Coupled wheels | 5ft. 0in. diameter. |
| Bogie wheels | 3ft. 0½in. diameter. |
| Fixed wheel base | 7ft. 6in. |
| Bogie wheel base | 5ft. 6in. |
| Total wheel base | 22ft. 0in. |
| Tubes | 220—1¾in. external diameter. |
| Heating surface—Tubes | 1,093.5sq.-ft. |
| " " Firebox | 113.5sq.-ft. |
| Total | 1,207.0sq.-ft. |
| Grate area | 18sq.-ft. |
| Working steam pressure | 165lb. per sq.-in. |
| Tank capacity | 1,200 gallons. |
| Bunker capacity | 2 tons coal. |
| Weight in working order | 53 tons 15 cwts. |

Built by Neilson.

In the same year Mr. Johnson designed some inside cylinder passenger engines (Fig. 139), which had the same size of boiler as the bogie tank engines.

They had a four-wheeled bogie in front, and four-coupled driving wheels. The tender was on six wheels, and carried the same amount of water as the bogie tenders previously described. Spring buffers are used between engine and tender. The principal dimensions are:—

| | |
|---|---|
| Cylinders | 18in. by 26in. |
| Coupled wheels | 6ft. 1in. diameter. |
| Bogie wheels | 3ft. 9½in. diameter. |
| Tender wheels | 6—4ft. 1in. diameter. |
| Wheel base of engine | 21ft. 9½in. |
| Wheel base of tender | 13ft. 0in. |
| Total wheel base of engine and tender | 43ft. 4½in. |
| Working pressure | 165 lb. per sq.-in. |
| Weight in working order—Engine | 43 tons 18 cwts. |
| " " " Tender | 35 tons 0 cwts. |
| Total | 78 tons 18 cwts. |

Built by Neilson and Co.

This is the present standard type of passenger engines on the Great North of Scotland Railway.

In addition to the engines of the Furness Railway, previously described, others deserve recognition, and it should be placed on record that the red-brown colour distinguishing the locomotives of this line has been the standard colour for a number of years. Some sixteen years or so back, the Midland Railway discarded green as the distinguishing colour for its engines, and adopted the red-brown shade of the Furness

Railway. Some people have imagined that the Furness Railway locomotives are painted in imitation of the Midland, but the facts show the opposite to be the case.

In 1870 a type of four-wheels-coupled passenger engines were introduced on the Furness Railway. The leading dimensions of these were:—

| | |
|---|---|
| Diameter of cylinders | 1ft. 4in. |
| Stroke | 1ft. 8in. |
| Diameter of coupled wheels | 5ft. 7½in. |
| Diameter of leading wheels | 3ft. 8in. |

CENTRE TO CENTRE OF WHEELS.

| | |
|---|---|
| Leading to driving | 6ft. 6in. |
| Driving to trailing | 7ft. 9in. |
| Total wheel base | 14ft. 3in. |
| Diameter of boiler (mean) | 3ft. 11in. |
| Length of barrel | 10ft. 0in. |
| Length of firebox (shell) | 4ft. 4in. |
| Number of tubes | 157—2in. external diameter |

HEATING SURFACE.

| | |
|---|---|
| Tubes | 839.5sq.-ft. |
| Firebox | 77.0sq.-ft. |
| Total | 916.5sq.-ft. |

TENDER (Four Wheels).

| | |
|---|---|
| Grate area | 11.5sq.-ft. |
| Diameter of wheels | 3ft. 8in. |
| Wheel base | 9ft. 6in. |
| Capacity of tank | 1,200 gallons. |
| Capacity coal | 3 tons. |
| Total wheel base, engine and tender | 120. |

WEIGHTS IN WORKING ORDER.

| | T. | c. | q. |
|---|---|---|---|
| Leading | 8 | 10 | 0 |
| Driving | 11 | 10 | 0 |
| Trailing | 10 | 5 | 0 |
| Total | 30 | 5 | 0 |
| Total weight of tender | 17 | 5 | 0 |
| Working pressure in lbs. per sq.-in. | 120. | | |

At this period the standard goods engines of the Furness Railway were six-wheels-coupled, of the following dimensions:—

| | |
|---|---|
| Diameter of cylinders | 1ft. 4in. |
| Stroke | 2ft. 0in. |
| Diameter of coupled wheels | 4ft. 7½in. |

CENTRE TO CENTRE OF WHEELS.

| | |
|---|---|
| Leading to driving | 6ft. 9in. |
| Driving to trailing | 8ft. 0in. |
| Total wheel base | 14ft. 9in. |
| Diameter of boiler (mean) | 3ft. 11in. |
| Length of barrel | 10ft. 4in. |
| Length of firebox (shell) | 4ft. 11¼in. |
| Number of tubes | 156—2in. external diameter. |

### HEATING SURFACE.

| | |
|---|---|
| Tubes | 871.27sq.-ft. |
| Firebox | 88.08sq.-ft. |
| Total | 959.35sq.-ft. |

### TENDER (Four Wheels).

| | |
|---|---|
| Grate area | 13.8sq.-ft. |
| Diameter of wheels | 3ft. 8in. |
| Wheel base | 9ft. 6in. |
| Capacity of tank | 1,600 gallons. |
| Capacity coal | 3 tons. |
| Total wheel base, engine and tender | 32ft. 7in. |

### WEIGHTS IN WORKING ORDER.

| | T. | C. | Q. |
|---|---|---|---|
| Leading | 10 | 11 | 0 |
| Driving | 11 | 10 | 0 |
| Trailing | 8 | 18 | 0 |
| Total | 30 | 19 | 0 |
| Total weight of tender | 19 | 10 | 0 |
| Working pressure in lbs. per sq.-in. | 120. | | |

The modern main line Furness Railway passenger engines have four wheels coupled of 6ft. diameter, with a leading bogie, the wheels of which are 3ft. 6in. diameter. The cylinders are inside 18in. diameter, with a 24in. stroke. The other dimensions are:—

### CENTRE TO CENTRE OF WHEELS.

| | |
|---|---|
| Centre of bogie to centre of driving axle | 9ft. 6¼in. |
| Centre of driving to trailing | 8ft. 6in. |
| Centres of bogie wheels | 5ft. 9in. |
| Total wheel base | 20ft. 11in. |
| Diameter of boiler (mean) | 4ft. 3in. |
| Length of barrel | 10ft. 3in. |
| Length of firebox (shell) | 5ft. 9in. |
| Number of tubes | 230—1¾in. external diameter. |

### HEATING SURFACE.

| | |
|---|---|
| Tubes | 1,109.0sq.-ft. |
| Firebox | 99.5sq.-ft. |
| Total | 1,208.5sq.ft. |

### TENDER (6 wheels).

| | |
|---|---|
| Grate area | 17sq.-ft. |
| Diameter of barrel | 3ft. 10in. |
| Wheel base | 12ft. 0in. |
| Capacity of tank | 2,500 gallons |
| Capacity coal | 4½ tons. |
| Total wheel base, engine and tender | 42ft 1in. |

### WEIGHTS IN WORKING ORDER.

| | T. | C. | Q. |
|---|---|---|---|
| Leading bogie | 13 | 12 | 0 |
| Driving | 14 | 10 | 0 |
| Trailing | 13 | 4 | 0 |
| Total | 41 | 6 | 0 |
| Total weight of tender | 28 | 5 | 0 |
| Working pressure in lbs. per sq.-in. | 150. | | |

This express class of passenger engines was introduced in 1896. When Mr. W. Pettigrew, M.Inst.C.E., who was, during the latter years of Mr. Adams's *régime*, practically the chief at Nine Elms Locomotive Works, was appointed locomotive superintendent at Barrow, to

FIG. 140.—PETTIGREW'S NEW GOODS ENGINE FOR THE FURNESS RAILWAY

succeed Mr. Mason, he got out designs for a new and powerful class of goods engines, which are now being delivered to the Furness Railway. Fig. 140 represents one of these engines, the leading dimensions of which are:—

| | |
|---|---|
| Diameter of cylinders | 18in. |
| Stroke | 26in. |
| Diameter of coupled wheels | 4ft. 8in. |
| Wheel base of engine | 5ft. 6in. |
| Diameter of boiler (inside) | 4ft. 4in. |
| Length of barrel | 10ft. 6in. |
| Length of fire-box (outside) | 6ft. 9in. |
| The boiler contains | 208 tubes, 1¾in. external diameter |

HEATING SURFACE.

| | |
|---|---|
| Tubes | 1,029sq.-ft. |
| Firebox | 105sq.-ft. |
| Total | 1,134sq.-ft. |
| Grate surface | 20.5sq.-ft. |
| Weight of engine in working order (about) | 38½ tons. |
| Working pressure | 150lbs. per sq.-in. |

TENDER.

| | |
|---|---|
| Diameter of wheels | 3ft. 10in. |
| Wheel box | 12ft. |
| Weight in working order (about) | 28¼ tons. |
| Capacity of tanks | 2,500 gallons. |
| Coal | 4 tons. |
| Total wheel base of engine and tender | 37ft. 11in. |
| Total weight in working order (about) | 66¾ tons. |

Fig. 141 represents one of the Highland Railway's 10-wheel main line engines, with outside cylinders. The six-coupled wheels make this design to be well adapted for the heavy traffic of the system, whilst the leading bogie gives sufficient facility for easily negotiating the curves of the Highland Railway.

The first newest class of express engines, designed by Mr. P. Drummond, is just delivered, and is very similar to those designed for the Highland Railway by Mr. D. Jones, the late locomotive superintendent, except that the new class has inside cylinders, whilst those built two years ago had outside cylinders. The dimensions of No. 1, "Ben-y-Gloe," just delivered, are: cylinders, 18¼in. by 26in. The coupled wheels 6ft. and the leading bogie wheels 3ft. 6in. diameter. Heating surface, 1,175 sq. ft. Steam pressure, 175lb. per sq. in. Weight, in working order: engine, 46 tons; tender, 37½ tons.

Mr. H. Pollitt's design of locomotive for working the express traffic over the London extension of the Great Central Railway has four-coupled wheels 7ft. diameter; cylinders, 18½in. by 26in., with piston valves; a Belpaire fire-box, and steam-pressure 170lbs. per sq. in. The tender holds 4,000 gallons of water and 5 tons of coal.

FIG. 141.—SIX-WHEELS-COUPLED BOGIE ENGINE, WITH OUTSIDE CYLINDERS, HIGHLAND RAILWAY

Before closing this account of locomotive evolution, some few details of modern Irish locomotives will be of interest.

Fig. 142 represents a four-coupled passenger engine of the Belfast and Northern Counties Railway. This engine is of the compound type, and is fired by petroleum on Holden's system.

"Jubilee" (Fig. 143) is also a compound express passenger engine of the same railway. Both these engines were designed by Mr. B. Malcolm, the Company's locomotive superintendent.

EVOLUTION OF THE STEAM LOCOMOTIVE 317

The modern passenger engines on the Great Northern Railway (Ireland) are of the four-coupled type, with a leading bogie, and are

FIG. 142.—LIQUID FUEL ENGINE, BELFAST AND NORTHERN COUNTIES RAILWAY

known as the "Rostrevor" class. The leading dimensions are as follows:—

### CYLINDERS.

| | |
|---|---:|
| Diameter of piston | 18¼in. |
| Stroke of piston | 24in. |
| Centre to centre | 2ft. 7in. |
| Steam ports | 14¼in. by 1¼in. |
| Exhaust ports | 14¼in. by 5in. |
| Outside lap | 1in. |
| Lead | 1-8in. |
| Maximum travel | 3⅞in. |

### WHEELS

| | |
|---|---:|
| Diameter of bogie wheels | 3ft. 1¼in |
| Diameter of driving wheels | 6ft. 7in. |
| Diameter of trailing wheels | 6ft. 7in. |
| Bogie wheel base | 5ft. 3in |
| From bogie wheel centre to trailing | 17ft. 9in. |
| Total wheel base | 20ft. 4¼in. |

### HEATING SURFACE.

| | |
|---|---:|
| In firebox | 109sq.-ft |
| Tubes | 1,013sq.-ft. |
| Total | 1,122sq.-ft |
| Grate area | 18½sq. ft. |
| Working pressure per sq. in.) | 160lb |

### WEIGHT.

| | In working order. | | |
|---|---:|---:|---:|
| | T. | C. | Q. |
| Bogie | 13 | 5 | 0 |
| Driving axle | 14 | 15 | 0 |
| Trailing | 14 | 0 | 0 |
| Total | 42 | 0 | 0 |

318   EVOLUTION OF THE STEAM LOCOMOTIVE

FIG. 143.—" JUBILEE," 4-WHEELS COUPLED COMPOUND LOCOMOTIVE, BELFAST AND NORTHERN COUNTIES RAILWAY

Fig. 144 represents one of the engines of the Great Northern (Ireland) Railway, as decorated to haul the Duke of York's train during his recent visit to Ireland.

FIG. 144.—" No. 73," STANDARD PASSENGER ENGINE, GREAT NORTHERN RAILWAY (IRELAND)

EVOLUTION OF THE STEAM LOCOMOTIVE     319

Fig. 145.— FOUR-COUPLED BOGIE EXPRESS ENGINE, GREAT SOUTHERN AND WESTERN RAILWAY

Fig. 145 is from a photograph of one of the standard passenger engines of the Great Southern and Western Railway. This engine was designed by Mr. R. Coey, the Company's locomotive superintendent. The coupled wheels are 6ft. 6in. diameter, the cylinders being 18in. diameter, with a stroke of 24in.

Our last illustration (Fig. 146) is produced from a photograph of "Peake," one of the "light" engines of the Cork and Muskerry Light Railway. Engines of this type are specially designed for working on "light" railways.

Fig. 146.—"PEAKE," A LOCOMOTIVE OF THE CORK AND MUSKERRY LIGHT RAILWAY

# INDEX.

[*N.B.—The letters B.G. denote a Broad Gauge locomotive.*]

## A

| | PAGE |
|---|---|
| Adams, Bridges, combination engines and carriages | 130, 133 |
| Adams, Bridges, radial axle-boxes | 209 |
| Adams, Bridges, system of intermediate driving shafts | 133 |
| Adams, Bridges, spring tyres | 211 |
| Adams, W., engines for the N.L. Ry. | 226 |
| Adams, W., engines for the L. & S.W. Ry. | 272 |
| "Æolus," B.G. | 71 |
| "Agenoria," | 27 |
| "Agilis," with double flanged wheels | 86 |
| "Ajax," B.G. | 73, 75 |
| "Albion" on the "Cambrian system" | 125 |
| Allan claims to have introduced "back-coupled" engines, 97; link motion | 97 |
| American engines for the Birmingham & Gloucester Railway | 87 |
| "Apollo," B.G. | 72 |
| "Areo-steam.'" engines | 234 |
| "Ariel," B.G. | 75 |
| "Armstrong" class, G.W.R. | 297 |
| Aspinall, J. A. F., locomotives for the Lancashire & Yorkshire Railway | 280 |
| Aston, W., engines for the Cambrian Railways | 264 |
| "Atlas," B.G. | 75 |
| "Atlas," M. & S.R. | 110 |

## B

| | PAGE |
|---|---|
| "Bacchus," B.G. | 72 |
| Back-coupled engines by Allan | 97 |
| Balanced locomotives | 84 |
| Beattie's engines, 162, 169 (coal-burning), 185, 194, 203, 207, 226, 231, 240 | |

| | PAGE |
|---|---|
| Belfast & Northern Counties Railway engines | 316 |
| Beyer's single iron plate frames | 97 |
| Beyer's "Atlas," for the M. & S.R. | 110 |
| Billinton, R. J., engines for L.B. & S.C.R. | 260 |
| Birmingham & Gloucester Ry., American engines on | 87 |
| Birmingham & Gloucester Ry., McConnell's engine for | 102 |
| "Black Prince," L. & N.W.R. | 247 |
| Blackett, Hedley, and Hackworth construct an engine | 10 |
| Blenkinsopp's, J., engine | 5 |
| "Blucher" | 14 |
| "Boat engines," B.G. | 73 |
| Bodmer's reciprocating engines | 100 |
| Bogie tenders | 241, 277, 310 |
| Bogie engines (early) | 56, 173 |
| Braithwaite & Ericsson's "Novelty" | 30 |
| Braithwaite & Ericsson's "William the IV." and "Queen Adelaide" | 46 |
| Bristol & Exeter Ry. locomotives, B.G. | 173 |
| Broad gauge engines (see G.W. & Bristol & Exeter Railways) | |
| "Brougham," S. & D.R. | 206 |
| Brunel, I. K., and broad gauge locomotives, 67, 75; Vale of Neath Ry. | 39 |
| (See also Great Western Railway engines) | |
| Brunton's "leg propelled" engine | 7 |
| "Bull Dog," G.W.R. | 302 |
| Burnett's tanks for the M. & S.J.W.R. | 233 |
| Bury, Edward, inventor of the inside cylinder locomotive | 40 |
| Bury, his first "Liverpool" | 40 |
| Bury. Authentic list of his first engines | 43 |

Y

322     EVOLUTION OF THE STEAM LOCOMOTIVE

| | PAGE |
|---|---|
| Bury, Contractor to the London & Birmingham Ry. | 82 |
| Bury, Engines on the Furness Ry. | 123, 179 |
| Bury, Extract from the Minute-books of the L. & M.R. relating to the "Liverpool" | 42 |
| Bury, "Liver," for the L. & M.R. | 52 |
| Bury, "Meteor," N. & C.R. | 62 |

## C

| | PAGE |
|---|---|
| Cambrian locomotive system | 125 |
| Caledonian Ry.:—Engine "No. 15," 152; 8ft. 2in. "single," 207; modern | 285 to 292 |
| "Caledonian," L. & M.R. | 54 |
| Canterbury & Whitstable Railway | 44 |
| Cambrian Railways engines | 209, 264 |
| "Canute," an early coal-burning engine | 186 |
| Chapman's chain locomotive | 6 |
| "Charles Dickens," L. & N.W.R. | 239 |
| Clark's smoke consuming engines | 191 |
| Coal-burning locomotives, 84; Chanter's system, 84; Dewrance's, 102; London & North Western, 167; Beattie, 185; Yorston, 188; Cudworth, 189; Clark, 191; Wilson, 191; Lee and Jacques, 192; Sinclair, 192; Douglas or Frodsham | 192 |
| Coey, R., engines for the G.S. & W.R. | 319 |
| Cork & Muskerry Light Railway | 319 |
| Combined engines and carriages | 130, 136, 224 |
| "Comet," Newcastle & Carlisle Railway | 60 |
| Compound locomotives | 169, 242, 249, 316 |
| Compressed air locomotive | 169 |
| "Cornwall" | 119 |
| Cork & Bandon Ry., Adams's light engines on | 140 |
| "Caithness," L. & N.W.R. | 205 |
| Cowan, W., goods engine | 225 |
| Cowlairs incline, 98; rope traction on | 100 |
| Crampton, T. R., locomotives, 75; on the 10ft.-wheel. B.G. | 112, 145, 159, 203 |
| Crewe Works erected | 97 |
| Cudworth, I., coal-burning engines | 189 |

| | PAGE |
|---|---|
| Cudworth, coal burning engines | 189 |
| "Cycloped" horse locomotive | 38 |
| Cylinder valves, fitted to Roberts's "Experiment" | 57 |

## D

| | PAGE |
|---|---|
| Davis & Metcalfe's exhaust-steam injector | 302 |
| Dean, W., locomotives for the Great Western Ry. | 294 |
| "Devonshire" class, G.W.R. | 297 |
| Disc wheels | 74, 75 |
| Dodd's engines for the Monkland and Kirkintilloch Ry. | 50 |
| Douglas, coal burning engine | 192 |
| Drummond, D., engines for the L. & S.W.R. | 276 |
| Drummond, D., engines for the Caledonian Ry. | 288 |
| "Dunalastair," Cal. Ry. | 285 |
| Dundee & Newtyle Ry. engines | 57 |
| "Duplex," a two-boiler engine | 158 |

## E

| | PAGE |
|---|---|
| Eastern Counties Ry., Hancock's locomotive for, 86; "Essex," 111; compressed air engine, 169; coal burning, 192; Sinclair's engines | 195 |
| (See also G.E.R.) | |
| "Eclipse," Dr. Church's tank engine | 82 |
| Eight-wheels-coupled engines, early | 195 |
| Eight-wheels-coupled engines, Webb's | 246 |
| Eight-wheel rolling stock, the first | 46 |
| "Enfield," combined engine and carriage | 133 |
| England's "Little England" locomotives | 141 |
| "Essex," E.C.R. | 111 |
| Exhaust steam blast (see Hackworth) | |
| Exhaust steam injector (Davies & Metcalfe's patent) | 302 |
| "Experiment" engine for the L. & M.R. | 56 |
| "Experiment," L. & N.W.R. | 243 |

# INDEX. 323

## F

"Fairfield" combined engine and carriage, B.G. ............... 131, 133
Fairlie's "double bogie" engines ..................... 224, 234
Festiniog Railway, Fairlie's engines on .......................... 223
Fell's steep gradient engines ...... 219
Fletcher's 4-wheel tank engine ... 201
"Folkestone," a Crampton engine for the S.E.R. ....................... 159
Four-cylinder engines, L. & N.W. Railway ................................ 248
Four-cylinder engines, S.W.R..... 276
Fowler, Sir J., "hot-brick" engine .......................... 200, 217
French locomotive on the Eastern Counties Ry. ................... 195, 207
Furness Ry. engines, 123, 179, 236, 312
"Fury" and "Firefly" classes, G.W.R., B.G. .......................... 90

## G

Galloway's incline climbing experiments ........................ 109
Gauge locomotive experiments ... 105
Geared-up engines, B.G. 77, 79, 147
Giffard's injector .................... 197
"Gladstone" class, L.B. & S.C. Railway ............................... 252
Glasgow & South-Western Ry. locomotives ......................... 241
"Globe," the first engine with a steam dome ........................... 47
"Goliath," Newcastle & Carlisle Railway ............................... 61
Gooch, Daniel (see G.W.R.)
Gooch, J. V., engines by ... 161, 162
Grand Junction Ry., opening ... 64
Grand Junction Ry. early locomotives ................................... 64
"Grasshopper," B.G. ............... 73
Gray's expansion gear ............. 93
"Great Britain," M'Connell's ..... 102
"Greater Britain," L. & N.W.R. 245
Great Central Ry., Pollitt's engines for ............................. 316
Great Eastern Ry., locomotives (see also Eastern Counties Ry. ......... 206, 217, 249, 255 to 259
Great Northern Railway engine, "215" ........................ 171
Great Northern Ry. engines, 171, 216, 236, 303 to 309
Great North of Scotland Ry. engines ........................ 225, 309 to 311

Great Northern (Ireland) Ry. engine ................................... 318
Great Southern & Western Ry. ... 319
"Great Western," B.G. ............ 106
Great Western Ry. locomotives, the original, 66; first trial of, 69; table of dimensions, 70; the 10ft. wheel engines, 73, 76; geared-up engines, 77, 79; table of mileage of original engines, 81; Gooch's first engines, 90; first engine built at Swindon, 105; "Great Western," 106; trial trips, 107, 108; Galloway's engine, 109; "Iron Duke," 113; first narrow gauge engines, 182; "Robin Hood," 184; Metropolitan Ry., engines for, 213; Dean's designs, 294 to 303
"Grosvenor," L.B. & S.C.R. ...... 242

## H

Hackworth, Timothy, first engine ...................................... 10
Hackworth, Timothy, and the Stockton and Darlington Ry. locomotives ............................. 21
Hackworth, Timothy, "Royal George" ................................. 24
Hackworth, Timothy, "Sanspareil" .................................... 32
Hackworth, Timothy, and the exhaust steam blast, 24; the secret stolen at Rainhill ......... 33
Hackworth, Timothy, "Globe" for the S. & D.R. .................. 47
Hackworth, Timothy, "Majestic" and Wilberforce "classes" for the S. & D.R. .................. 52, 53
Hackworth, Timothy, trunk or ram engine, 61; "Arrow" ... 61
Hackworth, Timothy, builds "Jenny Linds" ....................... 104
Hackworth, Timothy, "Sanspareil 2," 149; challenge to R. Stephenson concerning ......... 150
Harrison's patent engines, B.G.... 76
"Harvey Combe" ballast engine 60
Hancock's engine for the Eastern Counties Ry. ......................... 86
Haigh Foundry engines, B.G. ... 79
"Hawthorn" ........................... 157
Hawthorne's engines ...... 52, 58, 156
Hedley (see Blackett) ............... 10
Highland Railway locomotives ... 316

Historical locomotives sold by auction ................................ 51
Holden, J., liquid fuel locomotives ............................... 254, 316
Holden, J., engines for the G.E.R. 253
Holmes, M., engines for the N.B.R. ................................. 253, 277
"Hot-brick" locomotive, Fowler's, for Met. Ry. .................. 200, 217
Howe and the "link" motion, 96; 3-cylinder engine. ......... 105
"Hundred miles an hour!" B.G. 79
Hurricane 10ft. wheel engine, B.G. ..................................... 76, 79

**I**

Injector, Gifford's invention of ...197
"Iron Duke" BG. ...................... 113
Inside cylinder locomotive, "Liverpool," the first .................... 40
Inside cylinder locomotive—extract from the minute books of the L. & M.R. relating to same ......... 42
"Inspector," L.B. & S.C.R. ... 261
International Exhibition, 1851, locomotives at ..................... 156
"Invicta," Canterbury & Whitstable Ry. ............................. 44
Ivatt, H. A., engines for the G.N.R. .......................... 303 to 309

**J**

"Jason,'" B.G., Gooch's first goods engine ........................ 92
James and the link motion......... 96
"Jenny Lind" engines ...... 104, 115
"Jenny Sharps" ..................... 116
"Jinks's Babies" ..................... 234
Johnson, S. W., engines for the Midland Ry. ......................... 250

**K**

Kirtley, W., engines for the L.C. & D.R. ............................... 262
Kendall, W., 3-cylinder engine ... 231
Kennedy's, James, testimony regarding the first inside cylinder locomotive ............................. 42

"Lablache" .............................. 124
"Lambro" ................................ 95
Lancashire & Yorkshire Ry. engines .............................. 234, 280
L.B. & S.C.R. locomotives, 240, 242, 252, 260
L.C. & D.R. locomotives ... 203, 262

"Light locomotives," Samuels' 130; Adams's, 139; England's 141
Liquid fuel locomotives ...... 253, 285
"Little Wonder," Festiniog Ry. 224
"Little England" .................... 141
"Little Wonder," Samuels' combined engine and carriage ......... 130
"Liverpool," Crampton's engine for the L. & N.W.R. ............. 145
"Link" Motion, 96; Allan's .... 97
"Liver," Bury's, for L. & M.R.... 52
"Liverpool," the first engine with inside cylinders and crank axles 40
"Liverpool," description of ......... 44
Liverpool & Manchester Ry., early locomotives on, 45, 46, 50, 52, 85
Liverpool & Manchester Ry., opening of ........................... 46
Liverpool & Manchester Ry., 8-wheel passenger carriage at opening .............................. 46
Liverpool & Manchester Ry., Rainhill contest, 28; the competitors ............................... 30
"Locomotion," S. & D.R. ......... 20
London & Birmingham Ry., opening, 82; Bury's engines for ... 82
London, Brighton & S.C.R., Bodmer's engine on .................. 101
"Long boiler" engines, 94, 103, 111, 122, 137
London, Brighton & S.C. Ry., "Jenny Linds" ..................... 116
"London," Crampton's engine for the L. & N.W.R. .................. 113
"Lord of the Isles," B.G. ......... 115
L. & S.W.R. locomotives, 162, 169, 187, 194, 202, 207, 226, 231, 240, 272
L. & N.W.R. locomotives, 163, 122, 153, 155, 205, 238, 239, 243, 281
(See also London & Birmingham Ry.)

**M**

"Magnet," S. & D.R. ............... 54
"Majestic" class, S. & D.R. ...... 52
Malcolm, B., engines for the Belfast & Northern Counties Ry.... 316
Manson, engines for the G.N. of S. Ry. ................................. 309
"Mars," B.G. ............................ 73
McConnell's "Great Britain," 102; counterbalancing experiments, 122; "most powerful N.G. engine," 122; "Mac's Mangle," 153; "Bloomer's 155; "300," 163; "Caithness," 205

# INDEX.

McIntosh, J. F., locomotives for the Caledonian Ry. ............. 286
Metropolitan Ry., hot brick engine for, 200; B.G. engines on, 213; first engines ............ 214
"Meteor," L. & S.W. R. ............. 203
"Meteor," Bury's, for the N. & C. Ry. ................................. 62
Metallic piston packing, first used ...................................... 51
"Michael Longridge," for the S. & D. Ry. ............................ 64
Midland Ry., trials of "Jenny Sharps" and "Jenny Linds" on 116 to 118
Midland Ry., Johnson's engines 250
Monkland & Kirkintilloch Ry., first engines on the ............. 50
Murray's, M., engine [see Blenkinsopp] ................................ 5

## N

"Namur," on Crampton's system 112
Narrow gauge engines on the G.W.R., the first ................. 182
Neilson's type of goods engine... 180
Newcastle & Carlisle Ry., opening of ................................. 59
Newcastle & Carlisle Ry., "Goliath" locomotive .................. 61
Newcastle & Carlisle Ry., "Atlas" locomotive ................... 61
Newcastle & Carlisle Ry., "Tyne" locomotive .................. 61
Newcastle & Carlisle Ry., "Eden" locomotive ................. 62
Newcastle & Carlisle Ry., "Meteor" locomotive ............. 62
Norfolk Ry., light engines on ... 140
North British Ry. engines ... 253, 276
North Eastern Ry. locomotives, 249, 252
North London Ry. engines, 191, 226, 230
"No. 266," G.N.R. ............... 308
"No. 990," G.N.R. ............... 306
"Nunthorpe," S. & D.R. ......... 193

## O

"Old Coppernob," Furness Ry, the oldest engine now at work 123
Opening of the Canterbury and Whitstable Railway ............. 44
Opening of the Liverpool and Manchester Ry. ..................... 46
Opening of the Stockton & Darlington Ry. .......................... 47

Opening of the Newcastle & Carlisle Ry. .............................. 59
Opening of the Grand Junction Railway ............................ 64
Opening of the Great Western Ry. 72
Opening of the London & Birmingham Ry. ...................... 82
Opening of the London & Southampton Ry. ..................... 85
Opening of the East Kent Ry. ... 195
Opening of the Metropolitan Ry., B.G. ............................... 213
Opening of the Metropolitan & St. John's Wood Ry. ............. 233

## P

Pambour, on the early L. & M. Ry. engines ...................... 50
Casey's compressed air locomotive 169
"Patentee," Stephenson's 6-wheel passenger engine for the L. & M.R. .................................... 59
Paton's Cowlairs Incline engine... 98
Pearson's design for a double tank locomotive, 147; 9ft. singles... 173
Pettigrew, W., engines for the Furness Ry., ....................... 315
"Perseverance," at Rainhill ...... 37
"Planet," L. & M. Ry. ............. 49
"Plews," Y.N & B.R. ............... 144
Pollitt, H., engines for the Great Central Ry. ........................ 316
"Precedent" type, L. & N.W. R. 238
"Precursor" type, L. & N.W.R. 239
"Pretolea," G.E.R. ................. 255
"Premier," B.G. ..................... 105
"Problem," L. & N.W.R., first engine fitted with the injector... 197
Pryce, H. J., engines for the N.L.R. ................................. 230
"Puffing Billy" ...................... 12
"Python," L. & S.W.R. ........... 231

## Q

"Queen-Empress," L. & N.W.R. 246

## R

Rainhill locomotive contest, "Cycloped" at the ................... 38
Rainhill locomotive contest, Manumotive carriages at the ... 38
Rainhill locomotive contest, "Perseverance" at the ........... 37
Rainhill locomotive contest, "Rocket" at the .................... 35

Rainhill locomotive contest: "Sanspareil" at the .............. 33
Rainhill locomotive contest, "Novelty" at the .............. 30
Ramsbottom's water pick-up apparatus .............................. 198
"Red Star," B.G. .............. 136
Rennie's "Lambro" .............. 95
Ritchie's design for a locomotive 148
Roberts's "Experiment," L. & M.R., with cylinder valves ...... 56
Robertson's steam brake .......... 97
"Rocket," at Rainhill, 35; later history ........................... 37
"Rocket," Colburn's opinion of her ................................ 36
"Rocket," her tubular boiler invented by Booth .............. 36
"Rocket," awarded the Rainhill prize .......................... 36
Russia, Hackworth's trunk engine, the first locomotive in...... 61
"Royal George," first financially successful locomotive ........ 23
"Royal William" .............. 19

## S

Samuels' "Little Wonder" ...... 130
"Sanspareil," at Rainhill, 33; later history .................. 34
"Sanspareil 2" .................. 149
Sheffield & Manchester Ry., Bodmer's engines on........ 100, 101
Sheffield & Manchester Ry., "Atlas" ................................ 110
Short-stroke engines .......... 60, 61
Sinclair's smoke consuming engines, 192; "singles," 206; tanks ............................ 217
"Snake" ........................ 122
Steam organ, fitted to the "Tyne" 62
Steam tenders .................. 217
Steam blast (see also Hackworth) 24
Stephenson, Geo., & Hackworth, 11, 21, 25, 33
Stephenson's engines: first, 14; second, 15; third, 16; later types, 26, 27, 35, 43, 46, 49, 50, 51, 56, 59, 60, 64; long boiler, 103; 3-cylinder, 105; "A," 105; "White Horse of Kent," 111, double engine ...... 194
Stephenson's, Robt., valve gear, 94; link motion ................ 96
Stephenson's, Robt., challenge to, re "Sanspareil 2" .............. 150
Stewart. W., early locomotives by 13

Stirling, P., 8ft. 1in. "singles" for the G.N.R. .............. 236
Stirling, J., design for the G. & S.W.R. ........................ 241
Stirling, J., reversing apparatus... 241
Stirling, J., locomotives for the S.E.R. ........................ 266
Stockton & Darlington Ry., opening of .......................... 20
Stockton and Darlington Railway, "Royal George" .............. 23
Stockton & Darlington Ry. locomotives, 22, 27, 47, 52, 53, 54, 61, 64, 65, 193, 206 ...... 234
Stroudley's engines for the L.B. & S.C.R. .................. 242, 252
Sturrock, Archibald, apprenticed, 58; "215," G.N.R., 171; and Met. R., 214; condensing engines, 216; steam tenders ...... 217
"Soho" .......................... 86
South Eastern Ry., Bodmer's engines on ........................ 101
South Eastern Ry., "White Horse of Kent," 111; "Folkestone," 159; Sharp's engines, 161; Cudworth's, 189, 225; Watkin's, 241; Stirling's ...... 266
South Devon Ry. locomotives, B.G. ............................ 179
"Sunbeam," S. & D.R. .......... 64
"Swiftsure," Forrester's, engine for the L. & M.R. .............. 59

## T

Tayleur's short stroke locomotives 60
Tank engines, early ...... 137, 158
Taff Valley Ry. engines, 163, 202, 233, 292
Ten feet driving wheel engines, B.G. ...................... 69, 73, 76
"Teutonic," L. & N.W.R. ...... 244
"Thunderer," B.G. .............. 77
Three-cylinder engines, Stephenson & Howe's .............. 105
Three-cylinder engines, Kendall's 231
"Tiny," Crewe Works, engine ... 209
Tosh's goods engine .......... 182
Trevithick's, F., "Cornwall" ... 119
Trevithick, R., inventor of the steam locomotive .............. 1
Trevithick, R., his first railway engine ........................ 2
"Tyne," N. & C.R. .............. 62

# INDEX.

## V

Vale of Neath Railway (see Brunel).
Valve gears, "Experiment," 57; "Soho," 86; Gray's, 93; Dodds & Owen's, 93; Stephenson's, 94; link motion, 96; vertical, 123; rotatory, 181; Dubs' .................................. 199
"Venus," B.G. .................. 72, 136
"Vulcan," B.G. .................. 69, 70

## W

"Wallace," Dundee & Arbroath Railway ........................... 82
Water pick-up apparatus, Ramsbottom's ........................... 198
"Welsh Pony," Festiniog Ry. ... 223
Webb, F. W., engines for the L. & N.W.R., " Precedents," 238; "Precursors," 239; compounds, 243; "Greater Britain," 245; 8 wheel-coupled, 246; "Black Prince" .................... 248

Whishaw on Great Western Ry. B.G. engines (see Chapter VI., pages 66 to 81)
Williams and the "link" motion 96·
Wilson's, Ed., system of smoke consuming ........................ 191
"Wilberforce," S. & D.R. ...... 53
"William the 4th," and "Queen Adelaide" locomotives ......... 46
Winans' Manumotive carriages at Rainhill ........................... 38
Worsdell's compounds ............ 249
"Windcutter," locomotive ...... 274
Wood, N., on Great Western Ry. B.G. engines (see chapter VI., pages 66 to 81)
"Wrekin," ........................... 151

## Y

Yarrow's coal-burning engine ...... 190
Yorston's coal-burning engine ... 188
"Ysabel" ........................... 168

**RETURN TO → CIRCULATION DEPARTMENT**
202 Main Library

| LOAN PERIOD 1 **HOME USE** | 2 | 3 |
|---|---|---|
| 4 | 5 | 6 |

**ALL BOOKS MAY BE RECALLED AFTER 7 DAYS**
1-month loans may be renewed by calling 642-3405
6-month loans may be recharged by bringing books to Circulation Desk
Renewals and recharges may be made 4 days prior to due date

## DUE AS STAMPED BELOW

ICLF

OCT 10 1980

RECEIVED by

SEP 22 1980

CIRCULATION DEPT.

LIBRARY USE ONLY
JUL 0 1987
CIRCULATION DEPT.

AUG 08 1987

AUTO. DISC.

JUL 15 1987

UNIVERSITY OF CALIFORNIA, BERKELEY
FORM NO. DD6, 60m, 3/80      BERKELEY, CA 94720

www.ingramcontent.com/pod-product-compliance
Lightning Source LLC
Chambersburg PA
CBHW030007240426
43672CB00007B/852